Controlled Release Technologies: Methods, Theory, and Applications

Volume I

Editor

Agis F. Kydonieus, Ph.D.
Executive Vice President
Hercon Group
Health-Chem Corporation
New York, New York

CRC Press is an imprint of the
Taylor & Francis Group, an informa business

CRC Press
Taylor & Francis Group
6000 Broken Sound Parkway NW, Suite 300
Boca Raton, FL 33487-2742

Reissued 2019 by CRC Press

A Library of Congress record exists under LC control number:

Publisher's Note
The publisher has gone to great lengths to ensure the quality of this reprint but points out that some imperfections in the original copies may be apparent.

Disclaimer
The publisher has made every effort to trace copyright holders and welcomes correspondence from those they have been unable to contact.

ISBN 13: 978-0-367-25360-8 (hbk)
ISBN 13: 978-0-367-25364-6 (pbk)
ISBN 13: 978-0-429-28739-8 (ebk)

Visit the Taylor & Francis Web site at http://www.taylorandfrancis.com and the CRC Press Web site at http://www.crcpress.com

PREFACE

Delivery of chemical ingredients by controlled release processes occurs in nature. Examples of such processes include the delivery and control of the flow of food and waste across the exterior membrane of one-celled animals and the oxygenation of blood in mammals by the diffusion of oxygen through the alveolar walls. Development of controlled release systems for our modern technology may be considered an attempt to simulate nature's processes and thereby provide more efficient and more effective delivery of chemicals to intended targets. Information and knowledge gained in these exploits form the substance of the two volumes of this treatise.

Intensive research and development by the drug industry in the last 20 years has provided the scientific foundation for this fledging new science. Concern over the administration of high doses of conventional drugs was the impetus toward the development of controlled release oral drugs initially and implantable devices more recently. Applying much the same principles, scientists working in the pesticide field have developed new technologies and formulations to (1) extend the duration of effectiveness of pesticides without increasing rates of application, (2) reduce the hazard associated with the use of highly poisonous chemicals, (3) prolong the effective life of unstable, volatile, or hydrolyzable pesticides, (4) improve pest control efficiency, and (5) minimize pollution of the environment.

The technologies described in these two volumes depend almost exclusively on the use of polymers and polymer technology. Exceptional advancement in polymer science during the past several decades has made possible the creation of delivery systems which previously could not have been conceived. Described herein are dozens of polymers that comprise the elements of delivery systems that control the permeation and release of active ingredients.

The applications, advantages, and fundamental concepts of controlled release have been the subject of many symposia and of several books dealing with formulations: however, all known delivery systems have not been assembled for consideration as they are here.

Our emphasis in these volumes is on the description of the controlled release technologies, their theoretical basis, mechanisms of release, and their advantages and limitations. Many commercial applications, as well as experimental products and formulations, are discussed to illustrate the potential of each of these technologies. Young as this technology may be, many scientists are convinced that it has the momentum and the secure fundamental basis for solving the complex problems of the delivery of drugs to humans and of directing pesticides to targeted biota with minimum disruption of environmental values. Formulations have already been developed that have been proven commercially successful in many fields, including medicine, agriculture, forestry, public health, and products for the consumer and industry. At this juncture, our aim is that this volume will add to the current momentum to explore this already proven technology and that it will help product development scientists choose the proper controlled release technology for their specific product needs.

The fundamental concepts and theoretical background of controlled release processes are given in Volume 1, Chapter 1. Having introduced the subject matter in the introductory chapter, Chapters 2 through 6 and Volume II, Chapters 1 through 12 follow, describing in detail 17 different controlled release technologies; Chapter 13 describes briefly several other controlled release technologies and provides a 1977 to 1978 patent search of controlled release methods and applications. Each chapter was written by an author who either invented a specific controlled release technology or has had a major roll in advancing the state of its art. Representing all the major technological areas of controlled release, the subject matter is grouped as follows in the chapters indicated:

1. Volume I, Chapters 2 through 6: monolithic devices, membranes, porous and homogenous films, and laminated structures.
2. Volume II, Chapters 1 and 2: erodible, bioerodible, soluble, hydrolyzable, and biodegradable devices.
3. Chapter 3: retrograde chemical reaction systems.
4. Chapters 4 through 8: microencapsulation systems by coascervation, interfacial polymerization, air suspension, and centrifugal extrusion.
5. Chapters 9 through 12: other important controlled release technologies, such as hollow fibers, osmotic devices, starch xanthate matrices, and kraft lignin carriers.

As an introductory background and perspective for the novice or the recent practitioner in the field, the first chapter was included to introduce all the major controlled release technologies and to describe the basic components of controlled release devices and the release characteristics of the controlled release processes. New developments and improvements in existing technologies continue to appear. Described in Chapter 13 are some new and improved technologies and products, actually those found in a review of the Official Gazettes of the U.S. Patent and Trademark Office for 1977 and part of 1978. Several other controlled release technologies which were found in the literature have also been included in Chapter 13. Obviously, in the fast-expanding field of controlled release, several omissions must have occured despite our effort to present all significant developments known by the end of 1977, when most of the manuscripts were collected. Nevertheless, we hope that this effort will prove to be of value to scientists and product development engineers seeking up-to-date information in the field.

Several friends and associates should be given credit for their helpful suggestions and criticisms. Special thanks should go to Dr. Morton Beroza for scrutinizing parts of the manuscript and to Drs. Nate Cardarelli and Bill O'Neill for their guidance and encouragement. I am also indebted to the authors for their cooperation in adhering to strict manuscript specifications and to Ms. Camille Boxhill and Ms. Adriane Chisholm for their efforts in typing and assisting in the editorial endeavors. Finally, I would like to thank the management of Hercon Group and Health-Chem Corporation, who have been strong advocates of controlled release for many years and have given the editor all the support required to complete this undertaking.

Agis F. Kydonieus
September 1978

THE EDITOR

Agis Kydonieus, Ph.D., is Executive Vice President of Hercon Products Group, a division of the Health-Chem Corporation, New York, N.Y. and Adjunct Professor of Chemical Engineering, The Cooper Union, New York, N.Y.

Dr. Kydonieus graduated in 1959 from the University of Florida, Gainesville, Florida, with a BS degree in Chemical Engineering (summa cum laude) and obtained his Ph.D. degree from the same school in 1964.

Dr. Kydonieus is a member of the Board of Governors of The Controlled Release Society as well as a member of The American Institute of Chemical Engineers, The Entomological Society of America, American Chemical Society, Pest Control Association, and The Society of Plastics Engineers.

Dr. Kydonieus is the author of several dozen patents, publications and presentations in the fields of controlled release and biomedical devices.

CONTRIBUTORS

G. Graham Allan, Ph.D., D.Sc.
Professor of Fiber and Polymer Science
Department of Chemical Engineering
College of Forest Resources
University of Washington
Seattle, Washington

Joseph A. Bakan, B.S.
Director of Research
Capsular Products Division
Appleton Papers, Inc.
Dayton, Ohio

John W. Beer
Research Associate
Centro Agronomico Tropical de Investigaciony Ensenanza
Turriabla, Costa Rica

Thomas W. Brooks, Ph.D.
Technical Director
Conrel
Needham Heights, Massachusetts

Massimo Calanchi
Manager
Research and Development
Microcaps
Eurand, S.p.A.
Milan, Italy

Nate F. Cardarelli, M.S.
Director
Environmental Management Laboratory
University of Akron
Akron, Ohio

Michael J. Cousin, Ph.D.
Research Scientist
Battelle Columbus Laboratories
Columbus, Ohio

H. T. DelliColli, Ph.D.
Product Development Manager
Agrichemical Products
Westvaco Corporation
North Charleston, South Carolina

John T. Goodwin, Ph.D.
Vice President
Chemistry and Chemical Engineering Division
Southwest Research Institute
San Antonio, Texas

Harlan S. Hall, B.S.
President
Coating Place, Inc.
Verona, Wisconsin

Frank W. Harris, Ph.D.
Professor of Chemistry
Wright State University
Dayton, Ohio

Robert C. Koestler, Ph.D.
Project Leader—Microencapsulation
Pennwalt Corporation
King of Prussia, Pennsylvania

Mario Maccari, Ph.D.
Director
Research and Development
Eurand S.p.A.
Milan, Italy

Ruth Ann Mikels, B.S.
Research Assistant
College of Forest Resources
University of Washington
Seattle, Washington

Arthur S. Obermayer, Ph.D.
President
Moleculon Research Corporation
Cambridge, Massachusetts

William P. O'Neill, Ph.D., M.B.A.
Management and Technical Consultant
Poly-Planning Services
Los Altos, California

Ralph E. Pondell, B.S.
Vice President-Secretary
Coating Place, Inc.
Verona, Wisconsin

Alberto R. Quisumbing, Ph.D
Director Field Applications
Hercon Group
Herculite Products, Inc.
New York, New York

Theodore J. Roseman, Ph.D.
Senior Research Scientist
The Upjohn Company
Kalamazoo, Michigan

Mario F. Sartori, Ph.D.
Consultant
Department of Chemistry
University of Delaware
Newark, Delaware

Baruch Shasha, Ph.D.
Research Chemist
Northern Regional Research Center
Agricultural Research
Science and Education Administration
United States Department of Agriculture
Peoria, Illinois

George R. Somerville, B.S.
Director
Department of Applied Chemistry and Chemical Engineering
Southwest Research Institute
San Antonio, Texas

Felix T. Theeuwes, Sc.D.
Program Director
OROS Products
Principal Scientist
Alza Research
Palo Alto, California

Seymour Yolles, Ph.D.
Professor of Chemistry
University of Delaware
Newark, Delaware

TABLE OF CONTENTS

Volume I

Volume II

Chapter 1

FUNDAMENTAL CONCEPTS OF CONTROLLED RELEASE

Agis F. Kydonieus

TABLE OF CONTENTS

TABLE 1

Controlled Release—Examples Found in Nature

One-celled animals controlling the flow of food and waste materials across the exterior membrane
Activation of bacteria spores after long periods of inactivity by exposure to correct environmental conditions
Chameleon's control of skin color
Honey in beehive—released by eating container
Natural systemic insecticides, e.g., pyrethrum
CR creation of an adhesive spider web
Nitrogen fixation in soil by legumes
Insecticide (poison) injection by wasps
CR of "ink" by squid
Protective odor release by skunk
Ovulation cycle in animals
Controlled decomposition of wood and leaves to produce humus
CR of aroma and nectar at specific times by flowering plants
Natural biological equilibrium reactions
Temperature activated control of blood vessels in skin (blushing)
Pickling, fermentation, etc. processes activated by bacteria
Plants emitting odor to attract or drive off insects and animals (pitcher plant, skunk cabbage)
Barnacle adhesive

TABLE 2

Controlled Release—Examples Invented by Man

Artificial kidney utilizing microcapsules of charcoal to adsorb uremic waste products
Detergents packaged in hot-water-soluble films
Membrane blood oxygenators
Microcapsular dry food flavors released during mastication
Enteric-coated capsular pharmaceuticals released by pH differential
Controlled release systemic insecticides absorbed by plants
Glue sticks containing microcapsular adhesives released by pressure
Controlled release foods tailored to the specific needs of the plant
Insecticide microcapsular bait for control of fire ants
Dye markers that release the dye on water contact
Microencapsulated mercaptan for natural gas identification
Slow-release steroids for birth control applications
Environmentally eroded coatings
Dry flavor powders released on water contact or under heat
Microcapsular injectables possessing controlled release in body fluids
Microcapsular adhesives employing temperature release
Controlled release coatings activated by bacterial attack
Antifouling controlled release coatings
Controlled release insecticide strips
Microcapsular primers and adhesion promoters
Seeds encapsulated in polyvinyl alcohol tapes

Controlled release may be defined as a technique or method in which active chemicals are made available to a specified target at a rate and duration designed to accomplish an intended effect. A definition perhaps more acceptable to the chemist and engineer may be: controlled release is the permeation-moderated transfer of an active material from a reservoir to a target surface to maintain a predetermined concentration or emission level for a specified period of time.

Nature, naiveté, and ingenuity have provided a number of well-known examples of permeation-controlled processes. Some well-known natural processes depending on permeation include respiration, osmosis, and the "bloom" or "patina" on grapes and other fruit. Tables 1 and 2 summarize, respectively, examples of controlled release found in nature and some invented by man.[1]

I. RATIONALE FOR CONTROLLED RELEASE

During the last decade, controlled release technology has received increasing attention in the face of a growing awareness that substances ranging from drugs to agricultural chemicals are frequently excessively toxic and sometimes ineffective when administered or applied by conventional means. Thus, conventionally administered drugs in the form of pills, capsules, injectables, and ointments are introduced into the body as pulses that usually produce large fluctuations of drug concentrations in the bloodstream and tissues and consequently, unfavorable patterns of efficacy and toxicity. Conventional application of agricultural chemicals similarly provides an initial concentration far in excess of that required for immediate results in order to assure the presence of sufficient chemical for a practical time period; such overdosing wastes much of the chemical's potential and all too often causes toxicity problems for nontarget organisms.

The process of molecular diffusion through polymers and synthetic membranes has been used as an effective and reliable means of attaining not only the controlled release of drugs and pharmacologically active agents, but also of fertilizers, pesticides, and herbicides. Central to the development of controlled delivery systems is the synthesis of the principles of molecular transport in polymeric materials and those of pharmacokinetics and pharmacodynamics. In drug delivery, pharmacokinetics is an important consideration because target tissues are seldom directly accessible, and drugs must be transported from the portal of entry in the body through a variety of biological interfaces to reach the desired receptor site.

During this transport process, the drug can undergo severe biochemical degradation and thereby produce a delivery pattern at the receptor site that differs markedly from the pattern of drug release into the system. For agricultural chemicals, concentration, persistence, and transport in the soil are decreased by biodegradation, chemical degradation, photolysis, evaporation, surface runoff, and ground water leaching.[2]

A. Conventional Delivery vs. Controlled Release of Active Agents

1. Drug Delivery

Conventionally, active agents are most often administered to a system by nonspecific, periodic application. For example, in medical treatment, drugs are introduced at intervals by ingestion of pills or liquids or by injection. The drugs then circulate throughout much of the body, and the concentration of the active agent rises to high levels, system-wide, at least initially. Such responses are shown in Figure 1.[3] Both by injection and orally, the initially high concentrations may be toxic and cause side effects both to the target organ and neighboring structures. As time passes, the concentration diminishes, owing to natural metabolic processes, and a second dose must be administered to prevent the concentration from dropping below the minimum effective level. This situation is, of course, very inconvenient and difficult to monitor, and careful calculations of the amount of residual active agent must be made to avoid overdosing. The close attention required, together with the fact that large amounts of the drug are lost in the vicinity of the target organ, make this type of delivery inefficient and costly. In addition, side effects owing to drug misdirected to nontarget tissues are also possible.

Cowsar has discussed a hypothetical drug that is effective at 5 ± 2 mg/kg (below 3 mg/kg ineffective, above 7 mg/kg toxic) and has a half-life in vivo of 8 hr; his regimen calls for the patient to be treated for 10 to 14 days.[4] He found that an initial injection of 7 mg/kg followed by 32 subsequent injections of 5 mg/kg at 10 hr intervals was required. If 14-mg/kg injections were given to reduce the number of injections needed, an effective level could be maintained, but for 8 hr, the concentration of the drug was

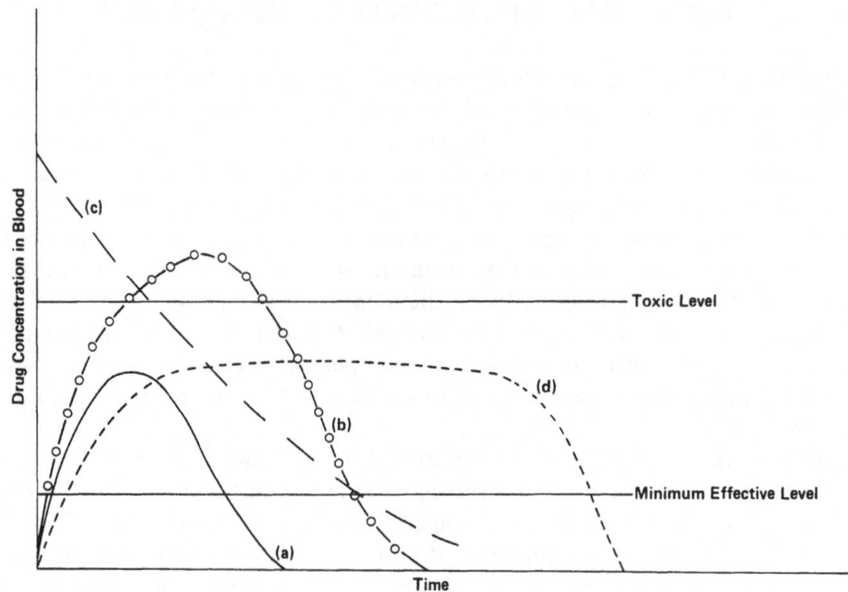

FIGURE 1. Typical drug level versus time profile. (a) Standard oral dose. (b) Oral overdose, (c) i.v. injection. (d) Controlled release ideal dose.

at a potentially toxic level. If a controlled release product were available, a single administration providing the 5 mg/kg would be needed.

In Figure 1, the ideal controlled release rate (d) is illustrated; i.e., a constant concentration, one that is effective but not toxic, is maintained for the desired time. Advantages of this system for therapeutic agents are (1) reproducible and prolonged constant delivery rate, (2) convenience of less frequent administrations, and (3) reduced side effects because the dose does not exceed the toxic level.

2. *Pesticide Delivery*

Although the above discussion focused primarily on drugs, the same problems are encountered in the application of agricultural chemicals. Conventionally, insecticides, fertilizers, and other pesticides are applied periodically to crops by broadcasting, spraying, etc. Again, initially very high and possibly toxic concentrations ensue with subsequent rapid diminution below the minimum effective level. The problem is magnified by recent requirements that agricultural chemicals must be biodegradable or, at least, nonpersistent in the environment. As a result, repeated applications are needed to maintain control.

The principal advantage of controlled release formulations is that they allow much less pesticide to be used more effectively for a given time interval, and they are specifically designed to counter the severe environmental processes that act to remove the pesticides applied conventionally. These processes include leaching and evaporation as well as photolytic, hydrolytic, and microbial degradation. In most instances, the rate of removal follows first order kinetics.[5-8] Thus, if M_e is the minimum effective level; M_∞, the amount of agent applied initially; and k_r, the rate constant; then t_e, the time during which an effective level of pesticide is present after a single application, would be

$$t_e = \frac{1}{k_r} \ln \frac{M_\infty}{M_e} \tag{1}$$

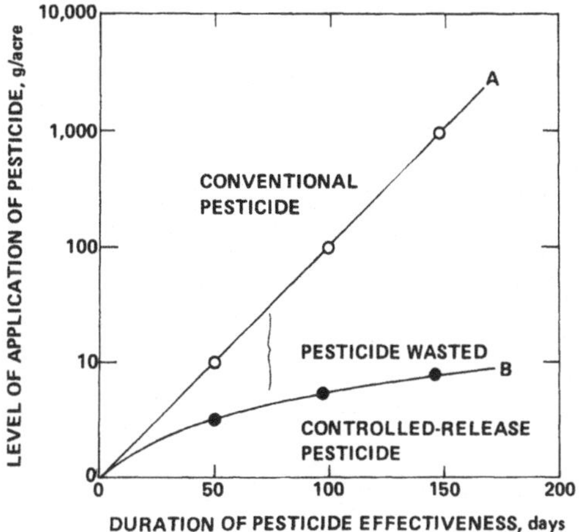

FIGURE 2. Relationships between the level of application and the duration of action for conventional and controlled release formulations.

It follows from Equation 1 that an increase in the effective duration of action of a conventionally applied pesticide would require that an exponentially greater quantity of the pesticide be applied. If, however, the pesticide could be maintained at the minimum effective level by a continuous supply from a controlled release system, then optimum performance of the insecticide would be realized, and this duration of action, t_e, would be:

$$t_e = \frac{M_\infty - M_e}{k_d M_e} \tag{2}$$

where k_d is the rate constant for pesticide delivery from the controlled release device.

Figure 2 shows the relationship between the level of application and the duration of action for conventional formulations with first-order release rate (Equation 1) and for controlled release formulations (Equation 2).[5,8] For Figure 2, the assumption is made that the half-life of the insecticide is 15 days, and the minimum effective level is 1 g/acre. The area between Curves A and B then represents the amount of pesticide that is being wasted. It is apparent that for a short duration of effectiveness, e.g., 1 week or less, the conventional method is adequately efficient. As the duration of effectiveness increases, the efficiency of the conventional system decreases exponentially as the logarithmic scale of Figure 2 indicates.

B. Advantages and Limitations

The advantages of controlled release indicated above are indeed great. However, controlled release systems can impart other important advantages to active agents that are sufficient to elevate many products to commercial successes. Table 3 lists a number of these.[9-12]

Though the advantages of controlled release are impressive, the merits of each application have to be examined individually, and the positive and negative effects weighed carefully before large expenditures for developmental work are committed. In other words, controlled release is not a panacea, and negative effects may, at times, more than offset advantages. Some of the disadvantages of controlled release or the areas that require a thorough appraisal include: (1) cost of controlled release preparation and processing, which may be substantially higher than the cost of standard for-

TABLE 3

Some Advantages of Controlled Release

Reduce mammalian toxicity of highly toxic substances
Extend duration of activity at equal level of active agent
Reduce evaporative losses and flammability of liquids
Reduce phytotoxicity
Protect pesticides from environmental degradation
Reduce leaching into the earth and transport into streams
Reduce pesticide contamination of the environment
Convert liquids to solids and flowable powders
Separate reactive components
Control the release of active agents
Mask the taste of bitter materials
Economical because less active material is needed
Convenience, including ease of handling
Avoid patient compliance problems in drug administration
Employ less total drug, thus minimize local side effects, systemic side effects, and the drug accumulation often encountered in chronic dosing
Improved drug efficiency in treatment of illness
Special effects, e.g., controlled release aspirin provides sufficient drug that on awaking, the arthritic patient has symptomatic relief despite no drug intake during the night

mulations; (2) fate of the polymer matrix and its effect on the environment; (3) fate of polymer additives, such as plasticizers, stabilizers, antioxidants, fillers, etc.; (4) environmental impact of the polymer degradation products following heat, hydrolysis, oxidation, solar radiation and biological degradation; (5) cost, time, and probability of success in securing government registration of the product, if this is required.

II. POLYMER SYSTEMS FOR CONTROLLED RELEASE

Once an application has been thoroughly investigated and found suitable, one must select (1) the controlled release technology that best fits the application, and (2) the basic physical form of device and the rate-controlling polymer matrix and active agent to be used.

A. Brief Description of Controlled Release Technologies

Table 4 categorizes the various controlled release technologies, including physical as well as chemical systems. A brief description of the technologies by category follows.

1. Reservoir Systems with Rate-Controlling Membrane

These include microcapsules and macrocapsules. Microencapsulation is a procedure that reproducibly applies a uniformly thin polymeric coating around small solid particles, droplets of liquid, or dispersions of solids in liquids, the size of the resulting capsules ranging dimensionally from a few tenths of a micron to several thousand microns.[9] Capsules greater than 2000 to 3000 μm are called macrocapsules. There are no real differences between micro- and macrocapsules as far as release characteristics or type of active agent that can be encapsulated are concerned. Many methods for microencapsulating active agents have been developed. These methods and the chapters describing them follow.

1. Phase separation methods
- Aqueous phase separation (complex coacervation), Volume II, Chapters 4 and 5
- Organic phase separation, Volume II, Chapter 13
- Meltable dispersion, Volume II, Chapter 13

TABLE 4

Categorization of Polymeric Systems for Controlled Release

- I. Physical Systems
  - A. Reservoir systems with rate-controlling membrane
    - 1. Microencapsulation
    - 2. Macroencapsulation
    - 3. Membrane systems
  - B. Reservoir systems without rate-controlling membrane
    - 1. Hollow fibers
    - 2. Poroplastic® and Sustrelle® Ultramicroporous Cellulose Triacetate
    - 3. Porous polymeric substrates and foams
  - C. Monolithic systems
    - 1. Physically dissolved in nonporous, polymeric, or elastomeric matrix
      - a. Nonerodible
      - b. Erodible
      - c. Environmental agent ingression
      - d. Degradable
    - 2. Physically dispersed in nonporous, polymeric, or elastomeric matrix
      - a. Nonerodible
      - b. Erodible
      - c. Environmental agent ingression
      - d. Degradable
  - D. Laminated structures
    - 1. Reservoir layer chemically similar to outer control layers
    - 2. Reservoir layer chemically dissimilar to outer control layers
  - E. Other physical methods
    - 1. Osmotic pumps
    - 2. Adsorption onto ion-exchange resins
- II. Chemical Systems
  - A. Chemical erosion of polymer matrix
    - 1. Heterogeneous
    - 2. Homogeneous
  - B. Biological erosion of polymer matrix
    - 1. Heterogeneous
    - 2. Homogeneous

- Fluidized-bed spray drying, Volume II, Chapter 7
- Spray drying, Volume II, Chapter 13
- Pan coating, Volume II, Chapter 13

2. Interfacial reactions

- Interfacial condensation polymerization, Volume II, Chapter 6
- *In situ* interfacial condensation polymerization, Volume II, Chapter 13
- Interfacial addition polymerization, Volume II, Chapter 13

3. Physical methods

- Multiorifice centrifugal, Volume II, Chapter 8
- Electrostatic, Volume II, Chapter 13

Although many of the microencapsulated products release their contents by permeation through the walls, other mechanisms are also very important, such as breakdown or dissolution of the walls and osmotic bursting.

2. Reservoir Systems without Rate-Controlling Membrane

These systems include hollow fibers (Volume II, Chapter 9), impregnation in porous plastics such as MPS porous PVC sheet, Millipore® filters, and Cellgard® porous polypropylene (Chapter 4), foams, and possibly hydrogels (Chapter 4 and Volume II, Chapter 9) and ultramicroporous cellulose triacetate (Chapter 6).

The simplest example is perhaps the hollow fibers which hold the active agent in their bore and release it by diffusion through the air layer above the agent. Systems utilizing impregnated porous plastics (PVC and Cellgard®, etc.) are more complex, but in all cases, the active agent is retained by capillary action, physically imbedded in the pores. Release also occurs by diffusion through the air layer above the liquid that fills the pores. Strictly speaking, most of these systems may be considered monolithic matrix systems, except that interaction of active agent and polymer is minimal.

3. Monolithic Systems

Probably the simplest and least expensive way to control the release of an active agent is to disperse it in an inert polymeric matrix. In polymeric systems, the active agent is physically blended with the polymer powder and then fused together by compression molding, injection molding, screw extrusion, calendering, or casting,[13] all of which are common processes in the plastics industry. This system of controlled release is discussed in Chapter 2.

Similarly, the active agent can be blended with elastomeric materials in the mixing step like any of the other additives, e.g., accelerators, reinforcing pigments, stabilizers, and processing aids.[14] This system of controlled release is discussed in Chapter 3.

In both of the above cases, the active agent dissolves in the polymeric or elastomeric matrix until saturation is reached. Additional active agent, if any, remains dispersed within the polymer matrix, and the system is physically dispersed. As the agent is removed from the surface of the monolithic device, more of the agent diffuses out from the interior to the surface in response to the decreased concentration gradient leading to the surface.

If the polymer used is soluble or degrades during its intended use, the monolithic device is erodible. In this case, the release of the active agent is by a combination of diffusion and liberation due to erosion.

A further clarification should be made here for pure degradable and erodible systems. Degradable systems can be defined as systems that release their contents by diffusion, osmotic bursting, leaching, and any other CR mechanism, but in addition have the very desirable property of chemically or biologically degrading after the useful life of the device has expired. Contamination of the environment into which a degradable device is placed is thus minimal.

Erodible systems can be considered a subdivision of degradable systems. They release their contents only by the liberation of the active agent due to the chemical or biological erosion of the matrix. Erodible and degradable systems are discussed in Volume II, Chapters 1 and 2.

If the polymer is plasticizable or swellable by an environmental agent such as water, the system involves ingression of an environmental agent into the device and plasticizes the polymeric matrix, thereby allowing physically bound active ingredient to diffuse out. This system which differs from the erodible one in that the matrix remains physically intact, includes the starch xanthate systems (Volume II, Chapter 11), hydrogels (Chapter 4 and Volume II, Chapter 13), and perhaps some of the lignin-modified controlled release systems (Volume II, Chapter 12).

4. Laminated Structures

In this system, at least two and usually three polymeric films are adhered or laminated together. The center layer of a three-layer laminate is the reservoir layer. It contains a large amount of the active agent and may be made of porous or nonporous polymeric material. The outer layers control the rate of release of the agent and are usually fabricated from a more rigid polymer than that of the reservoir. Should the reservoir layer and the outer films both be made of the same polymer, it is apparent that the system reverts to a monolithic matrix system. A discussion of its ramifications, as they relate to pharmaceutical and pesticidal systems, is presented in Chapters 4 and 5.

5. Other Physical Methods

Osmotic pumps comprise another system that is both novel and useful. In its simplest and most elegant design, the pump consists of a tablet containing the active agent and an osmotic "attractant", such as NaCl, surrounded by a semipermeable membrane with a small orifice. When the pump is placed in an aqueous environment, the osmotic pressure of the NaCl draws water to the device through the semipermeable membrane. Because the membrane coating is nonextensible, the NaCl-saturated solution (and active agent) inside the device is pumped out through the orifice as water is osmotically imbibed. This system is discussed in Volume II, Chapter 10.

Adsorption of active agents onto ion-exchange resins has been tried as a CR mechanism.[15] Thus, the adsorptive forces of these resins can decrease the release of an ionic species through an equilibrium favoring the resin's adsorption sites, but renewal of the medium can result in very fast release. In pharmaceutical applications, the release rate can depend upon pH and electrolyte concentration; e.g., in the GI tract, release can be higher in the stomach and lower during transit through the small intestine owing to pH differences.

6. Retrograde Chemical Reaction Systems

Controlled release of active agents may be achieved by a variety of chemical methods as well. In one approach, the active agent is bonded via covalent or ionic bonds to a preformed polymer, in accordance with the equation:

$$\text{Preformed polymer} + \text{Pesticide} \underset{\text{Environment}}{\overset{\text{Synthesis}}{\rightleftarrows}} \text{Polymeric pesticide}$$

This approach requires macromolecules with pendant functional groups capable of reacting with the active ingredients or their derivatives. The nature of the chemical bond may be varied to yield bonds with different rates of cleavage. The most common linkages are esters, anhydrides, and acetals.

Another approach involves the polymerization of difunctional monomeric pesticides. In these materials, the pesticide is incorporated directly in the polymer backbone. The pesticide is released by the chemical or biological depolymerization of the system in accordance with the equation:

$$\text{Monomer} + \text{Pesticide} \underset{\text{Environment}}{\overset{\text{Synthesis}}{\rightleftarrows}} \text{Polymeric pesticide}$$

The depolymerization reaction can be purely homogeneous, purely heterogeneous (i.e., surface reactions), or some combination of the two. A heteogeneous reaction might occur, for example, because the reaction products cannot leave the reaction site in the bulk of the polymer, locally reversing the reaction, or because one degradant (e.g., a microorganism) is excluded from the interior of the polymer. Volume II, Chapters 2 and 3 describe in detail the homogeneous and heterogeneous chemical reactions used in such controlled release systems.

B. Basic Components of Controlled Release Devices

The components of controlled release include (1) the active agent and (2) the polymer matrix or matrices that regulate release of the active agent.

1. Polymer Matrix

Advances in controlled release technology have been rapid because polymer science has become sophisticated enough to incorporate into polymers tailor-made properties for each controlled release application. The importance of polymer selection will be appreciated more if one considers the different design criteria that must be fulfilled:

1. Molecular weight, glass transition temperature, and chemical functionality of the polymer must allow the proper diffusion and release of the specific active agent.
2. The functionality of the polymer should be such that it will not chemically react with the active agent.
3. The polymer and its degradation products must be nontoxic to the environment, and, in medical applications, nontoxic or antagonistic to the host.
4. The polymer must not decompose in storage and generally not during the useful life of the device.
5. The polymer must be easily manufactured and fabricated into the desired product. It should allow incorporation of large amounts of active agent without excessively deteriorating its mechanical properties.
6. Finally, cost of the polymer should not be excessive and thereby cause the controlled release device to be noncompetitive.

A list of polymers that have been used in controlled release formulations is shown in Table 5.[16-21] These polymers are used as coatings in microencapsulation, films in laminated structures, slabs in monolithic systems, and flakes in many erodible devices.

2. Active Agents

Controlled release technology has been considered for a wide variety of applications. Broad product areas in which controlled release applications have been made are shown in Table 6. As already noted, the major effort in applying controlled release principles has been in the administration of pharmaceuticals and the application of pest control chemicals.[20]

Tables 7 and 8 present pharmaceutical, agricultural, and veterinary agents that have been used in experimental and commercialized controlled release devices.[19,20]

3. Physical Form of Controlled Release Devices

Controlled release devices have taken many and varied forms. They can be as small as a few microns, as is the case with some of the microcapsules, or as big as the standard pharmaceutical capsules or tablets. They can be in the form of spherical granules, strips, bandages, or 60-in-wide, 200-yard-long rolls, as is the case with the STAPH-CHEK® antibacterial antifungal laminated fabrics used for pillow covers and mattress ticking in hospitals, prisons, and nursing homes. Figure 3 illustrates some of the forms that controlled release products can take.[22]

III. RELEASE CHARACTERISTICS OF CONTROLLED RELEASE SYSTEMS

Of the different technologies listed in Table 4, all of the physical processes are in one way or another controlled by the diffusion of the active agent through a polymer barrier or by an inward diffusion of an environmental fluid in the case of "environmental agent ingression" devices and some homogeneous "retrograde chemical reaction" devices.

Possibly the only systems that are not controlled by diffusion are the "pure" erodible devices and some heterogeneous "retrograde chemical reaction" devices, but even in these systems, release is generally achieved through a combination of diffusion and erosion.

TABLE 5

Polymers Used in Controlled Release Devices

Natural polymers

Carboxymethylcellulose	Zein
Cellulose acetate phthalate	Nitrocellulose
Ethylcellulose	Propylhydroxycellulose
Gelatin	Shellac
Gum arabic	Succinylated gelatin
Starch	Waxes—paraffin
Bark	Proteins
Methylcellulose	Kraft lignin
Arabinogalactan	Natural rubber

Synthetic elastomers

Polybutadiene	Hydrin rubber
Polyisoprene	Chloroprene
Neoprene	Butyl rubber
Polysiloxane	Nitrile
Styrene-Butadiene rubber	Acrylonitrile
Silicone rubber	Ethylene-propylene-diene terpolymer

Synthetic polymers

Polyvinyl alcohol	Polyvinyl chloride
Polyethylene	Polyacrylate
Polypropylene	Polyacrylonitrile
Polystyrene	Chlorinated polyethylene
Polyacrylamide	Acetal copolymer
Polyether	Polyurethane
Polyester	Polyvinylpyrrolidone
Polyamide	Poly(*p*-xylylene)
Polyurea	Polymethylmethacrylate
Epoxy	Polyvinyl acetate
Ethylene vinyl acetate copolymer	Polyhydroxyethyl methacrylate
Polyvinylidene chloride	

TABLE 6

Broad Product Areas in Which Controlled Release has been Applied

Adhesives	Fuels	Perfumes
Antifoulants	Growth regulators	Photographic agents
Bacteria	Herbicides	Pigments
Blowing agents	Inks	Plasticizers
Catalysts	Insecticides	Propellants
Curing agents		Solvents
Detergents	Metals	Stabilizers
Drugs	Monomers	Viruses
Dyes	Oils	Vitamins
Flavors	Paints	
Foods		

The active agent passes through the polymeric barrier in the absence of pores or holes by a process of absorption, solution, and diffusion down a gradient of thermo-

TABLE 7

Pharmaceutical Active Agents Utilizing Controlled Release

Analgesics	Antitussives	Nutritional products
Anthelmintics	Cathartics	Potassium supplements
Antidotes	Diagnostic aids	Sedatives
Antiemetics	Diuretics	Sulfonamides
Antihistamines	Effervescents	Stimulants
Antimalarials	Enzymes	Sympathomimetics
Antimicrobials	Expectorants	Tranquilizers
Antipyretics	Hypnotics	Urinary Antiinfectives
Antiseptics	Microorganisms	Vitamins
Antituberculotics	Minerals	Xanthine derivatives

TABLE 8

Agricultural and Veterinary Active Agents Utilizing Controlled Release

Algicides	Fungicides	Nematicides
Analgesics	Germicides	Nutrients
Anthelmintics	Growth regulators	Repellents
Antimicrobials	Herbicides	Pheromones
Bactericides	Insecticides	Viruses
Chemosterilants	Insect diets	Vitamins
Disinfectants	Juvenile hormones	
Fumigants	Minerals	

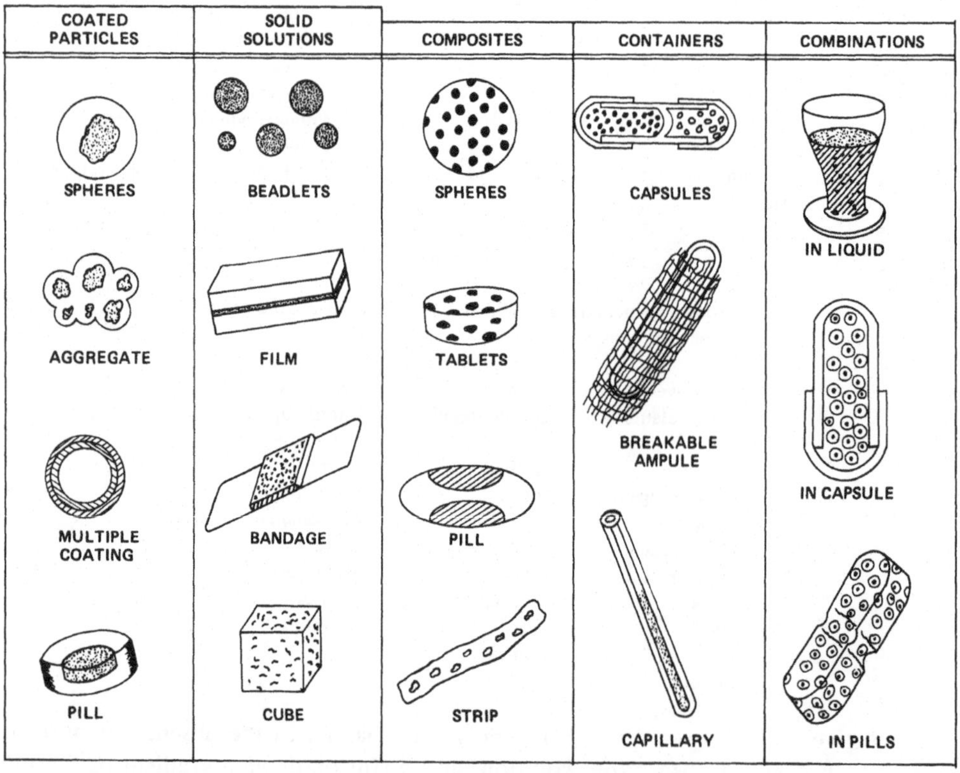

COATED PARTICLES	SOLID SOLUTIONS	COMPOSITES	CONTAINERS	COMBINATIONS
SPHERES	BEADLETS	SPHERES	CAPSULES	IN LIQUID
AGGREGATE	FILM	TABLETS	BREAKABLE AMPULE	IN CAPSULE
MULTIPLE COATING	BANDAGE	PILL	CAPILLARY	IN PILLS
PILL	CUBE	STRIP		

FIGURE 3. Forms of controlled release products.

dynamic activity until desorbed or removed. The transport of the active agent is governed by Fick's first law:

$$J = \frac{dM_t}{A\,dt} = \frac{-D\,dC_m}{dx} \qquad (3)$$

where J is the flux in g/cm²/sec, C_m is the concentration of active agent in the polymeric membrane in g/cm³, dC_m/d_x is the concentration gradient, D is the diffusion coefficient of the active agent in the polymeric membrane in cm²/sec, A is the surface area through which diffusion takes place in cm², M_t is the mass of agent released, and dM_t/dt is the steady-state release rate at time t.

Equation 3 can be integrated under the proper boundary conditions for each of the systems listed in Table 4 to obtain an equation giving the amount of agent released as a function of time. In many situations, however, the mathematics become rather complicated, and no explicit equations can be derived. The mathematics of diffusion have been discussed elsewhere and they will be reviewed by the authors in their respective chapters.[23-25] However, release characteristics encountered with the most common controlled release systems are mathematically described below.

A. Reservoir Systems with Rate-Controlling Membrane

When applied to these systems, Fick's law predicts that a steady state will be established with the release rate being constant and independent of time if an active agent is enclosed within an inert polymer membrane and concentration of the agent is maintained constant within the enclosure. The amount of active agent released per day is therefore constant for the life of the device, and

$$M_t = kt \qquad (4)$$

where k is a constant. The above applies for all geometries of the device, e.g., spheres, slabs, etc. This type of release is "zero order" and is shown as Curve I in Figure 4.

Microcapsules and macrocapsules are the major controlled release reservoir systems with rate-controlling membranes. Microcapsules are made with very thin walls, and, very often, the polymer phase is not homogeneous, but cracks and pores are observed when viewed under magnification. Therefore, in some field applications, microcapsules have been found to follow a "$\sqrt{t}$ order" release as described below.

Erodible reservoir systems—In capsules or microcapsules with erodible membranes, the mechanism of release can be the erosion and rupture of the barrier membrane. Various delivery patterns, including essentially constant release (zero order), can be achieved by blending microcapsules of appropriate wall thicknesses.

B. Reservoir Systems without Rate-Controlling Membrane

It can be shown that these devices should follow a rate of release proportional to $t^{-1/2}$. The amount of agent released is then proportional to $\sqrt{t}$ ($\sqrt{t}$ order), and given by the equation:

$$M_t = k\sqrt{t} \qquad (5)$$

This equation gives a parabolic curve as shown by Curve II in Figure 4. With this system, a large amount of the agent is released initially, and substantially smaller and decreasing amounts are released during the last half of the life of the device.

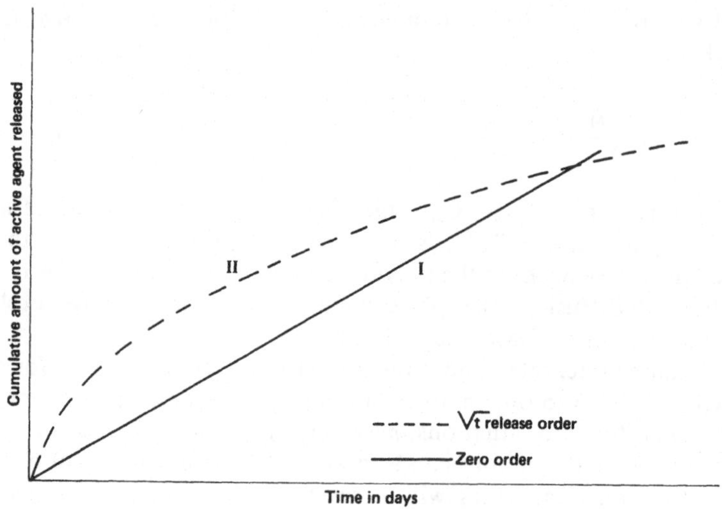

FIGURE 4. Representation of "zero order" and "√t order" release.

C. Monolithic Systems

Physically dissolved, nonerodible, polymeric or elastomeric matrix—Release rate in this system is proportional to $t^{-1/2}$, (same as Equation 5) until about 60% of the active agent is released. The release rate thereafter is related exponentially to time, i.e.,

$$\frac{dM_t}{dt} = k_1 e^{-k_2 t} \tag{6}$$

where k_1 and k_2 are constants.

Thus, the rate of release above 60% drops exponentially. This type of release, which is called "first order", is also observed in reservoir systems in which the solution of active agent within the enclosure is less than saturated.

Physically dispersed, nonerodible, polymeric or elastomeric matrix—The amount released in these systems is proportional to the $\sqrt{t}$ as long as the concentration of active agent present (dispersed and dissolved) is higher than the solubility of the agent in the matrix. Thus, the dispersed systems are similar to the dissolved systems, except that, instead of a decreased release rate after 60% of the chemical has been emitted, the relationship holds almost over the complete release curve.

Monolithic Erodible Systems—If one assumes that release of the active agent by diffusion is negligible in this system, the speed of erosion will control the release rate. Release by erosion is a surface-area-dependent phenomenon, and the release will be constant (zero order) as long as the surface area does not change during the erosion process. This is true for slab-shaped devices. Eroding cylinders and spheres give delivery rates that decrease with time (owing to decreasing surface area), even though the kinetic process providing the rate-determining step is, in fact, zero order.[26]

D. Laminated Structures

Two cases are easily discernible. First, if the distribution coefficient of the active agent between the reservoir layer and the barrier membrane is much smaller than unity, the system approximates "zero order" release (reservoir system with rate-controlling membrane), and the amount released per day is independent of time.[27] If the distribution coefficient is close to unity, the system approximates the $\sqrt{t}$ order release (monolithic, physically dispersed system).

E. Other Physical Methods—Osmotic Pumps

The mechanism of release is based on the osmotic pressure generated by the diffusion of water through the semipermeable membrane to form a saturated aqueous solution inside the device. The delivery rate of the solute by the pump is constant (zero order release) as long as excess solid is present inside the pump to form a saturated solution. The rate declines exponentially (first order) as soon as the solution inside the device drops below saturation.

F. Retrograde Chemical Reaction Systems

The heterogeneous systems, i.e., surface reactions, are similar in nature to the erodible matrix systems, and "zero order" release is obtained for slabs. Owing to geometry, i.e., changing of surface area as reaction proceeds, spherical or cylindrical systems have nonlinear release characteristics with time.

For homogeneous systems, the picture is much more complicated because reaction as well as diffusion play important roles. Cowsar indicates that for water-soluble systems,[28] the rate of release of pendant pesticide groups follows conventional first order kinetics, and systems based on depolymerization reactions may indeed be zero order, if the mechanism of release comprises unzipping of polymer chains.

IV. SOME COMMERCIAL APPLICATIONS

Though controlled release is still a young science, much has been accomplished as indicated by the number of controlled release products that have become commercially available in the last few years. In Table 9, data on a dozen or so products representing ten different technologies are shown. Some of the products have release characteristics approaching "zero order", such as those with some membranes and microcapsules. Others, such as the monolithic systems, laminated structures, and hollow fibers, have release characteristics roughly proportional to $\sqrt{t}$ or intermediate between zero order and $\sqrt{t}$. Perhaps it should be mentioned here that zero order is not necessarily better than the $\sqrt{t}$ order release. With many products, an initial burst of active agent followed by slower release for replenishment of degraded or otherwise consumed active agent is desirable. It turns out that systems with $\sqrt{t}$ release are generally less costly to manufacture than zero order systems.

In the commercial controlled release products shown in Table 9, as well as in those described elsewhere,[29] the scientists involved have undoubtedly selected and formulated, for their particular end use, the best system they could devise, including economic considerations as part of their deliberations.

TABLE 9

Some Commercial Controlled Release Products

Name	Active agent	Polymer	Form of device	CR technology	CR improvement	Company
Brocon® CR	Nasal Decongestant	Hydroxypropylmethylcellulose	Tablet	Erodible matrix	Prolonged relief	Forest Laboratories
Tox-Hid®	Rodenticide (Warfarin)	—	Erodible capsules	Air suspension Wurster process	Taste masking	WARF Institute
Ocusert®	Pilocarpine for glaucoma control	Ethylene vinyl acetate copolymer	Thin, elliptical device	Laminated structure	Convenience	Alza Corporation
No-Pest® Strip	Insecticide (DDVP) for fly control	Polyvinyl chloride	Rectangular slabs	Monolithic plastic matrix	Continuous control	Shell Chemical Company
Penncap M®	Insecticide (agricultural application)	Polyamide	Microcapsules	Interfacial polymerization	Reduces toxicity and prevents premature degradation	Pennwalt Corporation
Conrel® Microfibers	Pink bollworm pheromone (gossyplure)	Polyamide	Fiber	Hollow fiber	Increases duration of effectiveness	Conrel (Albany International Corp.)
Hercon Insectape®	Insecticide (Baygon®) for Roach control	Polyvinyl chloride and polyester	1″ × 4″ strip	Laminated structure	Increases duration of effectiveness (up to 6 months)	Herculite Products, Inc. (Health-Chem Corp.)
Incracide® E-51	Molluscicide (copper sulfate)	Rubber	Granules	Monolithic plastic matrix	Increases duration of effectiveness	International Copper Research Association
Dursban® 10 CR	Larvicide for mosquito control	Chlorinated polyethylene	Cubes (pellets)	Monolithic polymeric matrix	Minimizes toxicity to nontarget species	Dow Chemical
No Foul®	Marine antifoulant	Neoprene rubber	Sheets	Monolithic elastomeric matrix	Increases duration of effectiveness	B. F. Goodrich
Precise®	Fertilizer (plant food) for home and garden use	—	Microcapsules	—	Increases duration of effectiveness (3 to 6 months)	Minnesota Mining & Mfg. Co. (3M)
Altosid® SR-10	Juvenile hormone (methoprene)	Polyamide	Microcapsules	—	Prevents premature degradation	Zoecon Corporation
Equi-Palazone®	Anti-inflammatory agent	—	Microcapsules	Phase separation	Taste masking	Arnold's Veterinary Products
Cap-Cyc®	Plant growth regulator (chlormequat)	—	Microcapsules	—	Increases duration of effectiveness	Minnesota Mining & Mfg. Co. (3M)

Zodiac®	Insecticide (Baygon®) for flea and tick control	Polyvinyl chloride	Collar	Monolithic plastic matrix	Increases duration of effectiveness	Zoecon Corporation
Progestasert®	Intrauterine contraceptive (progesterone hormone)	Ethylene vinyl acetate copolymer	T-shaped device	Membrane	Eliminates side effects	Alza Corporation
Biomet® SRM	Molluscicide (Organotin)	Natural rubber	Pellets	Monolithic elastomeric matrix	Increases duration of effectiveness	M&T Chemical Co.
Ecopro™ 1700	Insecticide (Temephos)	Ethylene propylene copolymer	Granules	Monolithic plastic matrix	Increases duration of effectiveness	Environmental Chemicals, Inc.
14ACE-B	Herbicide (2,4-D)	Natural rubber	Pellets	Monolithic elastomeric matrix	Increases duration of effectiveness	Creative Biology Lab., Inc.

REFERENCES

1. **Fanger, G. O.**, General background and history of controlled release, in *Proceedings of Controlled Release Pesticide Symposium*, Cardarelli, N. F., Ed., University of Akron, Ohio, 1974, 1.3.
2. **Chardrasekaran, S. K.**, Theory of controlled delivery systems, in *Proceedings of Controlled Release Pesticide Symposium*, Goulding, R. L., Ed., Oregon State University, Corvallis, 1977, 382.
3. **Robinson, J. R.**, Controlled release pharmaceutical systems, in *Chemical Marketing and Economics Reprints*, Long, F. W., O'Neill, W. P., and Stewart, R. D., Eds., American Chemical Society, San Francisco, 1976, 212.
4. **Cowsar, D. R.**, Introduction to controlled release, in *Controlled Release of Biologically Active Agents*, Vol. 47, Tanquary, A. C. and Lacey, R. E., Eds., Plenum Press, New York, 1974, 1.
5. **Lewis, D. H. and Cowsar, D. R.**, Principles of controlled release pesticides, in *Controlled Release Pesticides*, Scher, H. B., Ed., ACS Symposium Series 53, American Chemical Society, Washington, D.C., 1977, 1.
6. **Fanger, G. O.**, General background and history of controlled release, in *Proceedings of Controlled Release Pesticide Symposium*, Cardarelli, N. F., Ed., University of Akron, Ohio, 1974, 1.18.
7. **Baker, R. W., and Lonsdale, H. K.**, Principles of controlled release, in *Proceedings of Controlled Release Pesticide Symposium*, Harris, F. W., Ed., Wright State University, Dayton, Ohio, 1975, 10.
8. **Neogi, A. M., and Allan, G. G.**, Controlled-release pesticides: concepts and realization, in *Advances in Experimental Medicine and Biology*, Vol. 47, Tanquary, A. C. and Lacey, R. E., Eds., Plenum Press, New York, 1974, 195.
9. **Bakan, J. A.**, Microencapsulation of pesticides and other agricultural materials, in *Proceedings of Controlled Release Pesticide Symposium*, Harris, F.W., Ed., Wright State University, Dayton, Ohio, 1975, 76.
10. **Fanger, G. O.**, General background and history of controlled release, in *Proceedings of Controlled Release Pesticide Symposium*, Cardarelli, N. F., Ed., University of Akron, Ohio, 1974, 1.18.
11. **Scher, H. B.**, Microencapsulated pesticides, in *Controlled Release Pesticides*, Scher, H. B., Ed., ACS Symposium Series 53, American Chemical Society, Washington, D.C., 1977, 126.
12. **Robinson, J. R.**, Controlled-release pharmaceutical systems, in *Chemical Marketing and Economics Reprints*, Long, F. W., O'Neill, W. P., and Stewart, R. D., Eds., American Chemical Society, San Francisco, 1976, 212.
13. **Harris, F. W.**, Preparation of plastic, controlled release, pesticide formulations, in *Proceedings of Controlled Release Pesticide Symposium*, Cardarelli, N. F., Ed., University of Akron, Ohio, 1976, 1.33.
14. **Cardarelli, N. F.**, Compounding methods for controlled release elastomers, in *Proceedings of Controlled Release Pesticide Symposium*, Cardarelli, N. F., Ed., University of Akron, Ohio, 1976, 1.44.
15. **Ritschel, W. A.**, Peroral solid dosage forms with prolonged action, in *Drug Design*, Vol. 4, Ariens, E. J., Ed., Academic Press, New York, 37.
16. **Paul, D. R.**, Polymers in controlled release technology, in *Controlled Release Polymeric Formulations*, Paul, D. R. and Harris, F. W., Eds., ACS Symposium Series 33, American Chemical Society, Washington, D.C., 1976, 9.
17. **Scher, H. B.**, Microencapsulated pesticides, in *Controlled Release Pesticides*, Scher, H. B., Ed., ACS Symposium Series 53, American Chemical Society, Washington, D.C., 1977, 127.
18. **Cardarelli, N. F. and Kanakkanatt, S. V.**, Matrix factors affecting the controlled release of pesticides from elastomers, in *Controlled Release Pesticides*, Scher, H. B., Ed., ACS Symposium Series 53, American Chemical Society, Washington, D.C., 1977, 64.
19. **Zweig, G.**, Environmental aspects of controlled release pesticide formulations, in *Controlled Release Pesticides*, Scher, H. B., Ed., ACS Symposium Series 53, American Chemical Society, Washington, D.C., 1977, 45.
20. **Bakan, J. A.**, Microcapsule drug delivery systems, in *Polymers in Medicine and Surgery*, Kronenthal, R. L., Oser, Z., and Martin, E., Eds., Plenum Press, New York, 1975, 213.
21. **Neogi, A. N. and Allan, G. G.**, Controlled-release pesticides: concepts and realization, in *Controlled Release of Biologically Active Agents*, Vol. 47, Tanquary, A. C. and Lacey, R. E., Eds., Plenum Press, New York, 1974, 210.
22. **Fanger, G. O.**, General background and history of controlled release, in *Proceedings of Controlled Release Pesticide Symposium*, Cardarelli, N. F., Ed., University of Akron, Ohio, 1974, 1.7.
23. **Baker, R. W. and Lonsdale, H. K.**, Membrane-controlled delivery systems, in *Proceedings of Controlled Release Pesticide Symposium*, Cardarelli, N. F., Ed., University of Akron, Ohio, 1974, 40.1.
24. **Crank, J. and Park, G. S., Eds.**, *Diffusion in Polymers*, Academic Press, London, 1968.
25. **Crank, J.**, *The Mathematics of Diffusion*, Oxford University Press, London, 1956.
26. **Hopfenberg, H. B.**, Controlled release from erodible slabs, cylinders, and spheres, in *Controlled Release Polymeric Formulations*, Paul, D. R. and Harris, F. W., Eds., ACS Symposium Series 33, American Chemical Society, Washington, D.C., 1976, 26.

27. **Kydonieus, A. F.,** The effect of some variables on the controlled release of chemicals from polymeric membranes, in *Controlled Release Pesticides,* Scher, H. B., Ed., ACS Symposium Series 53, American Chemical Society, Washington, D.C., 1977, 152.
28. **Lewis, D. H., and Cowsar, D. R.,** Principles of controlled release pesticides, in *Controlled Release Pesticides,* Scher, H. B., Ed., ACS Symposium Series 53, American Chemical Society, Washington, D.C., 1977, 11.
29. **Cardarelli, N. F.,** *Controlled Release Pesticide Formulations,* CRC Press, Cleveland, 1976.

Chapter 2

MONOLITHIC POLYMER DEVICES

Theodore J. Roseman and Nate F. Cardarelli

PRELUDE

Monolithic devices provide one method to deliver substances at controlled rates of release. The solute is incorporated into a polymeric matrix, and migration to the surface occurs either by diffusion through pores within the matrix structure or by diffusion through the polymeric phase itself. Devices of this type provide release patterns which are characteristically different from other types of controlled release delivery systems.

This chapter demonstrates the application of monolithic devices to a wide spectrum of solutes. Release mechanisms are reviewed with special emphasis on the use of the physical model approach to describe the kinetics of solute transport. Tractable mathematical expressions are presented which allow for the *a priori* design and evaluation of monolithic systems. For convenience, the chapter is divided into two separate sections. The first section focuses on the mechanisms of solute release, factors influencing the release process, and pharmaceutical applications. The second part deals with plastic monolithic dispensers used in the agricultural and pesticide fields. The special case of controlled release from monolithic elastomers is not included in this chapter, but is explicitly described in Chapter 3.

SECTION 1

Theodore J. Roseman

TABLE OF CONTENTS

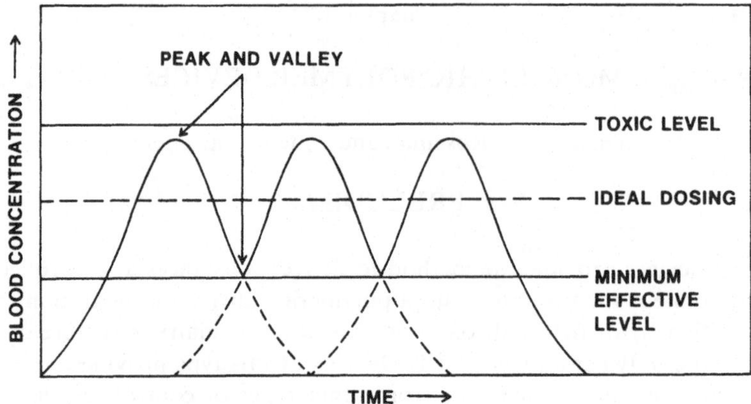

FIGURE 1. Hypothetical blood level pattern from a conventional multiple dosing schedule, and the idealized pattern from a controlled release system.

I. INTRODUCTION

The recognition that drugs can be released from polymeric devices has resulted in a proliferation of research in the area of controlled delivery of bioactive agents.[1,2] The ability of polymer matrices to meter drugs at controlled and reproducible rates for extended time periods provides significant benefits over conventional methods of drug delivery. Ideal drug administration requires that a constant level of the medicament be maintained throughout the course of therapy. This is rarely achieved in practice due to peak and valley blood concentrations which result from multiple dosing regimens. A schematic comparison of the two types of drug delivery is illustrated in Figure 1.

The advantages of controlled release devices are (1) reduced side effects due to the optimization of the blood concentration-time profile, (2) greater patient compliance due to the elimination of multiple dosing schedules, (3) reversibility of drug delivery which would allow for removal of the drug source if needed, and 4) self-administration. The route of administration will dictate the degree to which these advantages apply. Implantable devices, for examples, are not self-administered, and oral systems certainly are not reversible. However, delivery of drugs to accessible areas of the body, e.g., skin, vagina, rectum, eye, etc., represent excellent routes to totally capitalize on the concept of controlled release devices.

There are several designs of controlled release dosage forms, each of which provides a distinctive release profile and requires unique methodology for manufacture. Among the arsenal of delivery modules discussed in this book are reservoir, osmotic, erodible, laminated, and monolithic devices, as well as microencapsulated systems. The selection of a particular device depends upon a number of factors, such as the required release rate and duration, the lipophilic character of the drug and its chemical stability, geometry, and the route of administration. This chapter is devoted to a description of monolithic devices of pharmaceutical interest with special emphasis on the kinetics of drug release and the factors that influence it. A monolithic device is defined as a solid matrix that contains either (1) dissolved solute or (2) dispersed solute in equilibrium with dissolved solute. The solute is released by diffusion through the continuum of the matrix or pores within its structure. The term monolith[3] is used to distinguish it from other types of devices.

II. THEORETICAL PRINCIPLES

The derivation of mathematical expressions which describe the release of solutes from matrix systems are based upon Fick's laws of diffusion.[4-7] Fick's first law states that the flux (J) or rate of solute transfer across a plane of unit area is

$$J\,(\text{Rate}) = -D\frac{dC}{dx} \tag{1}$$

where dC/dx is the concentration gradient or the change in concentration (C) with respect to distance (x), and D is the diffusion coefficient. Fick's second law is given by

$$\frac{\partial C}{\partial t} = D\frac{\partial^2 C}{\partial x^2} \tag{2}$$

where the rate of change of concentration with time (t) at a particular level is proportional to the rate of change of the concentration gradient at that level.[7] Although , in actuality the driving force for diffusion is the difference in thermodynamic activity of the solute, it is assumed that solutions are sufficiently dilute so that the concentration approximates the activity.

A. Homogeneous Monolith (Dissolved Solute)

A solution to Equation 2 for a monolith of finite dimensions (slab geometry) containing only dissolved solute at an initial concentration of C_o is[6]

$$C = \frac{4C_o}{\pi}\sum_{m=o}^{\infty}\frac{1}{2m+1}\sin\frac{(2m+1)\pi x}{h} \exp\left[\frac{-D_s(2m+1)^2\pi^2 t}{h^2}\right] \tag{3}$$

where h is the thickness, the subscript s refers to the polymer phase, and the concentration on the desorbing side is zero. Concentration-distance profiles are illustrated in Figure 2 for various times. The fraction of solute released (F) as a function of time is found by differentiating C with respect to x, substituting into Equation 1, and integrating from zero time to time, t:

$$F = \frac{Q_t}{Q_\infty} = \left[1 - \frac{8}{\pi^2}\sum_{m=o}^{\infty}\frac{1}{(2m+1)^2} \exp\left\{\frac{-D_s(2m+1)^2\pi^2 t}{h^2}\right\}\right] \tag{4}$$

where Q_t is the amount released per unit area at time t, and Q_∞ is the amount desorbed per unit area at infinite time. Alternatively, the fraction desorbed is given below.[3,8]

$$F = \frac{Q_t}{Q_\infty} = 4\left(\frac{D_s t}{h^2}\right)^{1/2}\left[\pi^{-1/2} + 2\sum_{n=1}^{\infty}(-1)^n \text{ierfc}\frac{nh}{2(D_s t)^{1/2}}\right] \tag{5}$$

A form of Equation 4 was deduced by Higuchi[9] to be applicable to the release of drugs from ointment vehicles.[10]

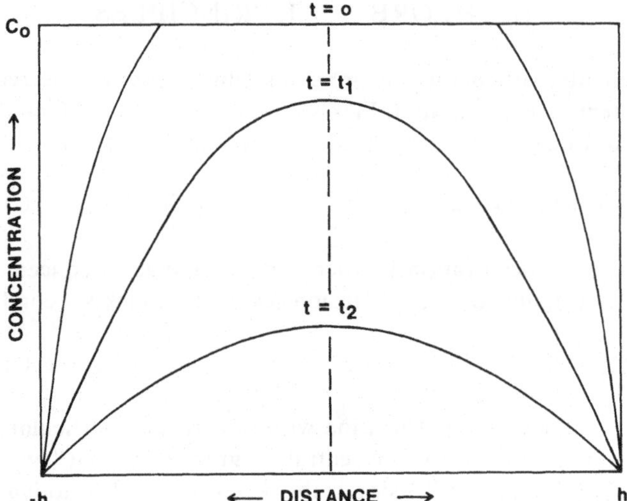

FIGURE 2. Concentration distance profiles in a monolithic slab containing dissolved solute. Release occurs from two faces, and the concentration at the boundaries is maintained at zero. The curves are drawn for three times such that $t_2 > t_1$.[6]

TABLE 1

Equations Describing the Release Kinetics of a Solute Dissolved in a Monolithic Device

	Early time approximation		Late time approximation	
Device	Fraction released[a]	Release rate[b]	Fraction released[a]	Release rate[b]
Slab[c]	$4\left(\frac{D_st}{\pi h^2}\right)^{1/2}$	$2\left(\frac{D_s}{\pi h^2 t}\right)^{1/2}$	$1-\frac{8}{\pi^2}\exp\left(\frac{-\pi^2 D_st}{h^2}\right)$	$\frac{8D_s}{h^2}\exp\left(\frac{-\pi^2 D_st}{h^2}\right)$
Cylinder[d,e]	$4\left(\frac{D_st}{r^2\pi}\right)^{1/2}-\frac{D_st}{r^2}$	$2\left(\frac{D_s}{r^2\pi t}\right)^{1/2}-\frac{D_s}{r^2}$	$1-\frac{4}{(2.405)^2}\exp\left[\frac{-(2.405)^2 D_st}{r^2}\right]$	$\frac{4D_s}{r^2}\exp\left[\frac{-(2.405)^2 D_st}{r^2}\right]$
Sphere[d,e]	$6\left(\frac{D_st}{r^2\pi}\right)^{1/2}-\frac{3D_st}{r^2}$	$3\left(\frac{D_s}{r^2\pi t}\right)^{1/2}-\frac{3D_s}{r^2}$	$1-\frac{6}{\pi^2}\exp\left(\frac{-\pi^2 D_st}{r^2}\right)$	$\frac{6D_s}{r^2}\exp\left(\frac{-\pi^2 D_st}{r^2}\right)$

[a] Fraction released (F) equals Q_t/Q_∞.

[b] Release rate equals $d(Q_t/Q_\infty)$.

[c] Short time approximation is valid for up to 60% release while long time approximation is valid when 40% or more release has occured.

[d] Short time approximation is valid for up to 40% release while long time approximation is valid for greater than 60% release.

[e] In the case of the cylinder and sphere, surface area changes with time. To obtain the cumulative amount released (Q'_t) or the release rate (dQ'_t/dt), allow Q_∞ to be the total amount released at infinite time.

From Baker, R. W. and Lonsdale, H. K., Controlled release: mechanisms and rates, in *Advances in Experimental Medicine and Biology*, Vol. 47, Tanquary, A. C. and Lacey, R. E., Eds., Plenum Press, New York, 1974, 15.

Equations 4 and 5 are not conveniently handled, and two simple approximations can be utilized. These are summarized in Table 1 (from Baker and Lonsdale[3]) for both the fraction of solute released and the release rate as a function of time. Consideration is given to three geometries of general interest, i.e., the slab, cylinder, and sphere of radius r. In the case of the slab, the release pattern is characterized by a square root

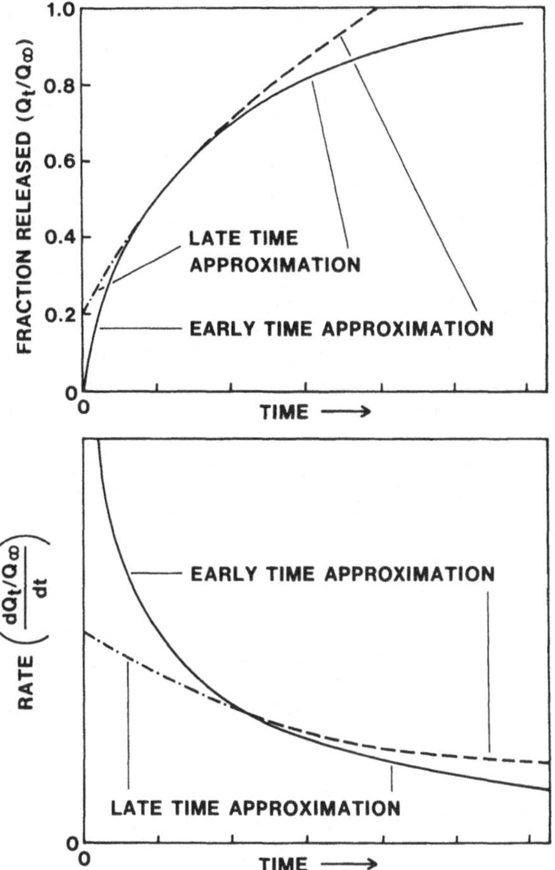

FIGURE 3. Theoretical release curves from a monolithic slab (dissolved solute) illustrating the early and late time approximations. The continuous line represents the complete profile. (From Baker, R. W. and Lonsdale, H. K., Controlled release: mechanisms and rates in *Advances in Experimental Medicine and Biology,* Vol. 47, Tanquary, A. C. and Lacey, R. E., Eds., Plenum Press, New York, 1974, 15.)

of time dependence at early times and an exponential dependence on time during the terminal phase of solute release. This is illustrated in Figure 3. The initial release period for the slab is analogous to the desorption kinetics from a semi-infinite medium where:[5]

$$Q_t = 2\,C_o \left(\frac{D_s t}{\pi}\right)^{1/2} \tag{6}$$

and

$$\frac{dQ_t}{dt} = C_o \left(\frac{D_s}{\pi t}\right)^{1/2} \tag{7}$$

Examples of the release profiles for the three geometries are given in Figure 4. In this analysis, it is assumed that height of the cylinder is large compared to its radius, i.e.,

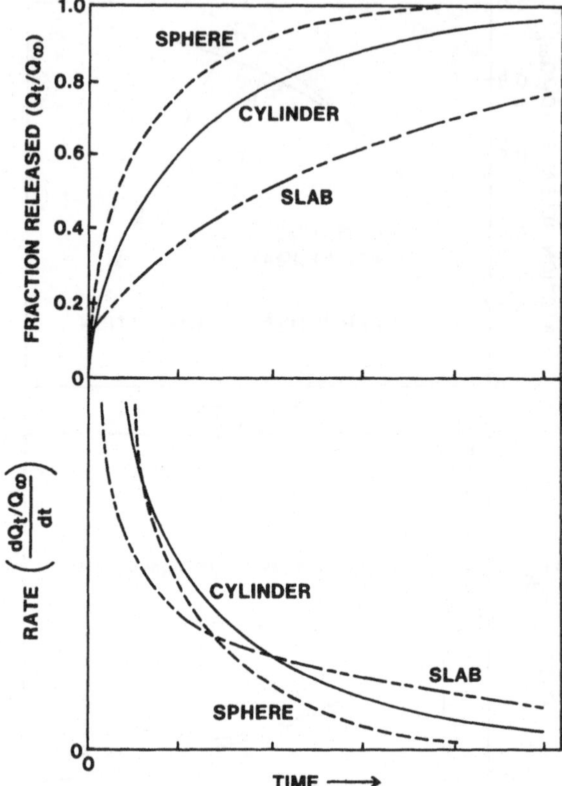

FIGURE 4. Theoretical release curves from monolithic devices (dissolved solute) of different geometries. The continuous line represents the complete profile. (From Baker, R. W. and Lonsdale, H. K., Controlled release: mechanisms and rates, in *Advances in Experimental Medicine and Biology*, Vol. 47, Tanquary, A. C. and Lacey, R. E., Eds., Plenum Press, New York, 1974, 15.

negligible radial diffusion, and that for the slab, edge effects are negligible. When this is not the situation, more complex mathematical expressions are required.[11,12]

B. Homogeneous Monoliths (Dispersed Solute)

Monoliths of this type contain finely divided solute particles which are uniformly dispersed or suspended within the matrix phase. The dispersed solute is in equilibrium with dissolved solute, and the total concentration is designated as A. In an analogous fashion to the case of monoliths containing only dissolved drug, diffusion occurs only through the matrix phase. The derivation of the release rate expression, however, relies on Fick's first law (Equation 1), and the kinetics of solute release is different. The analysis of this system was originally presented by Higuchi for the release of drugs dispersed in a stationary matrix, e.g., semisolid ointment.[13]

A schematic illustration of the physical model is shown in Figure 5, depicting the dependence of the concentration gradient on time. Assuming that (1) a pseudo-steady state exists, (2) the drug particles are small compared to the average distance of diffusion, (3) the diffusion coefficient is constant, and (4) a perfect sink condition exists in the external media (i.e., concentration equals zero), the following equations were derived*.

* For simplification, the volume fraction (ϕ) and tortuosity (τ) of the matrix are assumed to be unity. To account for these factors, D, is simply replaced with an effective diffusion coefficient (D_e) where $D_e = D_e\phi_1/\tau$.

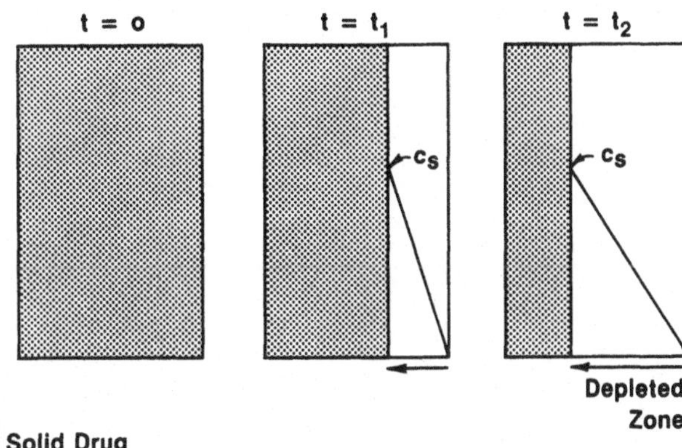

FIGURE 5. Schematic representation of the pseudo-steady state concentration profiles for the release of solute from a single face of a monolith containing dispersed drug. The lines are drawn for three times such that $t_2 > t_1$.[13]

The change in the amount released per unit area, dQ_t, corresponding to a change in the thickness of the depleted zone, $d\ell$, is

$$dQ_t = A d\ell - \frac{C_s}{2} d\ell \quad (8)$$

where C_s is the solubility of solute in the matrix phase. According to Equation 1,

$$dQ_t = \frac{D_s C_s}{\ell} dt \quad (9)$$

and equating Equations 8 and 9, integrating, and solving for yields:

$$\ell = \left(\frac{2D_s C_s t}{A - C_s/2}\right)^{1/2} \quad (10)$$

The depleted zone, ℓ, recedes as function of $t^{1/2}$. Substituting Equation 10 into the integrated form of Equation 8 gives:

$$Q_t = \left[2C_s D_s \left(A - \frac{C_s}{2}\right) t\right]^{1/2} \quad (11)$$

When $A >> C_s$, Equation 11 reduces to:

$$Q_t = \left(2AC_s D_s t\right)^{1/2} \quad (12)$$

The fraction of solute released (F) from the slab is

$$F = \left(\frac{2C_s D_s t}{Ah^2}\right)^{1/2} = k\, t^{1/2} \quad (13)$$

where h is the thickness, and k is a constant equal to $(2C_sD_s/Ah^2)^{1/2}$. The corresponding equations for the cylinder[14] and sphere[15] respectively are

$$\left[\frac{F}{4} + \left(\frac{1-F}{4}\right) \ln(1-F)\right]^{1/2} = k' t^{1/2} \quad (14)$$

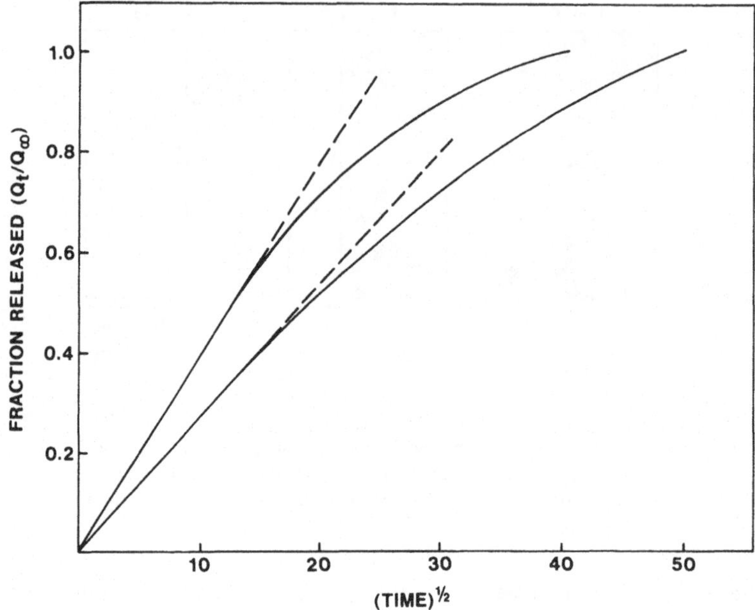

FIGURE 6. Fraction of solute released as a function of the square root of time for monoliths containing dispersed drug. Dashed lines represent the linear dependence of a slab normalized for the area of the cylinder or sphere. The upper curve is the sphere and the lower curve is the cylinder.

$$\left[\frac{1 - (1 - F)^{2/3} - 2/3\,F}{2}\right]^{1/2} = k't^{1/2} \tag{15}$$

where $k' = (C_sD_s/Ar^2)^{1/2}$, and r is the radius. Graphs of the fraction released vs. the square root of time are shown in Figure 6 for k′ equal to unity. The dashed lines represent Equation 13 with the appropriate normalization for geometry.* For up to 50% of solute release, the slab is a good approximation for either the cylinder or sphere. The rate of release from a slab is

$$\frac{dQ_t}{dt} = \left(\frac{AC_sD_s}{2t}\right)^{1/2} \tag{16}$$

Comparison of the release expressions during early times for dissolved (Equations 6 and 7) and dispersed (Equations 12 and 16) solute in a matrix show certain similarities and differences. Identical dependencies on time and diffusivity are predicted, but there is a different relationship of concentration. In the dissolved case, release is proportional to the initial concentration (C_0), while release is dependent on the square root of the total concentration ($A^{1/2}$) when dispersed drug is present.

The pseudo-steady state approximation in the development of Equation 11 requires that A be greater than C_s. Paul and Mc Spadden[16] have provided a rigorous derivation which relies on Fick's second law. Starting with Equation 2, the following expression resulted:

$$Q_t \simeq [2D_sC_s(A - C_s)\,t]^{1/2} \tag{17}$$

which was the limiting form of an error function series expansion for $A >> C_s$. Equations 11 and 17 differ only in that the coefficient of C_s in Equation 17 is unity instead

* The area of the sphere and cylinder were $4\pi r^2$ and $2\pi rh_c$, respectively. Unit height (h_c) was assumed for the cylinder.

of one half. However, when $A >> C_s$, Equations 11 and 17 both reduce to the same result, supporting the validity of the pseudo-steady state assumption. In the limit $A \rightarrow C_s$, the exact analysis yields Equation 6, while the pseudo-steady state expression, Equation 11, gives a result which is 11% lower.

C. Granular or Porous Monoliths

Release of solute from a granular matrix occurs through connecting capillaries or pores. The steps in the derivation of the release rate expression are identical to those presented for the homogeneous monolith in the preceding section.[15] The resulting terms in the equations, however, refer to values in the elution medium rather than the matrix phase. From Equation 11, we have:

$$Q_t = [C_a D_a \frac{\epsilon}{\tau} (2A - \epsilon C_a)\, t]^{1/2} \qquad (18)$$

where C_a and D_a are the solubility and diffusion coefficient, respectively, in the leaching solvent, ε is the porosity, and τ is the tortuosity of the matrix. The tortuosity constant is required to account for the circuitous channels through which the solute diffuses.

Specific cases have been considered and equations derived which account for (1) equilibrium solute binding to the matrix phase,[17] (2) release of two noninteracting drugs,[18] and (3) release of mutual interacting drugs.[19] When the initial porosity of the matrix is small or the fraction of the matrix volume occupied by the solute is relatively large, i.e., $\varepsilon = A/\varrho$ where ϱ is the density of the solute, then Equation 18 yields:

$$Q_t = A \left[\frac{D_a C_a}{\tau \rho} \left(2 - \frac{C_a}{\rho} \right) t \right]^{1/2} \qquad (19)$$

In contrast to the homogeneous monolith containing dispersed drug, the release from a porous monolith is expected to be linearly dependent upon A.

D. Monoliths and Boundary Layers

Diffusion of solute from the surface of a solid into the surrounding milieu has been theorized to involve transport across a region adjacent to the solid. The solute concentration at the surface is maintained at saturation, and diffusion occurs across an unstirred region termed the boundary diffusion layer or Nernst diffusion layer.[20] This diffusion layer offers an additional resistance to mass transfer and can influence the kinetics of solute release from monolithic devices. A schematic representation of a diffusion layer, h_a, in series with a monolith containing dispersed drug is depicted in Figure 7.[21] With the assumptions given in II.B, the rate of transport across a plane of unit area according to Equation 1 is

$$\frac{dQ_t}{dt} = \frac{D_s}{\ell} (C_s - C'_s) \qquad (20)$$

where C'_s is the concentration in the matrix at x = o. The rate of transport across the diffusion layer (h_a) is given by

$$\frac{dQ_t}{dt} = \frac{D_a}{h_a} (C'_a - C_B) \qquad (21)$$

where C'_a is the concentration in water at x = o, and C_B is the concentration at x =

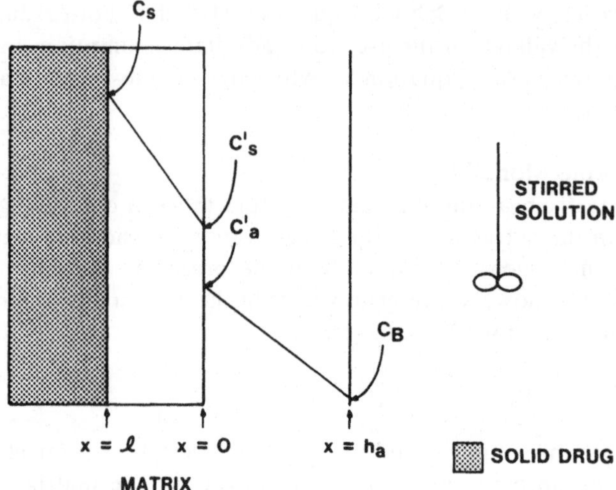

FIGURE 7. Hypothetical diagram for the matrix diffusion layer model.[21]

h_a. The solute must pass from the matrix phase to the eluting solvent, and the concentrations on either side of x = o are related by the partition coefficient, K:

$$K = \frac{C_s}{C_a} = \frac{C'_s}{C'_a} \tag{22}$$

Equating Equations 20 and 21 and solving for the unknown C'_s yields:

$$\frac{dQ_t}{dt} = \frac{D_s}{\ell}\left[C_s - \frac{C_s D_s h_a + D_a \ell C_B}{\frac{D_a \ell}{K} + D_s h_a}\right] \tag{23}$$

Equation 23 relates the rate of release to the time dependent zone of depletion (ℓ). However, the rate of release per unit area is also

$$\frac{dQ_t}{dt} = A\frac{d\ell}{dt} - \frac{C_s}{2}\frac{d\ell}{dt} \tag{24}$$

Equating Equations 23 and 24, integrating, and assuming perfect sink conditions, i.e., C_B = o, gives:

$$\ell^2 + 2\frac{D_s h_a K \ell}{D_a} = \frac{2 D_s C_s t}{A - C_s/2} \tag{25}$$

When A >> C_s, the cumulative amount released per unit area is

$$Q_t = \frac{-D_s h_a K A}{D_a} + \left[\left(\frac{D_s h_a K A}{D_a}\right)^2 + 2 A D_s C_s t\right]^{1/2} \tag{26}$$

and the rate becomes:

$$\frac{dQ_t}{dt} = \frac{\alpha C_s}{2\,(\beta^2 K^2 + \alpha\, C_s t)^{1/2}} \tag{27}$$

where $\alpha = 2AD_s$ and $\beta = D_s h_a A/D_a$. For small values of β^2K^2, Equations 26 and 27 reduce to the previously derived equations, i.e., Equations 12 and 16 which follow square root of time relationships. This is commonly referred to as matrix-controlled kinetics. However, a linear dependency of Q_t on time results when β^2K^2 is large (boundary layer-controlled kinetics)

$$Q_t = \frac{C_a D_a}{h_a}\, t \tag{28}$$

and the corresponding rate expression is zero order:

$$\frac{dQ_t}{dt} = \frac{C_a D_a}{h_a} \tag{29}$$

Regardless of the value of β^2K^2, Equations 28 and 29 will be operative at early times. The matrix-boundary diffusion layer model represents a generalized description of the kinetics of solute release from monolithic devices when diffusion from the surface offers a significant resistance to the mass transport process. Figure 8 shows the delivery rate-time of behavior predicted by Equation 27. The graph is characterized by a transition from zero order release to a square root of time dependency, and its ramifications are discussed below.

Release expressions for the cylinder and sphere are not explicit, but the total amount desorbed at any time Q'_t is related to time by the following equations:*

$$Q'_t = \pi h_c A\,(r^2 - r'^2) \tag{30}$$

$$\frac{r'^2}{2} \ln \frac{r'}{r} + {}^1\!/\!_4\,(r^2 - r'^2) + \frac{D_s h_a K}{2D_a r}(r^2 - r'^2) = \frac{C_s D_s t}{A} \tag{31}$$

where r' is the radius of the undepleted region, and h_c is the height of the cylinder. The corresponding equations for the sphere are

$$Q'_t = \frac{4}{3}\,\pi A\,(r^3 - r'^3) \tag{32}$$

$$r^3 - 3rr'^2 + 2r'^3 - \frac{2D_s h_a K}{D_a r}(r'^3 - r^3) = \frac{6rD_s C_s t}{A} \tag{33}$$

The governing rate equation for release from a monolithic device (semi-infinite media) containing only dissolved drug is[16]

$$\frac{dQ_t}{dt} = \frac{C_0}{\gamma}\left(\exp \frac{t}{\gamma^2 D_s}\right) \operatorname{erfc} \left(\frac{t}{\gamma^2 D_s}\right)^{1/2} \tag{34}$$

where γ is $h_a K/D_a$. The limiting cases of this expression result when (1) γ is large (or t is small):

$$\frac{dQ_t}{dt} = \frac{C_0}{\gamma} = \frac{C'_a D_a}{h_a} \tag{35}$$

where C'_a is the solute concentration in the receptor phase at the matrix-solvent interface, and when (2) γ is small (or t is large). In this latter case, Equation 34 reverts to

* The thickness of the diffusion layer is assumed to be small compared to the radius of the monolith.

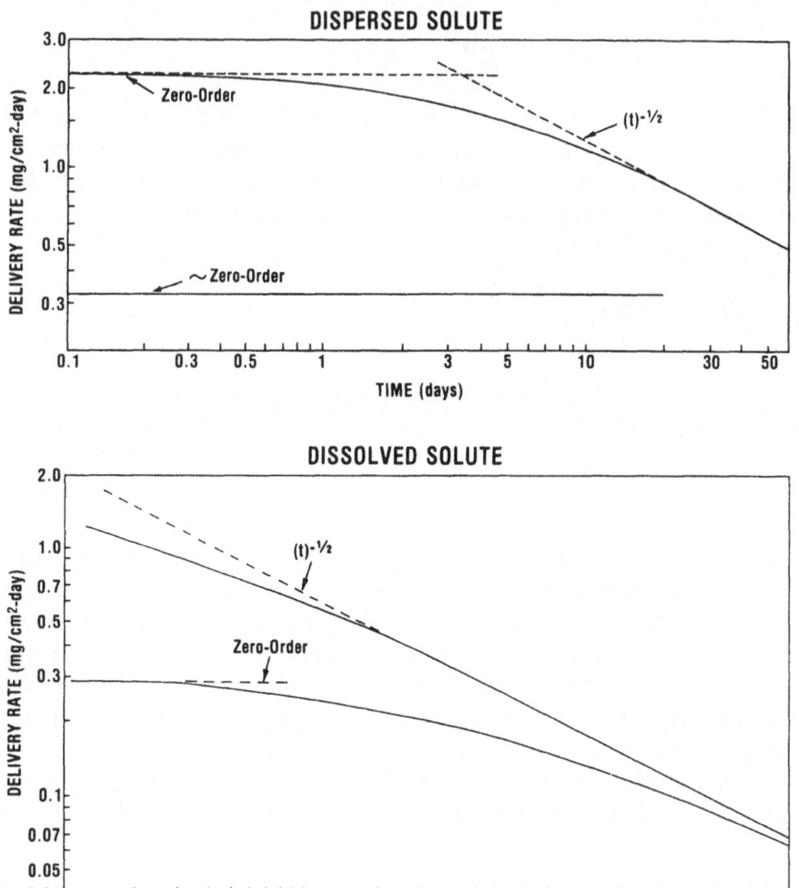

FIGURE 8. Release rate-time behavior for the matrix-diffusion layer model where the solute is the hexyl ester of *p*-aminobenzoic acid, and it is dispersed (Figure 8a) or dissolved (Figure 8b) in the monolith. In each case, the upper curve represents an in vitro situation, while the lower curve is the expected in vivo release. This latter case corresponds to a tenfold increase in the resistance (h_aK/D_a) in the diffusion layer.[40]

Equation 7. The amount released, Q_t, with respect to time, t, is simply obtained by integrating the rate equations.

In Figure 8, the transition from diffusion layer control to matrix control is graphically illustrated for a monolith containing only dissolved solute, e.g., $A = C_s$, and a monolith containing dispersed solute, i.e., $A > C_s$. The hexyl ester of *p*-aminobenzoic acid was chosen as a model compound. Constant release (zero order) is observed during early times due to boundary layer control. At long times, matrix control dominates, and the delivery rate follows a $t^{-1/2}$ dependence. When drug is dissolved in the matrix, the contribution of the diffusion layer resistance is less pronounced. This is evidenced by the observation that constant release is maintained for a shorter time period. Figure 9 also shows that an increase in the resistance (h_aK/D_a) in the boundary layer extends the zero-order release phase and reduces the release rate.

The duration of the zero-order kinetics will depend upon the value of A when $A >> C_s$. For $A \leqslant C_s$ the delivery rate curves merely shift proportionally with the initial loading dose, and the duration of constant release is independent of concentration.

TABLE 2

Parameters Which Influence the Kinetics of Solute Release from Monolithic Devices

Dependent upon solute	Independent of solute
Solubility	Concentration
Partition coefficient	Geometry
Diffusion coefficient	Tortuosity
	Porosity
	Volume fraction
	Diffusion layer

III. FACTORS INFLUENCING THE KINETICS OF SOLUTE RELEASE

Monolithic devices are usually prepared by either incorporating the solute within the monomeric material before polymerization and molding, extrusion, injection molding or by film casting.[11,14,21-31] Alternatively, the matrix may be soaked in a solution of known solute concentration until the designated loading dose is achieved. Granular or porous matrices are manufactured utilizing tableting technology and have been prepared by direct compression of the mixture of the solute in a plastic matrix or by adding it to melted wax, granulating, and then compression.[32-36] The geometry of the device is controlled by the dimensions of the mold, die, or film.

The parameters which influence the release profiles of solutes from monolithic devices are summarized in Table 2. These are categorized as solute-dependent or solute-independent properties. The former are related to the physical-chemical nature of the solute in the polymer, while the latter are system variables. Either set can be varied to optimize the delivery rate. Structural modifications of the drug molecule provide a useful method of controlling the physical-chemical properties. In the field of pharmaceutics, however, the chemical composition of the polymer is usually varied to achieve the desired release rate, since a drug has already been selected which possesses the required potency and biological activity. The following sections describe how these factors influence drug release.

A. Concentration

The influence of concentration on the release profiles of steroids from silicone rubber has been widely studied.[14,28,37,38] For monolithic devices containing dispersed drug, release rates are proportional to the square root of concentration ($A^{1/2}$). This dependence is shown in Figure 9 for medroxyprogesterone acetate and is in agreement with the prediction of Equation 12. Photographs of the time-dependent zone of depletion (ℓ) are presented in Figure 10. The well-defined depleted regions provide experimental evidence to support the theoretical principles which were discussed previously. Zones of depletion were microscopically measured, and their thicknesses are proportional to $A^{-1/2}$ as given in Equation 10. It has been observed that the shape of Q_t vs. $t^{1/2}$ plots are also dependent upon A. As concentration is ascended, greater periods of nonlinearity are observed at early times (Figure 11). Eventually, the curves become linear with the slopes being proportional to $A^{1/2}$. Initial curvature was also noted for a series of progestins,[14] ethynodiol diacetate,[39] and esters of *p*-aminobenzoic acids.[40] Due to the relatively high partition coeficients for these compounds, it was shown that the boundary diffusion layer influences the kinetics of drug release. The higher values of A result in smaller zones of depletion at any given time, and hence, a larger resistance to diffusion resides in the boundary diffusion layer. Release profiles are expected to be independent of A as shown in Equations 28 and 29, when diffusion layer control

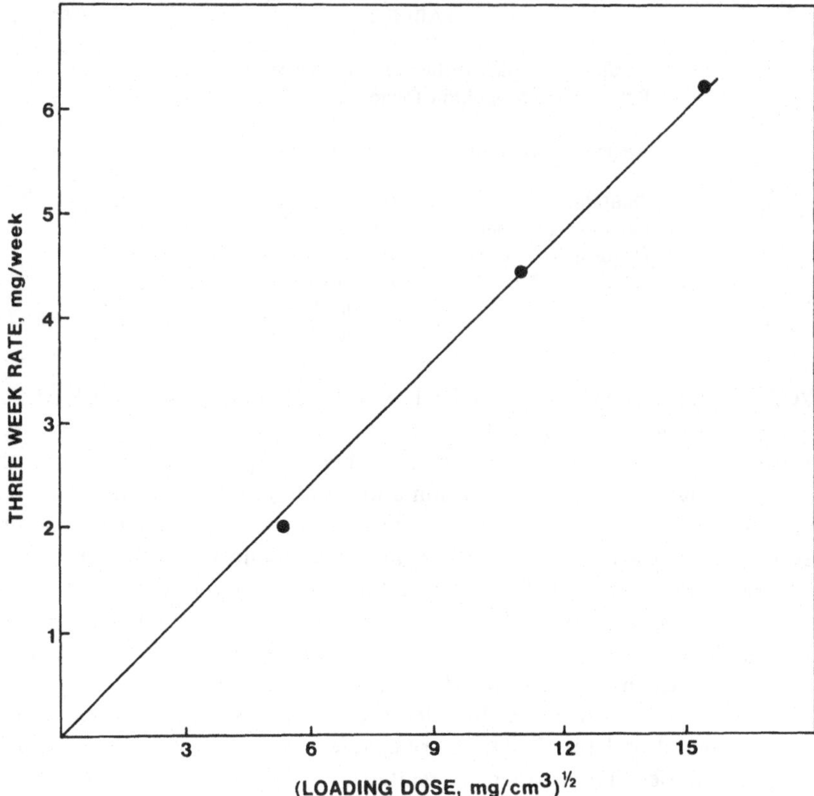

FIGURE 9. Dependence of the release rate of medroxyprogesterone acetate on the square root of the initial concentration in a monolith containing dispersed drug.

is operative. The effect of concentration on the permeability of the matrix was considered by Baker and Lonsdale.[41] At high loadings, diffusion through the polymer phase, as well as pores, resulting from dissolved drug particles was postulated. Equations 12 and 16 were modified to account for the possibility:

$$Q_t = \left[2AD_sC_st\left(\frac{1 + 2A/p}{1 - A/p}\right)\right]^{1/2} \tag{36}$$

$$\frac{dQ_t}{dt} = \left[\frac{AD_sC_s\,(1 + 2A/p)}{2t\;(1 - A/p)}\right]^{1/2} \tag{37}$$

The influence of drug loading on matrix permeability (D_sC_s) is illustrated in Figure 12. A substantial increase in the permeability of chloramphenicol is noted, while the permeability of hydrocortisone increases to a lesser degree with concentration. This difference is consistent with a higher permeability of chloramphenicol through the pores due to its greater water solubility.

Release of drug from monoliths containing dissolved drug is expected to be linearly related to the initial concentration, C_o. Pharmaceutical devices of this type have not been as widely studied as dispersion systems. The concept, however, has been employed in topical dosage forms[9,10,42,43] and in pesticide delivery systems.[44]

The relationship between drug release and concentration (A) for porous monoliths was studied using plastic,[32,45,46] wax,[33,47] methyl acrylate-methyl methacrylate copolymers,[35] and hydroxypropyl methyl cellulose.[34,48] Release increases with concentration, but the linear dependence predicted by Equation 19 is not always noted. This finding

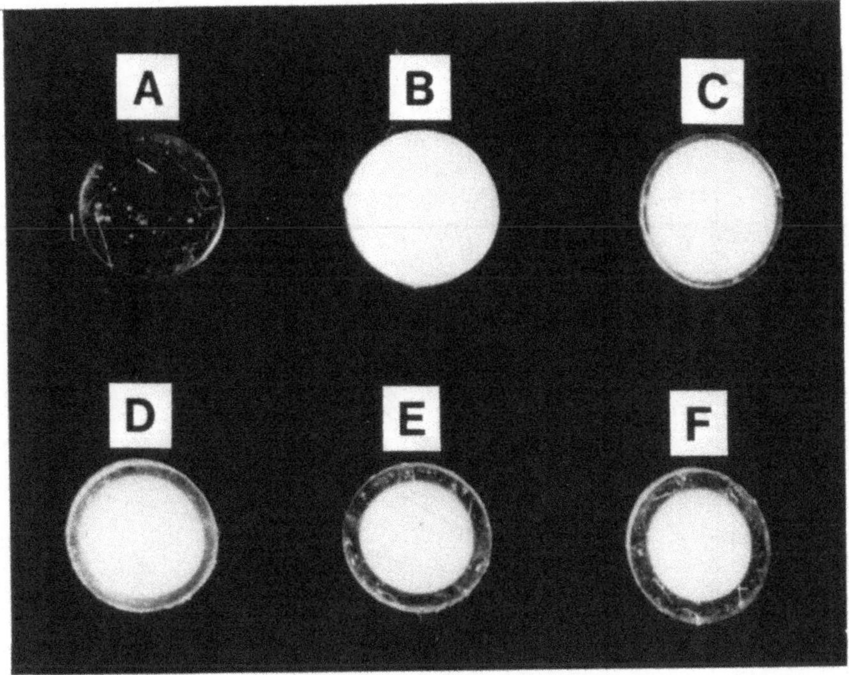

FIGURE 10. Photograph of cross-sectioned slices of medroxyprogesterone acetate-silicone rubber cylinders depicting zones of depletion (clear region) as a function of time of dissolution. Key: A_4, placebo: B, drug-filled, initial; C, 1 week; D, 2 weeks; E, 3 weeks; F, 4 weeks.[21]

suggests that porosity may not be a direct function of A or that another variable(s) was altered in the equation when concentration was changed.[32]

B. Diffusion Coefficients

The magnitude of the diffusion coefficient is dictated by the energy as well as the geometry of the system. The size and shape of the diffusing molecule, the degree of polymer crystallinity, and polymer chain-chain interactions and flexibility will affect the polymers ability to form a hole large enough to accommodate the penetrating molecule.[49-51] The diffusion coefficient of solutes in liquids are given by the Stokes-Einstein equation:[52,53]

$$D_a = \frac{RT}{4\pi\eta N}\left(\frac{4\pi N}{3V}\right)^{1/3} \tag{38}$$

for small particles and:

$$D_a = \frac{RT}{6\pi\eta N}\left(\frac{4\pi N}{3V}\right)^{1/3} \tag{39}$$

for large particles, where R is the Boltzman constant, η is the viscosity, N is Avogadro's number, T is the absolute temperature, and V is the partial molar volume. Diffusion coefficients in polymers are difficult to predict *a priori* in a reliable fashion, but can be empirically related to molecular size or weight:

$$\log D_s = -S_V \log V + k_V = -S_M \log M + k_M \tag{40}$$

where S_V, k_V, S_M, and k_M are constants, and M is the molecular weight. The dependence of the diffusion coefficient on molecular size is much greater in polymers than in water,

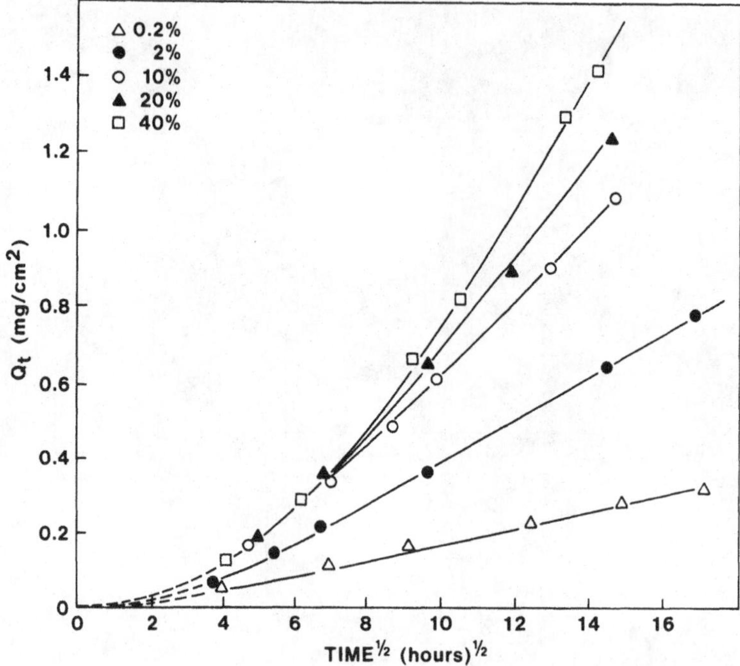

FIGURE 11. Release of chlormadinone acetate from silicone rubber monoliths containing dispersed drug as a function of the square root of time. (From Haleblian, J., Runkel, R., Mueller, N., Christopherson, J., and Ng, K., *J. Pharm. Sci.*, 60, 541, 1971. With permission.)

and this is illustrated in Figure 13. Limited data are available on diffusion coefficients of drugs in polymers. They, however, can be readily determined by the lag-time method of Barrer.[54] Diffusion coefficients of some selected compounds are given in Table 3. The wide range of values indicates that the diffusion coefficient is a major factor in determining the rate of solute delivery. Based upon the derivation presented in the Theoretical Principles Section, release curves are expected to be proportional to the square root of the diffusion coefficient. Concentration-dependent diffusion coefficients have been considered elsewhere.[5]

C. Partition Coefficients

The polymer-water partition coefficient comes into play when transfer from the polymer phase to the eluting solvent occurs. In the case of membrane transport, the product of the partition coefficient and diffusion coefficient defines the permeability constant. The partition coefficient is an additive property of the functional groups present in a molecule;[55] and in contrast to the diffusion coefficient it is extremely sensitive to slight changes in molecular structure. For example, the addition of a single hydroxyl group to progesterone forming 17α-hydroxyprogesterone results in 50-fold reduction in the silicone rubber-water partition coefficient. Values for the partition coefficient of some steroids in silicone rubber are listed in Table 3.

The effect of the partition coefficient on the mass transport of steroids and a homologous series of esters of *p*-aminobenzoic acid from monoliths containing dispersed drug has been investigated.[14,21,39,40,56-58] Employing Equation 26 or 27, Roseman and Yalkowsky[40] demonstrated that the overall release kinetics are biphasic, with the early portion being invariant with time (zero order) and the later phase being dependent upon the square root of time. The magnitude of the partition coefficient (K) is a con-

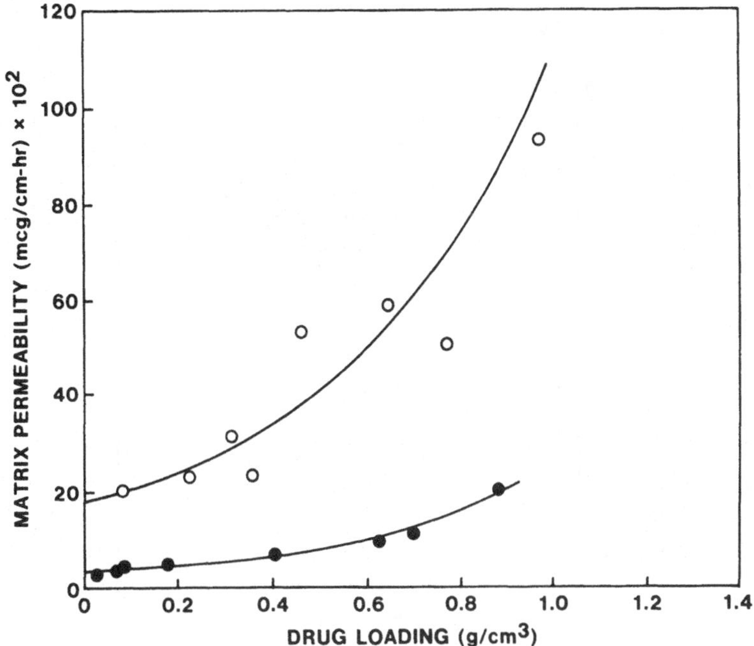

FIGURE 12. Plot of matrix permeability (C_sD_s) vs. initial drug concentration in ethylene-vinyl acetate monoliths containing dispersed drug. Key: O, chloramphenicol; •, hydrocortisone alcohol. (From Baker, R. W. and Lonsdale, H. K., *Proc. Int. Controlled Release Pesticide Symp.*, Cardarelli, N. F., Ed., University of Akron, Ohio, September 16 to 18, 1974.

trolling factor in determining the duration of the zero-order period, as seen in Equations 28 and 29 when K^2 is large and Equations 12 and 16 when K^2 is small. Esters of *p*-aminobenzoic acid are ideal compounds to demonstrate the dual nature of the release profile. The value of K is systematically related to the number of carbons (n) in the alkyl chain by:[59]

$$\log K_n = \log K_0 + \sigma n \tag{41}$$

where K_0 and K_n are the partition coefficients of the reference and higher homolog, respectively, and σ is a constant, characteristic of the membrane and environmental solvent. Considering a variation of n from one (methyl) to seven (heptyl) for example, results in about a four order of magnitude increase in K. Since it is not expected that these moderate changes in molecular weight with chain length affect diffusion coefficients to any great extent and the other parameters in Equations 26 and 27 are fixed by the experimental design, i.e., α and β are constant, the release profile is only dependent upon the value of K. Plots of Q_t vs. $t^{1/2}$ are shown in Figure 14. For small values of n, a linear relationship is observed. As n increases to above four (i.e., butyl), the duration of the nonlinear period increases dramatically. In fact, the hexyl and heptyl show a linear dependence of Q_t on t and not $t^{1/2}$. The influence of alkyl chain length on release rate is shown in Figure 15 for three times values. The solid lines represent the matrix control rate, Equation 16, and the dashed line signifies the diffusion layer control rate, Equation 29, which is the limitation imposed by the aqueous solubility of the drug. A transition region from matrix control to diffusion layer control is indicated by the dotted line. At long times ($t \rightarrow \infty$), matrix control is operative,

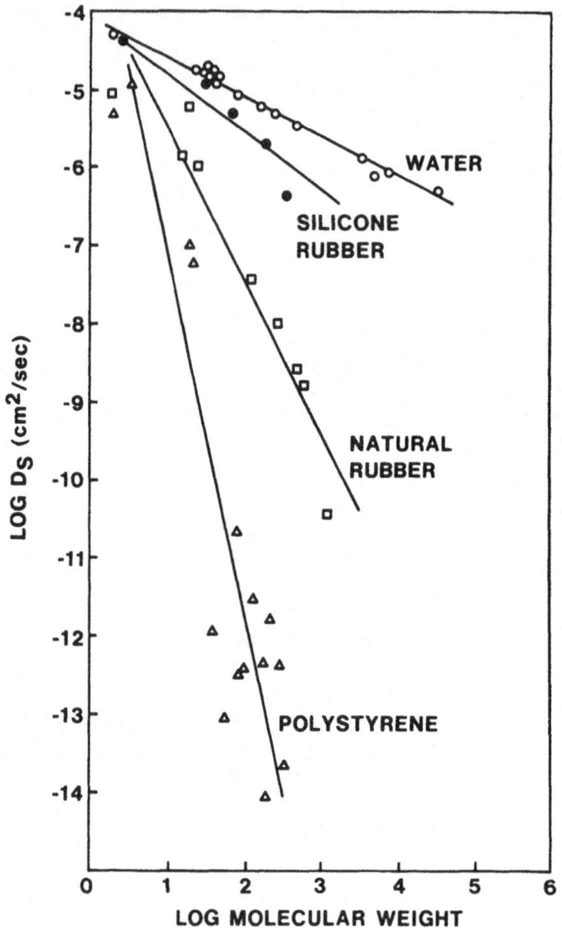

FIGURE 13. Dependence of diffusion coefficients on molecular weight. (From Baker, R. W. and Lonsdale, H. K., Controlled release: mechanisms and rates, in *Advances in Experimental Medicine and Biology,* Vol. 47, Tanquary, A. C. and Lacey, R. E., Eds., Plenum Press, New York, 1974, 15.

while at short times ($t \rightarrow 0$), diffusion layer control dominates regardless of the value of n. However, as n increases so does the partition coefficient, and longer periods of diffusion layer control result.

D. Solubility

The solubility of the solute in the polymer or elution media depends upon the intermolecular forces between the solute and solvent. Relative extents of solubility can be estimated from the solubility parameter (δ) which is a measure of the cohesive energy densities of the same molecules.[60] The smaller the difference of δ's between a solute and the solvent, the greater the solubility of the solute. Release of drugs from monolithic devices containing excess drug is proportional to the square root of their solubility in the polymer, that is, $(C_s)^{1/2}$. The linear portion of the curves in Figure 14 follow this dependence. This behavior was also experimentally verified for steroids where plots of $(Q_t/t)^{1/2}$ vs. $C_s^{1/2}$ were linear.[14] The solubility values of some steroids in silicone rubber are given in Table 3.

TABLE 3

Physicochemical Constants of Selected Solutes in Polymers

Solute	Polymer	Diffusion coefficient (cm^3/sec)	Solubility (mg/cm^3)	Partition coefficient (polymer/water)	Ref.
Acetophenone[a]	Polyethylene	3.55×10^{-8}	—	3.16	
p-Aminoacetophenone	Silicone rubber	2.44×10^{-6}	0.317	0.0321	65
Androstenedione	Silicone rubber	14.8×10^{-7}	0.365	7.4	105
Benzaldehyde[a]	Polyethylene	3.39×10^{-8}	—	3.74	
Benzoic acid[a]	Polyethylene	5.29×10^{-10}	—	6.25	
Chlormadinone acetate	Silicone rubber	3.03×10^{-7}	0.082	82	28
Delmadinone acetate	Silicone rubber	0.38×10^{-7}	0.850	140	100
Deoxycorticosterone acetate	Silicone rubber	4.94×10^{-7}	0.105	—	90
Estriol[b]	Polyurethane ether	2×10^{-9}	2	133	
Estrone	Silicone rubber	2.4×10^{-7}	0.324	8.0	105
Ethyl-*p*-aminobenzoate[c]	Silicone rubber	2.67×10^{-6}, 1.78×10^{-6}	1.68	0.966	
Ethylnodiol diacetate[d]	Silicone rubber	3.79×10^{-7}	1.48	108	
Fluphenazine	Polymethylmethacrylate	1.74×10^{-17}	—	—	108
Fluphenazine	Polyvinyl acetate	1.05×10^{-12}	—	—	108
Fluphenazine enanthate	Polymethylmethacrylate	1.12×10^{-17}	—	—	108
Fluphenazine enanthate	Polyvinylacetate	1.82×10^{-12}	—	—	108
Hydrocortisone	Silicone rubber	4.5×10^{-7}	0.014	0.05	58
Hydrocortisone	Polycarpolactone	1.58×10^{-10}	—	—	11
Hydrocortisone	Ethylene-vinyl acetate	1.18×10^{-11}	—	—	11
Hydrocortisone	Polyvinyl acetate terpolymer	4.31×10^{-12}	—	—	11
Hydroquinone	Polymethylmethacrylate	5.75×10^{-15}	—	—	108
17α-hydroxyprogesterone	Silicone rubber	5.65×10^{-7}	0.0072	0.89	14
Medroxyprogesterone acetate	Silicone rubber	4.17×10^{-7}	0.0874	26.9	14
4-Methylacetophenone[a]	Polyethylene	1.79×10^{-8}	—	12.5	
4-Methylbenzaldehyde[a]	Polyethylene	1.37×10^{-8}	—	25.6	
6α-methyl-11β-hydroxyprogesterone	Silicone rubber	2.84×10^{-7}	—	—	14
4-nitro-aniline	Polymethylmethacrylate	3.02×10^{-15}	—	—	108
4-nitro-aniline	Polyvinyl acetate	3.02×10^{-11}	—	—	108
Norprogesterone	Silicone rubber	18.5×10^{-7}	0.631	22.4	105
Procaine	Polymethylmethacrylate	1.35×10^{-15}	—	—	108

TABLE 3 (continued)

Physicochemical Constants of Selected Solutes in Polymers

Solute	Polymer	Diffusion coefficient (cm^2/sec)	Solubility (mg/cm^3)	Partition coefficient (polymer/water)	Ref.
4-Methylacetophenone[a]	Polyethylene	1.79×10^{-8}	—	12.5	
Procaine	Polyvinyl acetate	1.45×10^{-11}	—	—	108
Progesterone[e]	Silicone rubber	5.78×10^{-7}, 6.4×10^{-7}	0.513, 0.763	45.0, 59.4	
Promethazine	Polymethylmethacrylate	1.41×10^{-17}	—	—	108
Promethazine	Polyvinyl acetate	1.45×10^{-12}	—	—	108
Pyrimethamine	Silicone rubber	1.10×10^{-10}	—	—	11
Salicylic acid	Polymethylmethacrylate	9.55×10^{-15}	—	—	108
Salicylic acid	Polyvinyl acetate	4.37×10^{-11}	—	—	108
Trinitrophenol	Polymethylmethacrylate	3.55×10^{-16}	—	—	108
Trinitrophenol	Polyvinyl acetate	7.59×10^{-12}	—	—	108
Steroid-type compounds[f]	Silicone rubber (polydimethylsiloxane)	6×10^{-7}	—	—	
	Low-density Polyethylene	1×10^{-9}	—	—	
	Poly(co-ethylene-vinyl acetate) 9%	4×10^{-9}	—	—	
	Poly(co-ethylene-vinyl acetate) 18%	3×10^{-9}	—	—	
	Poly(co-ethylene-vinyl acetate) 40%)	5×10^{-9}	—	—	
	Poly(co-polytetramethylene ether glycol-diphenylmethane di-isocyanate)	9×10^{-10}	—	—	

Note: Temperature is 37°C unless stated otherwise.

[a] From Reference 106, at 40°C.
[b] From Reference 107 at 30°C; density of polymer assumed to be unity.
[c] From Reference 65, second value listed for the diffusion coefficient from Reference 64.
[d] From Reference 37, solubility in polymer determined in liquid silicone.
[e] First value listed is from Reference 14, while the second value is from Reference 105.
[f] From Reference 62; percent is w/w composition of vinyl acetate.

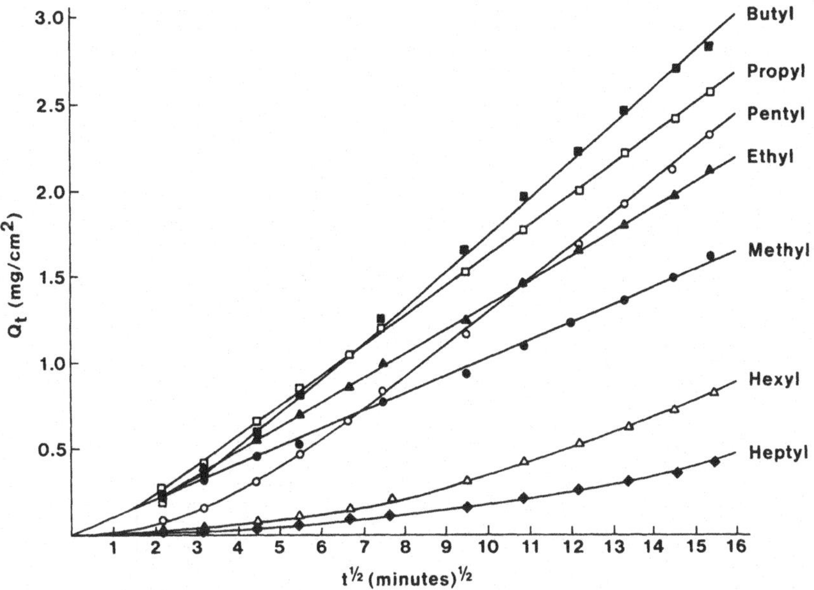

FIGURE 14. Plots of Q_t vs. $t^{1/2}$ for esters of *p*-aminobenzoic acid from silicone rubber monoliths containing 5% dispersed solute.[40]

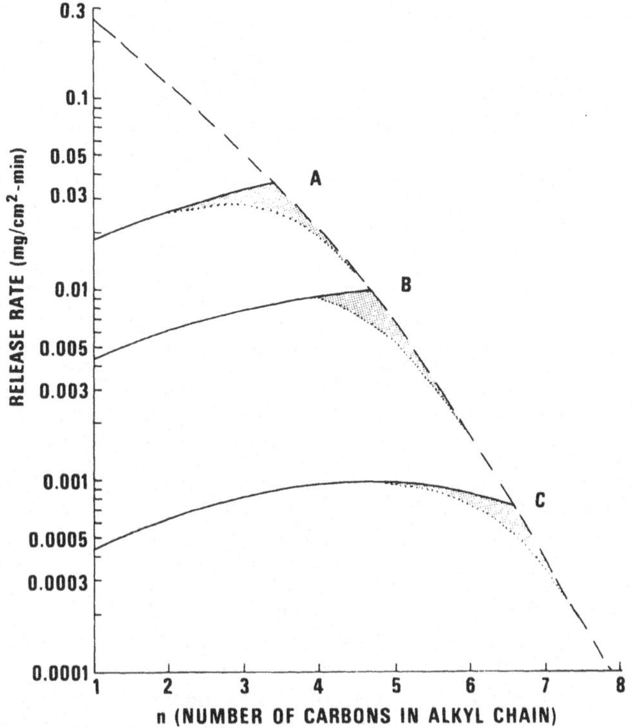

FIGURE 15. In vitro theoretical release rates of esters of *p*-aminobenzoic acid from silicone rubber monoliths containing dispersed solute as a function of chain length; A, 10 min; B, 3 hrs; C, 12.5 days. Key: —-, diffusion layer control rate (Equation 29), matrix control rate (Equation 16), and ..., expected rate (Equation 27).[56]

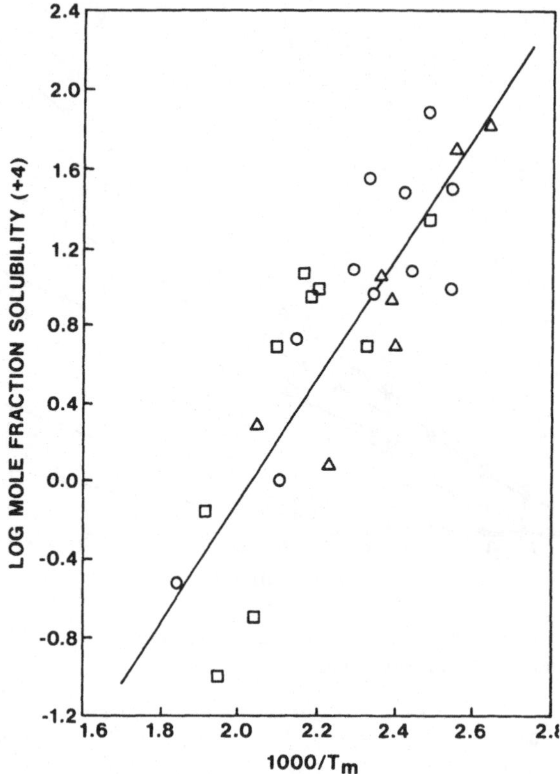

FIGURE 16. Relationship between the mole fraction solubility of steroids and melting point, T_m. Key: ○, testosterone derivatives; □, progesterone derivatives; △, estradiol derivatives. (From Chien, Y. W., *Controlled Release Polymeric Formulations,* Paul, D. R. and Harris, F. W., Eds., ACS Symposium Series 33, American Chemical Society, Washington, D.C., 1976, 53.

The thermodynamic relationships for the dependency of the solute's solubility in the polymer on its melting point (T_m) was presented by Chien.[61] Semilogarithmic plots of the mole fraction solubility vs. the reciprocal value of T_m were linear, and equations were presented that statistically related these parameters for different classes of steroids. Based upon the data in Figure 16, C_s can be predicted from a knowledge of a steroids melting point. The thermodynamic approach was cleverly applied to the prediction of steroid permeability rates across polymer membranes as a function of melting point.[62] This was expressed by the following equation

$$J_{max}\, h \exp(1 + \chi) = \frac{-\Delta S_f}{R} \left[(T_m/T) - 1\right] + \ln \rho_d D_s \qquad (42)$$

where J_{max} is the flux from a saturated solution across a membrane of thickness, h, χ is a solute-polymer interaction constant, and ΔS_f is the entropy of fusion. The interaction constant depends upon the solubility parameter and is given by

$$\chi = \frac{V_d}{RT} (\delta_d - \delta_p)^2 \qquad (43)$$

where the subscript d and p refer to the solute and polymer, respectively. Michaels et al.[62] have pointed out that if for a class of steroids in a particular polymer, ΔS_f, D_s, δ_d

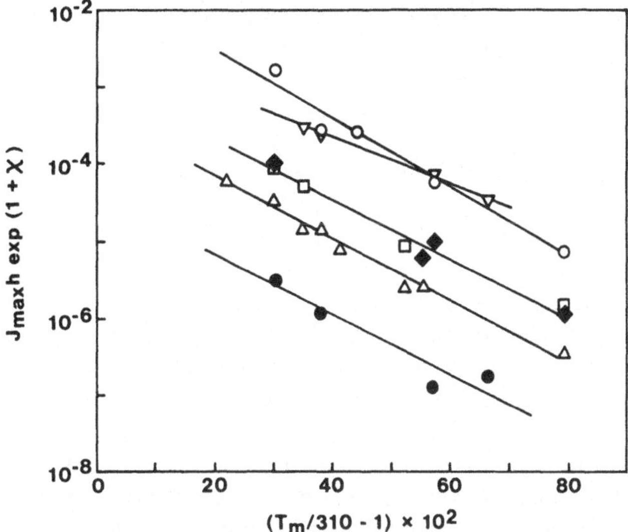

FIGURE 17. Dependence of steroid permeability in various polymers as a function of melting point, T_m. Key: ○, polydimethylsiloxane; ●, low density polyethylene; △, poly(co-ethylenevinyl acetate) 9% w/w vinyl acetate; □, poly (co-ethylene-vinyl acetate) 18% w/w vinyl acetate; ∇, poly(co-ethylene-vinyl acetate) 40% w/w vinyl acetate; ■, poly(co-polytetramethylene ether glycol-diphenylmethane diisocyanate. (From Michaels, A. S., Wong, P. S. L., Prather, R., and Gale, R. M., *AIChE*, 21, 1073, 1975. With permission.)

are constant, then plots of J_{max} h exp $(1 + \chi)$ vs. $[(T_m/T) - 1]$ should be linear. This is demonstrated in Figure 17 for a variety of steroidal compounds in five polymers. With a knowledge of the melting point and interaction constant, which is determined from the solubility parameters, J_{max} h can be calculated. An estimate of the release rate of a particular steroid from a monolithic device can then be described by:

$$\frac{dQ_t}{dt} = \left(\frac{J_{max} hA}{2t}\right)^{1/2} \qquad (44)$$

Considering the vast number of drug-polymer combinations available for controlled-release systems, this approach provides a useful method for the *a priori* prediction of release rates.

E. Additional Factors

Filler materials, such as silica, zinc oxide, carbon black, etc., are incorporated into polymers to modify physical properties and improve handling characteristics.[63] Assuming that the filler is inert with regards to drug transport, its presence in monolithic devices reduces the volume fraction for diffusion and increases tortuosity. Diffusion coefficients determined by the lag-time method will be decreased if the solute adsorbs onto filler particles. The degree to which this occurs depends upon the binding capacity and extent of absorption.[53] Adsorption effects have been observed during the transport of organic solutes across silicone rubber membranes.[14,64,65] Once the adsorption sites were saturated, the steady-state rates remained relatively constant. The presence of filler in a silicone rubber monolith reduces the release of medroxyprogesterone acetate.[14] However, increased permeabilities have been reported for the diffusion of gases in filled elastomers. This was attributed to the nonwetting of filler particles by the polymer, resulting in void spaces which were filled by the gas.[66]

Surfactants increase solute release from porous monoliths by the wetting of channels for subsequent diffusion, thereby effectively increasing the porosity.[45,67,68] And it is expected that the rate of solvent penetration is an important factor which can affect the drug release pattern. Additionally, the composition of the inert matrix also influences release profiles from these systems. For example, polyvinyl chloride releases sodium salicylate four to six times faster than polyethylene, and a sigmoidal dependence of Q_t on $t^{1/2}$ is observed.[46] The effect of tortuosity in altering release kinetics was shown for the release of sulfanilamide from a wax matrix, where extremely large values were reported.[47] Increased tortuosity values were also related to the slower release of drugs from a methyl acrylate-methyl methacrylate copolymer which was exposed to acetone vapor.[69] Different types of materials are employed to release variety of substances at a sustained rate, and many of these follow square root of time kinetics. A comprehensive listing of the polymers and drugs can be found in Reference 70.

In general, the polymers mentioned thus far are relatively nonpolar, and their use with highly water soluble compounds or macromolecules is limited. Monolithic devices have been prepared using hydrophilic polymers, such as cross-linked polyacrylamide and polyvinylpyrrolidone, hydroxyethylmethacrylate, polyvinyl alcohol, and ethylene-vinyl acetate copolymers to release proteins and hormones.[71,72] Release was dependent upon the molecular weight of the diffusant as well as the design of the monolith. Hydrophilic polymers are also used to deliver smaller organic molecules, such as the narcotic antagonist cyclazocine,[73] steroids,[27,74] and inorganic fluoride for caries prevention.[75]

The theoretical principles section for drug delivery from monolithic devices was developed for the situations where the controlling step is in the matrix or the boundary diffusion layer (h_a). In the latter situation, transport rates will be inversely related to the thickness of this unstirred region (Equation 29). An example of the dependence of release on h_a is shown in Figure 18. Diffusion layer thicknesses are related to the hydrodynamics in the eluting solvent and decrease as agitation is increased.[20] In biological systems where in vivo release was slower than in vitro, diffusion layers were purported to be a major factor in explaining the differences.[76] The degree to which it influences the release profile, however, is also a function of the partition coefficient, diffusion coefficients, and loading dose which were discussed previously. Because the kinetics of drug release is affected by the magnitude of the diffusion layer resistance, it is an important consideration in the design and evaluation of controlled release delivery systems. Ideally, release rates should be matrix controlled and independent of external environmental conditions.

The overall model considering drug delivery and absorption from a dispersion monolith is a four step process: (1) dissolution of drug particles within the polymer, (2) diffusion of dissolved drug through the polymer phase, (3) partitioning into the eluting solvent with subsequent diffusion across the boundary diffusion layer, and (4) absorption across the biological membrane. These steps are schematically illustrated in Figure 19 for a vaginal device. In general, micronized drug is incorporated into the matrix to assure rapid drug dissolution, as well as uniformity of content and reproducible release patterns. The agreement of theory and data is supportive of a rapid dissolution step under these conditions. Increasing particle size has been shown to reduce the release rates of chlormadinone acetate from silicone rubber[28] and other solutes from ointments.[77] Figure 20 demonstrates this effect for the release of the butyl ester of *p*-aminobenzoic acid from silicone rubber, where a rank correlation of release with the surface area of the particle within the matrix is noted. A mathematical model which includes dissolution controlled release was derived by Ayres and Lindstrom.[78,79]

Employing the physical model depicted in Figure 19, Flynn et al.[58] treated absorption

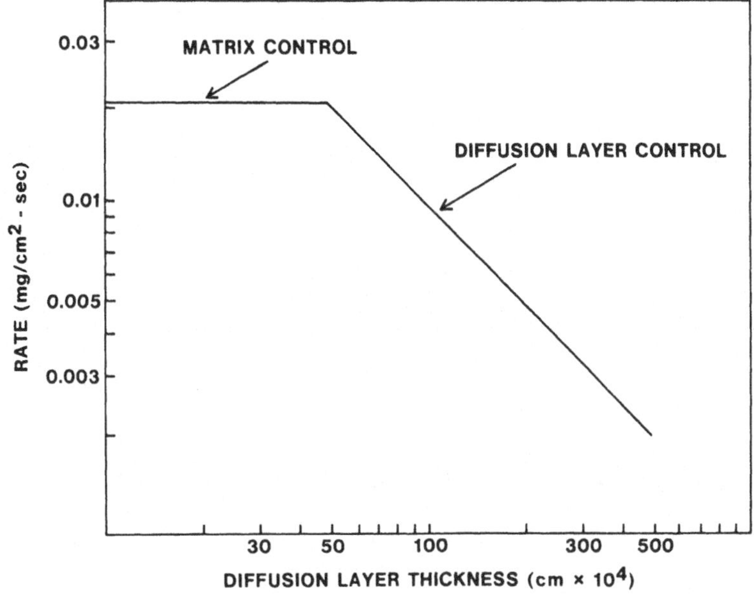

FIGURE 18. Dependence of release rate on diffusion layer thickness for a monolith containing dispersed drug.

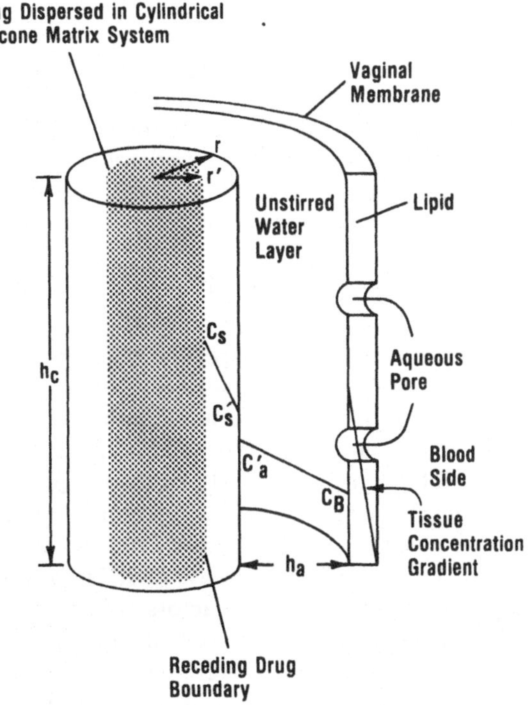

FIGURE 19. Schematic representation of a monolith containing dispersed drug interfaced with a vaginal membrane. (From Flynn, G. L., Ho, N. F. H., Hwang, S., Owada, E., Molokhia, A., Behl, C. R., Higuchi, W. I., Yotsuyanagi, T., Shah, Y., and Park, J, *Controlled Release Polymeric Formulations*, Paul, D. R. and Harris, F. W., Eds., ACS Symposium Series 33, American Chemical Society, Washington, D.C., 1976, 87.

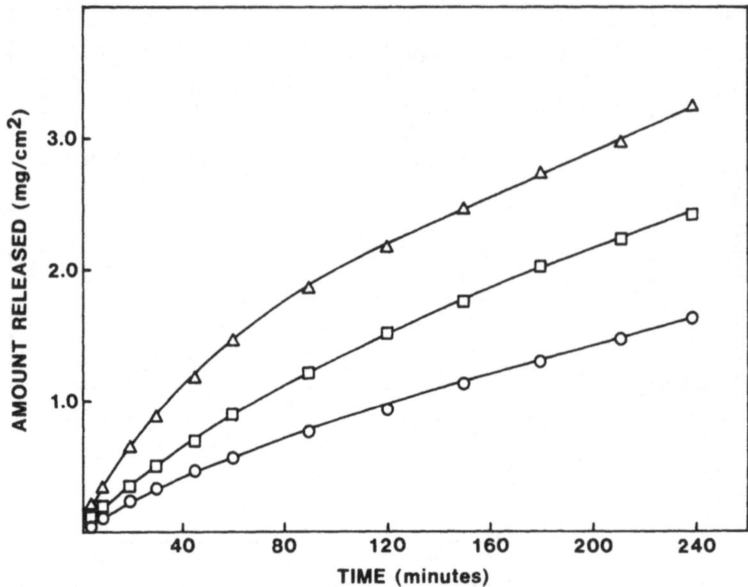

FIGURE 20. Release of the butyl ester of *p*-aminobenzoic acid from silicone rubber monoliths containing dispersed drug as a function of particle surface area. Key: △, 0.64 m^2/g; □, 0.098 m^2/g; ○, 0.032 m^2/g.

as an additional barrier in series with the matrix and diffusion layer. The absorption step was factored into two transport terms which included both drug permeability through aqueous pores and the lipid region of the membrane. Although no single model has been presented which considers the complete process from dissolution to absorption, studies of this type provide a firm physicochemical foundation for interfacing drug delivery with biological parameters.

IV. PHARMACEUTICAL APPLICATIONS OF MONOLITHIC DEVICES

Controlled release delivery systems are particularly useful in administering therapeutic agents which are pharmacologically active at low doses and exhibit relatively short biological half lives. As the drug is metered at a continuous rate, rapid drug metabolism prevents harmful concentration build-ups within the body. Matrix materials must be nontoxic, biocompatible with tissues, and possess the proper permeability characteristics. The dimensional constrains in designing a monolithic device depend upon the route of administration and the targeted release rate. Within reasonable limitations, geometry can be adjusted to optimize the release rate. The selection of a drug-polymer combination depends upon the numerous factors which were discussed previously. Drugs can be administered which are structurally distinct and exhibit a wide spectrum of biological activities. Porous monoliths were developed to be ingested orally (i.e., tablets), while homogeneous monoliths of a variety of shapes have been implanted or placed within animal or human body cavities, such as the uterus or vagina. This section describes some representative examples of monoliths which were designed to deliver therapeutic agents either systemically or to the target organ. The reader is referred to the original articles to obtain details regarding efficacy and other pertinent biological parameters.

Homogeneous monoliths are shown in Figure 21. These types of systems have been used for a variety of drugs[26-31,80-93] and complement other controlled release devices

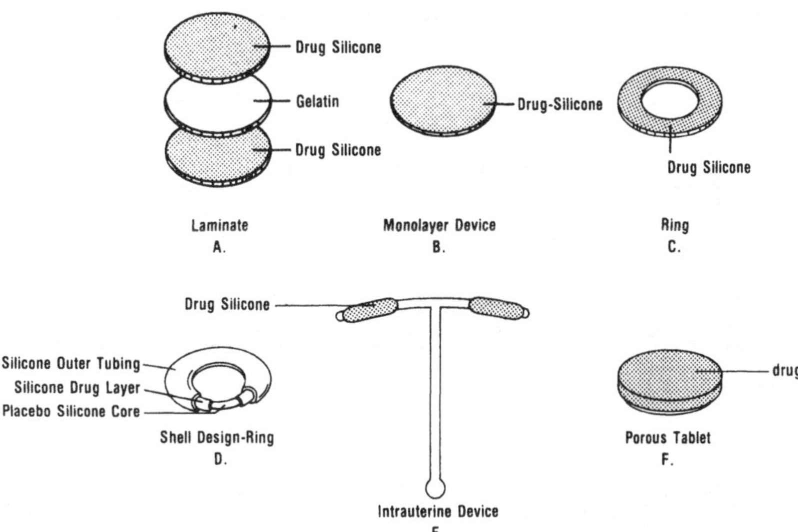

FIGURE 21. Some examples of controlled release monolithic delivery systems of pharmaceutical interest. (A to C from Reference 29; D from Reference 92; E from Reference 88.)

which were used in animals and clinical trials. For instance, drug-filled capsules are frequently employed to provide a continuous dosing of drug.[94-102] The capsular delivery system is one example of a reservoir-type device which has been considered elsewhere in this book.

A vaginal ring containing the contraceptive agent medroxyprogesterone acetate supports the concept of a continuous dosing pattern from a monolithic device. Compared to conventional tablets which are taken daily, a single ring provides reasonably uniform blood levels and eliminates the peak and valley concentrations associated with tablets (Figure 22). Interestingly, this constancy of concentration results from a device which is expected to release drug according to a square root of time dependence and suggests that the diffusion layer control mechanism is operative. Other steroids were incorporated into vaginal rings for clinical evaluation. Serum levels of drug were prolonged, but did not show the degree of constancy which was attained with medroxyprogesterone acetate. These differences are most likely related to changes in the matrix permeability rates among the steroids. To overcome this problem, modifications in the basic ring design were made. Burton et al.[93] prepared rings which contained a time-invariant diffusional barrier which controlled the release of steroid. Reasonably constant release was achieved by this method.

In an effort to eliminate irregular break-through bleeding patterns and maintain a high level of contraceptive efficacy, Mishell[92] studied the clinical performance of a multilayered vaginal ring (Figure 21) containing two hormones, a progestogen and estrogen. Clinical acceptance of rings with the combination was superior to single therapy with the progestogen alone. Fertility regulation with intrauterine devices containing steroids has also proven to be an effective method. The advantage of this approach is that the drug exerts its effect mainly at the receptor organ, thereby reducing side effects associated with systemic administration. An example of a monolithic matrix containing norgestrel attached to an intrauterine device is depicted in Figure 21.[88]

Sustained uterine stimulation and abortifacient activity of 15(S)-15-methyl prostaglandin F2 Alpha methyl ester vaginal laminates and rings were demonstrated in monkeys.[29] Monolayer devices were tested clinically and were shown to have potential as a

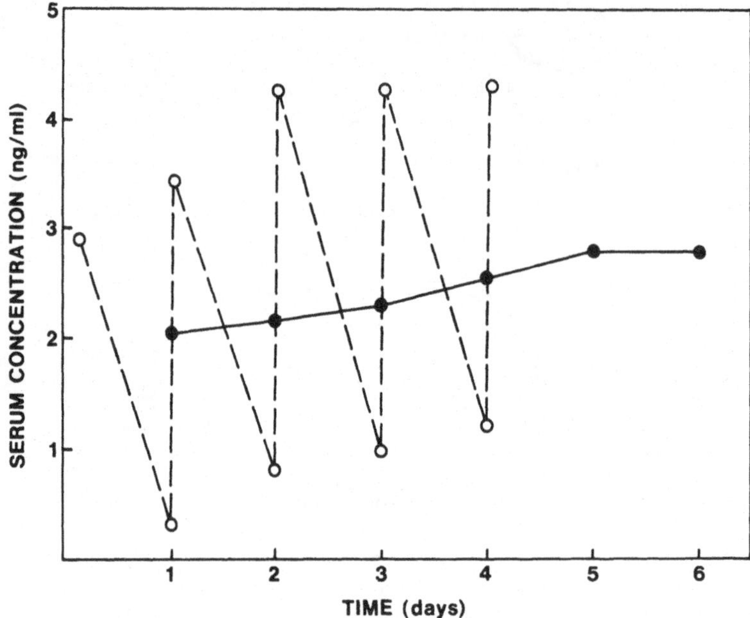

FIGURE 22. Human serum blood levels of medroxyprogesterone acetate from oral tablets[109] and a vaginal ring.[40] Values are averaged from different patients. Key: O, tablet; ●, ring.

means of releasing the prostaglandin continuously with a single insertion of a device.[103] Such an approach circumvents the intermittent dosing schedule required with vaginal suppositories. In contrast to studies using steroids, these devices were designed for short-term administration.

Porous or granular monoliths which require a leaching solvent to penetrate the matrix are employed in the development of sustained release tablets. After oral ingestion, digestive fluids within the gastrointestinal tract diffuse into the tablet and release drug for subsequent absorption. The remaining skeleton of the carrier is excreted intact in the feces. A more gradual release of drug is achieved, and the dosing regimen is reduced. Drugs administered with this dosage form include, methamphetamine hydrochloride, ferrous sulfate, hexocyclium methylsulfate, lithium, alprenolol, potassium chloride, prednisolone, and hyoscyamine, and some of these are available commercially.[70,104]

V. SUMMARY

Conventional methods of drug delivery are associated with nonconstancy of blood levels. The blood level profile is characterized by concentration peaks and valleys which result from multiple dosing regimens. Controlled release systems are designed to reduce the dosing intervals and provide a uniform drug concentration within the body throughout the period of therapy. This approach reduces patient exposure to potential side effects by maintaining the blood concentration at the therapeutic level and below the toxic level.

Monolithic devices represent one type of controlled release delivery system for therapeutic agents. A monolith contains either dissolved or dispersed drug within a polymeric carrier. Upon contact with environmental fluids, the release process is activated and drug diffuses through the polymer or channels within its structure to the surface, where absorption takes place. Mathematical expressions are presented which describe the kinetics of release as a function of the design and geometry of the monolith. Experimental evidence supports the diffusional concepts which are utilized in the derivation of the release rate expressions.

Factors which influence the observed release kinetics are reviewed in terms of the physical chemical properties of the diffusant and system variables. Special consideration is given to the importance of the boundary diffusion layer and how it affects the drug release mechanism. Pharmaceutical applications of monoliths are discussed with respect to their design and evaluation.

SYMBOLS*

A, total concentration in matrix (mg/cm^3)
C, concentration (mg/cm^3)
C_a, solubility in eluting solvent (mg/cm^3)
C'_a, concentration in the solvent at the polymer-solvent interface (mg/cm^3)
C_B, concentration in the eluting solvent (mg/cm^3)
C_o, initial concentration in the polymer (mg/cm^3)
C_s, solubility in the polymer (mg/cm^3)
C'_s, concentration in the polymer at the polymer-solvent interface (mg/cm^3)
D, diffusion coefficient (cm^2/sec)
D_a, diffusion coefficient in the eluting solvent (cm^2/sec)
D_e, effective diffusion coefficient in the polymer, $D_s\phi\tau$ (cm^2/sec)
D_s, diffusion coefficient in the polymer (cm^2/sec)
F, fraction released
h, thickness (cm)
h_a, diffusion layer thickness (cm)
h_c, height of a cylinder (cm)
J, steady-state flux (mg/cm^2-sec)
J'_{max} maximum flux (mg/cm^2-sec)
k, $(2C_sD_s/Ah^2)^{1/2}$ $(sec)^{-1/2}$
k', $(C_sD_s/Ar^2)^{1/2}$ $(sec)^{-1/2}$
k_v, k_M constants (cm^2/sec)
K, partition coefficient (concentration in the polymer/concentration in the eluting solvent)
K_n, partition coefficient of homolog of chain length n
K_o, partition coefficient of the reference homolog
l, zone of depletion thickness (cm)
M, molecular weight (g)
n, number of carbons in alkyl chain
N, Avogadro's number
Q_t, amount released per unit area at time, t (mg/cm^2)
Q_∞, amount released per unit area at time, infinity (mg/cm^2)
Q_t, total amount released at time, t (mg)
r, radius (cm)
r', radius of undepleted polymer (cm)

* A uniform notation is presented for the units of the symbols. Selection of the units throughout the text was arbitrary, but consistency was maintained within a given set of data.

R, Boltzman gas constant (cal/mol −°K)
ΔS_f, entropy of fusion (cal/mol °K)
S_M, constant (cm²/sec − g)
S_v, constant (mol/cm −sec)
t, time (sec)
T, absolute temperature (°K)
T_m, melting point (°K)
V, partial molar volume (cm³/mol)
V_d, molar volume (cm³/mol)
x, distance (cm)

Greek Letters

α, $2AD_s$ (mg/cm −sec)
β, $D_s h_a A/D_u$ (mg/cm²)
γ, $h_u K/D_u$ (sec/cm)
ε, porosity
δ_d, δ_p, solubility parameters of solute and polymer, respectively $(cal/cm^3)^{1/2}$
η, viscosity (g/cm −sec)
ϱ, density (mg/cm³)
σ, constant
τ, tortuosity
ϕ_1, volume fraction of matrix
χ, interaction parameter between the solute and the polymer

REFERENCES

1. **Tanquary, A. C. and Lacey, R. E., Eds.,** Controlled release of biologically active agents, *Advances inExperimental Medicine and Biology,* Vol. 47, Plenum Press, New York, 1974.
2. **Paul, D. R. and Harris, F. W., Eds.,** *Controlled Release Polymeric Formulations,* ACS Symposium Series 33, American Chemical Society, Washington, D.C., 1976.
3. **Baker, R. W. and Lonsdale, H. K.,** Controlled release: mechanisms and rates, in *Advances in Experimental Medicine and Biology,* Vol. 47, Tanquary, A. C. and Lacey, R. E., Eds., Plenum Press, New York, 1974, 15.
4. **Barrer, R. M.,** *Diffusion In and Through Solids,* Cambridge University Press, London, 1951.
5. **Crank, J.,** *The Mathematics of Diffusion,* Oxford University Press, New York, 1956.
6. **Jost, W.,** *Diffusion in Solids, Liquids and Gases,* Academic Press, New York, 1960.
7. **Jacobs, M. H.,** *Diffusion Processes,* Springer-Verlag, New York, 1967.
8. **Collins, R. L. and Doglia, S.,** Theory of controlled release of biologically active substances, *Arch. Environ. Contam. Toxicol.,* 1, 325, 1973.
9. **Higuchi, T.,** Physical chemical analysis of percutaneous absorption process from creams and ointments, *J. Soc. Cosmet. Chem.,* 11, 85, 1960.
10. **Higuchi, W. I.,** Analysis of data on the medicament release from ointments, *J. Pharm. Sci.,* 51, 802, 1962.
11. **Fu, J. C., Hagemeir, C., Moyer, D. L., and Ng, E. W.,** A unified mathematical model for diffusion from drug-polymer composite tablets, *J. Biomed. Mater. Res.,* 10, 743, 1976.
12. **Cobby, J., Mayersohn, M., and Walker, G. C.,** Influence of shape factors on kinetics of drugs release from matrix tablets. 1. Theoretical, *J. Pharm. Sci.,* 63, 725, 1974.
13. **Higuchi, T.,** Rate of release of medicaments from ointment bases containing drugs in suspension, *J. Pharm. Sci.,* 50, 874, 1961.
14. **Roseman, T. J.,** Release of steroids from a silicone polymer, *J. Pharm. Sci.,* 61, 46, 1972.
15. **Higuchi, T.,** Mechanism of sustained-action medication, *J. Pharm. Sci.,* 52, 1145, 1963.

16. **Paul, D. R. and McSpadden, S. K.**, Diffusional release of a solute from a polymer matrix, *J. Membr. Sci.*, 1, 33, 1976.
17. **Desai, S. J., Singh, P., Simonelli, A. P., and Higuchi, W. I.**, Investigation of factors influencing release of solid drug dispersed in inert matrices. II. Quantitation of procedures, *J. Pharm. Sci.*, 55, 1224, 1966.
18. **Singh, P., Desai, S J., Simonelli, A. P., and Higuchi, W. I.**, Release rates of solid drug mixtures dispersed in inert matrices. I. Non-interacting drug mixtures, *J. Pharm. Sci.*, 56, 1542, 1967.
19. **Singh, P., Desai, S. J., Simonelli, A. P., and Higuchi, W. I.**, Release rates of solid drug mixtures dispersed in inert matrices. II. Mutually interacting drug mixtures, *J. Pharm. Sci.*, 56, 1548, 1967.
20. **Levich, V.**, *Physicochemical Hydrodynamics*, Prentice-Hall, Englewood Cliffs, N.J., 1962, Chap. 2.
21. **Roseman, T. J. and Higuchi, W. I.**, Release of medroxyprogesterone acetate from a silicone polymer, *J. Pharm. Sci.*, 59, 353, 1970.
22. **Fites, A. L., Banker, G. S., and Smolen, V. F.**, Controlled drug release through polymeric films, *J. Pharm. Sci.*, 59, 610, 1970.
23. **Donbrow, M. and Friedman, M.**, Timed release from polymeric films containing drugs and kinetics of drug release, *J. Pharm. Sci.*, 64, 76, 1975.
24. **Borodkin, S. and Tucker, F. E.**, Linear drug release from laminated hydroxypropyl cellulose — polyvinyl acetate films, *J. Pharm. Sci.*, 64, 1289, 1975.
25. **Sciarra, J. J. and Patel, S. P.**, *In vitro* release of therapeutically active ingredients from polymer matrices, *J. Pharm. Sci.*, 65, 1519, 1976.
26. **Kalkwarf, D. R., Sikov, M. R., Smith, L., and Gordon, R.**, Release of progesterone from polyethylene devices *in vitro* and in experimental animals, *Contraception*, 6, 423, 1972.
27. **Chien, Y. W. and Lau, E. P. K.**, Controlled drug release from polymeric delivery devices. IV. *In vitro-in vivo* correlation of subcutaneous release of Norgestomet from hydrophilic implants, *J. Pharm. Sci.*, 65, 488, 1976.
28. **Haleblian, J., Runkel, R., Mueller, N., Christopherson, J., and Ng, K.**, Steroid release from silicone elastomer containing excess drug in suspension, *J. Pharm. Sci.*, 60, 541, 1971.
29. **Spilman, C. H., Beuving, D. C., Forbes, A. D., Roseman, T. J., and Larion, L. J.**, Evaluation of vaginal delivery systems containing 15(S)15-methyl $PGF_{2\alpha}$ methyl ester, *Prostaglandins*, 12(Suppl.) 1, 1976.
30. **Ormsbee, H. S., III and Ryan, C. F.**, Production of hypertension with desoxycorticosterone acetate-impregnated silicone rubber implants, *J. Pharm. Sci.*, 62, 255, 1973.
31. **Neil, G. L., Scheidt, L. G., Kuentzel, S. L., and Moxley, T. E.**, Effectiveness of antitumor agents administered subcutaneously to L1210 leukemic mice in silicone rubber devices, *Chemotherapy (Basel)*, 18, 27, 1973.
32. **Desai, S. J., Simonelli, A. P., and Higuchi, W. I.**, Investigation of factors influencing release of solid drug dispersed in inert matrices, *J. Pharm. Sci.*, 54, 1459, 1965.
33. **Schwartz, J. B., Simonelli, A. P., and Higuchi, W. I.**, Drug release from wax matrices. I. Analysis of data with first-order kinetics and with the diffusion controlled model, *J. Pharm. Sci.*, 57, 274, 1968.
34. **Lapidus, H. and Lordi, N. G.**, Some factors affecting the release of a water soluble drug from a compressed hydrophilic matrix, *J. Pharm. Sci.*, 55, 840, 1966.
35. **Farhadieh, B., Borodkin, S., and Buddenhagen, J. D.**, Drug release from methyl acrylate-methyl methacrylate copolymer matrix. I. Kinetics of release, *J. Pharm. Sci.*, 60, 209, 1971.
36. **Sjögren, J. and Fryklöf, L. E.**, Duretter — a new type of oral sustained action preparation, *Sartryck ur Farmacevtisk Revy*, 59, 171, 1960.
37. **Chien, Y. W., Lambert, H. J., and Grant, D. E.**, Controlled drug release from polymeric devices. I. Technique for rapid *in vitro* release studies, *J. Pharm. Sci.*, 63, 365, 1974.
38. **Winkler, V. W., Borodkin, S., Webel, S. K., and Mannebach, J. T.**, *In vitro* and *in vivo* considerations of a novel matrix-controlled bovine progesterone-releasing intravaginal device, *J. Pharm. Sci.*, 66, 816, 1977.
39. **Chien, Y. W. and Lambert, H. J.**, Controlled drug release from polymeric delivery devices. II. Differentiation between partition controlled and matrix controlled drug release mechanisms, *J. Pharm. Sci.*, 63, 515, 1974.
40. **Roseman, T. J. and Yalkowsky, S. H.**, Importance of solute paritioning on the kinetics of drug release from matrix systems, in *Controlled Release Polymeric Formulations*, Paul, D. R. and Harris, F. W., Eds., ACS Symposium Series 33, American Chemical Society, Washington, D.C., 33, 1976, 33.
41. **Baker, R. W. and Lonsdale, H. K.**, Membrane controlled delivery systems, in *Proc. Int. Controlled Release Pesticide Symp.*, Cardarelli, N. F., Ed., University of Aron, Ohio, September 16 to 18, 1974.

42. **Ayres, J. W. and Lasker, P. A.**, Evaluation of mathematical models for diffusion from semisolids, *J. Pharm. Sci.*, 63, 351, 1974.
43. **Carelli, V., DiColo, G., Nannipieri, E., Saettone, M. F., and Serafini, M. F.**, Release of drugs from ointment bases, III. Influence of the membrane and receiving phase on *in vitro* release of progesterone from two different vehicles, *Farmaco. Ed. Prat.*, 32, 591, 1977.
44. **Baker, R. W. and Lonsdale, H. K.**, Controlled delivery — an emerging use for membranes, *Chem. Technol.*, 5, 668, 1975.
45. **Desai, S. J., Singh, P., Simonelli, A. P., and Higuchi, W. I.**, Investigation of factors influencing release of solid drug dispersed in inert matrices. III. Quantitative studies involving the polyethylene plastic matrix, *J. Pharm. Sci.*, 55, 1230, 1966.
46. **Desai, S. J., Singh, P., Simonelli, A. P., and Higuchi, W. I.**, Investigation of factors influencing release of solid drug dispersed in inert matrices. IV. Some studies involving the polyvinyl choride matrix, *J. Pharm. Sci.*, 55, 1235, 1966.
47. **Schwartz, J. B., Simonelli, A. P., and Higuchi, W. I.**, Drug release from wax matrices. II. Application of a mixture theory to the sulfanilamide-wax system, *J. Pharm. Sci.*, 57, 278, 1968.
48. **Lapidus, H. and Lordi, N. G.**, Drug release from compressed hydrophilic matrices, *J. Pharm. Sci.*, 57, 1292, 1968.
49. **Michaels, A. S. and Bixler, H. J.**, Membrane permeation: theory and practice, in *Progress in Separation and Purification*, Vol. 1, Perry, E. S., Ed., Interscience, New York, 1968, 143.
50. **Tuwiner, S. B.**, *Diffusion and Membrane Technology*, Van Nostrand Reinhold, New York, 1962.
51. **Crank, J. and Park, G. S., Eds.**, *Diffusion in Polymers*, Academic Press, New York, 1968.
52. **Edward, J. T.**, Molecular volumes and the Stokes-Einstein equation, *J. Chem. Educ.*, 47, 261, 1970.
53. **Flynn, G. L., Yalkowsky, S. H., and Roseman, T. J.**, Mass transport phenomena and models: theoretical concepts, *J. Pharm. Sci.*, 63, 479, 1974.
54. **Barrer, R. M.**, Permeation, diffusion and solution of gases in organic polymers, *Trans. Faraday Soc.*, 35, 628, 1939.
55. **Leo, A., Hansch, C., and Elkins, D.**, Partition coefficients and their uses, *Chem. Rev.*, 71, 525, 1971.
56. **Roseman, T. J. and Yalkowsky, S. H.**, Influence of solute properties on release of *p*-aminobenzoic acid esters from silicone rubber: theoretical considerations, *J. Pharm. Sci.*, 63, 1639, 1974.
57. **Chien, Y. W., Lambert, H. J., and Lin, T. K.**, Solution-solubility dependency of controlled release of dug from polymer matrix: mathematical analysis, *J. Pharm. Sci.*, 64, 1643, 1975.
58. **Flynn, G. L., Ho, N. F. H., Hwang, S., Owada, E., Molokhia, A., Behl, C. R., Higuchi, W. I., Yotsuyanagi, T., Shah, Y., and Park, J.**, Interfacing matrix release and membrane absorption — analysis of steroid absorption from a vaginal device in the rabbit doe, in *Controlled Release Polymeric Formulations*, Paul, D. R., and Harris, F. W., Eds., ACS Symposium Series 33, American Chemical Society, Washington, D.C., 1976, 87.
59. **Yalkowsky, S. H. and Flynn, G. L.**, Transport of alkyl homologs across synthetic and biological membranes: a new model for chain length-activity relationships, *J. Pharm. Sci.*, 62, 210, 1973.
60. **Hildebrand, J. H. and Scott, R. L.**, *Solubility of Nonelectrolytes*, Van Nostrand Reinhold, New York, 1950.
61. **Chien, Y. W.**, Thermodynamics of controlled drug release from polymeric delivery devices, in *Controlled Release Polymeric Formulations*, Paul, D. R. and Harris, F. W., Eds., ACS Symposium series 33, American Chemical Society, Washington, D.C., -976, 53.
62. **Michaels, A. S., Wong, P. S. L., Prather, R., and Gale, R. M.**, A thermodynamic method of predicting the transport of steroids in polymer matrices, *AIChE J.*, 21, 1073, 1975.
63. **Morton, M., Ed.**, *Rubber Technology*, Van Nostrand Reinhold, New York, 1973.
64. **Most, C. F., Jr.**, Some filler effects on diffusion in silicone rubber, *J. Appl. Polym. Sci.*, 14, 1019, 1970.
65. **Flynn, G. L. and Roseman, T. J.**, Membrane diffusion. II. Influence of physical adsorption on molecular flux through heterogeneous dimethylpolysiloxane barriers, *J. Pharm. Sci.*, 60, 1788, 1971.
66. **Crank, J. and Park, G. S., Eds.**, *Diffusion in Polymers*, Academic Press, New York, 1968, 202.
67. **Singh, P., Desai, S. J., Simonelli, A. P., and Higuchi, W. I.**, Role of wetting on the rate of drug release from inert matrices, *J. Pharm. Sci.*, 57, 217, 1968.
68. **Dakkuri, A., Schroeder, H. G., and DeLuca, P. P.**, Sustained release from inert wax matrixes. II. Effect of surfactants on tripelennamine hydrochloride release, *J. Pharm. Sci.*, 67, 354, 1978.
69. **Farhadieh, B., Borodkin, S., and Buddenhagen, J. D.**, Drug release from methyl acrylate-methyl methacrylate copolymer matrix. II. Control of release rate by exposure to acetone vapor, *J. Pharm. Sci.*, 60, 212, 1971.
70. **Rowe, R. C.**, Sustained release plastic matrix tablets, *Manuf. Chem. Aerosol News*, 46, 23, 1975.
71. **Davis, B. K.**, Diffusion in polymer gel implants, *Proc. Natl. Acad. Sci. U.S.A.*, 71, 3120, 1974.
72. **Folkman, J.**, Controlled drug release from polymers, *Hosp. Pract.*, 13, 127, 1978.

73. **Abrahams, R. A. and Ronel, S. H.,** Biocompatible implants for the sustained zero-order release of narcotic antagonists, *J. Biomed. Mater. Res.,* 9, 355, 1975.
74. **Anderson, J. M., Koinis, T., Nelson, T., Horst, M., and Love, D. S.,** The slow release of hydrocortisone sodium succinate from poly (2-hydroxyethyl methacrylate) membranes, in *Hydrogels for Medical and Related Applications,* Andrade, J. D., Ed., ACS Symposium Series 31, American Chemical Society, Washington, D.C., 1976, 167.
75. **Cowsar, D. R., Tarwater, O. R., and Tanquary, A. C.,** Controlled release of fluoride from hydrogels for dental applications, in *Hydrogels for Medical and Releated Applications,* Andrade, J. D., Ed., ACS Symposium Series 31, American Chemical Society, Washington, D.C., 1976, 180.
76. **Roseman, T. J.,** Silicone rubber: a drug delivery system for contraceptive steroids, in *Advances in Experimental Medicine and Biology,* Vol. 47, Tanquary, A. C. and Lacey, R. E., Eds., Plenum Press, New York, 1974, 99.
77. **Konning, G. H. and Mital, H. C.,** Sheabutter V: effect of particle size on release of medicament from ointment, *J. Pharm. Sci.,* 67, 374, 1978.
78. **Ayres, J. W. and Lindstrom, F. T.,** Diffusion model for drug release from suspensions. I. Theoretical considerations, *J. Pharm. Sci.,* 66, 654, 1977.
79. **Lindstrom, F. T. and Ayres, J. W.,** Diffusion model for drug release from suspensions. II. Release to a perfect sink, *J. Pharm. Sci.,* 66, 662, 1977.
80. **Long, D. M., Sehgal, L. R., DeRios, M. E., Rios, M. V., Szanto, P. B., and Forrest, R.,** Depot cancer chemotherapy through polymer membranes, *Rev. Surg.,* 30, 229, 1973.
81. **Fu, J. C., Kale, A. K., and Moyer, D. L.,** Drug-incorporated silicone discs as sustained release capsule. I. Chloroqine diphosphate, *J. Biomed. Mater. Res.,* 7, 71, 1973.
82. **Shippy, R. L., Hwang, S. T., and Bunge, R. G.,** Controlled release of testosterone using silcone rubber, *J. Biomed Mater. Res.,* 7, 95, 1973.
83. **Henzl, M. R., Mishell, D. R., Jr., Velazquez, J. G., and Leitch, W. E.,** Basic studies for prolonged progestogen administration by vaginal devices, *Am. J. Obstet. Gynecol.,* 117, 102, 1973.
84. **Mishell, D. R., Jr., Lumkin, M., and Jackanicz, T.,** Initial clinical studies of intravaginal rings containing norethindrone and norgestrel, *Contraception,* 12, 253, 1975.
85. **Doyle, L. L., Stryker, J. C., Clewe, T. H., and Harrison, R. A.,** Selection of steroids for incorporation into silastic intrauterine devices, *J. Steroid Biochem.,* 6, 885, 1975.
86. **Johansson, E. D. B., Luukkainen, T., Vartiainen, E., and Victor, A.,** The effect of progestin R 2323 released from vaginal rings on ovarian function, *Contraception,* 12, 299, 1975.
87. **Akinla, O., Lähteenmaki, P., and Jackanicz, T. M.,** Intravaginal contraception with synthetic progestin, R 2323, *Contraception,* 14, 671, 1976.
88. **Nilsson, C. G., Johansson, E. D. B., and Luukkainen, T.,** A *d*-norgestrel-releasing IUD, *Contraception,* 13, 503, 1976.
89. **Thiery, M., Vadekerckhove, D., Dhont, M., Vermeulen, A., and Decoster, J. M.,** The medroxyprogesterone acetate intravaginal silastic ring as a contraceptive device, *Contraception,* 13, 605, 1976.
90. **Chien, Y. W., Lambert, H. J., and Rozek, L. F.,** Controlled release of deoxycorticosterone acetate from matrix-type silicone devices: *in vitro-in vivo* correlation and prolonged hypertension animal model for cardiovascular studies, in *Controlled Release Polymeric Formulations,* ACS Symposium Series 33, American Chemical Society, Paul, D. R., and Harris, F. W., Eds., American Chemical Society, Washington, D.C., 1976, 72.
91. **Mishell, D. R., Jr., Roy, S., Moore, D. E., Brenner, P. F., Page, M. A., Gentzschein, E., and Fisk, P. D.,** Clinical performnces and endocrine profiles with contraceptive vaginal rings containing *d*-norgestrel, *Contraception,* 16, 625, 1977.
92. **Mishell, D. R., Jr., Moore, D. E., Roy, S., Brenner, P. F., and Page, M. A.,** Clinical performance and endocrine profiles with contraceptive vaginal rings containing a combination of estradiol and *d*-norgestrel, *Am. J. Obstet. Gynecol.,* 130, 55, 1978.
93. **Burton, F. G., Skiens, W. E., Gordon, N. R., Veal, J. T., Kalkwarf, D. R., and Duncan, G. W.,** Fabrication and testing of vaginal contraceptive devices designed for release of prespecified dose levels of steroids, *Contraception,* 17, 221, 1978.
94. **Folkman, J. and Long, D. M.,** The use of silicone rubber as a carrier for prolonged drug therapy, *J. Surg. Res.,* 4, 139, 1964.
95. **Dziuk, P. J. and Cook, B.,** Passage of steroids through silicone rubber, *Endocrinology,* 78, 208, 1966.
96. **Ermini, M., Carpino, F., Russo, M., and Benagiano, G.,** Studies on sustained contraceptive effects with subcutaneous polydimethylsiloxane implants, III. Factors affecting steroid diffusion *in vivo* and *in vitro, Acta Endocrinol. (Copenhagen),* 73, 360, 1973.
97. **Kulkarni, B. D., Avila, T. D., Pharriss, B. B., and Scommegna, A.,** Release of ^{3}H-progesterone from polymeric systems in the rhesus monkey, *Contraception,* 8, 299, 1973.

98. **Rosenblum, M. L., Bowie, D. L., and Walker, M. D.,** Diffusion *in vitro* and *in vivo* of 1-(2-chloroethyl)-3-(trans-4-methylcyclohexyl)-1-nitrosourea from silicone rubber capsules, a potentially new mode of chemotherapy administration, *Cancer Res.*, 33, 906, 1973.
99. **Scommegna, A., Avila, T., Luna, M., Rao, R., and Dmowski, W. P.,** Fertility control by intrauterine release of progesterone, *Obstet. Gynecol.*, 43, 769, 1974.
100. **Kent, J. S.,** Controlled release of delmadinone acetate from silicone rubber tubing: *in vitro-in vivo* correlations to diffusion model, in *Controlled Release Polymeric Formulations,* ACS Symposium Series 33, Paul, D. R. and Harris F. W., Eds., American Chemical Society, Washington, D.C., 1976, 157.
101. **Bahgat, M. R. and Atkinson, L. E.,** Contraceptive steroid administration by subdermal implants: serum concentrations of R-2323, estrogen, and progesterone in rhesus monkeys, *Contraception,* 15, 335, 1977.
102. **Diaz, S., Pavez, M., Quinteros, E., Robertson, D. N., and Croxatto, H. B.,** Clinical trial with subdermal implants of the progestin R-2323, *Contraception,* 16, 155, 1977.
103. **Ramwell, P. W., Ed.,** *Prostaglandins,* 12(Suppl.), 1976.
104. **Osol, A. and Hoover, J. E., Eds.,** *Remington's Pharmaceutical Sciences,* Mack Publishing, Easton, Pa., 1975, 1625.
105. **Lacey, R. E. and Cowsar, D. R.,** Factors affecting the release of steroids from silicones, in *Advances in Experimental Medicine and Biology,* Vol. 47, Tanquary, A. C. and Lacey, R. E., Eds., Penum Press, New York, 1974, 117.
106. **Gonzales, M. A., Nematollahi, J., Guess, W. L., and Autian, J.,** Diffusion, permeation, and solubility of selected agents in and through polyethylene, *J. Pharm. Sci.*, 56, 1288, 1967.
107. **Baker, R. W., Tuttle, M. E., Lonsdale, H. K., and Ayres, J. W.,** Development of an estriol-releasing intrauterine device, *J. Pharm. Sci.*, 68, 20, 1979.
108. **Vezin, W. R. and Florence, A. T.,** Diffusion of small molecules in amorphous polymers, *J. Pharm. Pharmacol.*, 29(Suppl.), 44, 1977.
109. **Hiroi, M., Stanczyk, F. Z., Goebelsmann, U., Brenner, P. F., Lumkin, M. E., and Mishell, D. R., Jr.,** Radioimmunoassay of serum medroxyprogesterone acetate (Provera®) in women following oral and intravaginal administration, *Steroids,* 26, 373, 1975.

SECTION 2

Nate F. Cardarelli

TABLE OF CONTENTS

I. INTRODUCTION

Synthetic polymers are conveniently classified as elastomers ("rubbers") and plastomers ("plastics"). Elastomers possess relatively weak interchain forces, giving rise to their elastic property. Plastics, in contrast, have moderate interchain forces so that if such molecules are extended through an applied stress, that orientation is retained after the external force is removed. Plastics are categorized as "thermoplastic" and "thermosetting". Thermoplastics can be repeatedly softened and reshaped through the use of heat, whereas thermosetting polymers once melted and cooled undergo extensive crosslinking so that remelting can not be performed without degradation. An intermediate group of materials possessing both elastic and plastic properties also exist and at one time were designated as "elastoplastics". Ethylene-propylene-diene terpolymers, polyethylene vinyl acetate, polyvinyl acetate, certain modified acrylics, and others fell in this class.

Elastomers and several elastoplastics are discussed in the next chapter in terms of their usage as base matrices in controlled release formulations. This section will be concerned mainly with plastics as binding matrices for pesticides and other additives wherein the agent is monolithically dispersed.

II. GENERAL CONSIDERATIONS

The solubility of a material in a polymer is determined by intermolecular forces. Where such forces are weak, as with the elastomers, a wide variety of substances will dissolve therein, the degree depending upon the respective solubility parameter. The stronger interchain forces characteristic of plastics generally negate the possibility of dissolving a desired agent in that type of matrix. Consequently, a diffusion-dissolution mode of agent release probably can not be developed.

In order to establish a uniform and continuous release of agent molecules from the surface of a given plastic dispenser, obviously (1) the agent must be incorporable in that plastic, and (2) the agent molecules must in some manner migrate through the matrix and towards a depleting surface. Migration of a solid through a solid or a liquid through a solid can occur through an interstitial or vacancy mechanism.[1]

Polymeric materials contain unoccupied space, a free volume, within the matrix. The degree and nature of this free volume has been extensively used in the interpretation of polymer properties, including small molecule diffusion.[2] An increase in free volume in a given material increases the mobility of polymer chain segments which gives rise to an increase in the diffusion coefficient, at least for relatively small molecules.

Processed plastics contain voids of varying magnitude, but are basically dependent upon processing conditions and incorporated additives. These voids are dimensionally larger and not to be confused with free volume. For simplicity we will term the larger voids and their interconnective channels "pores" and the network, "porosity". Free volume is basically a thermodynamic or kinetic quantity, whereas porosity is not.

The effects of free volume and micropore structure on transport dynamics has been elucidated by Rogers[3] and Crank.[4]

Utilization of plastic matrices rather than elastomeric materials in controlled release applications is economically attractive for many uses. Raw materials cost is usually less, and more importantly, processing costs tend to be much less expensive. The vulcanization step usually required with elastomers and inherent problems involved in comminution, such as the need for cryogenic cooling, are mainly responsible for the associated high product costs.

III. RELEASE MECHANISMS

Conceptually, agent release from a plastic material can be effected through several processes: leaching, wherein the ingress of a liquid element, such as water, partitions the agent which then migrates outward following the pore structure; volatilization, wherein the agent molecule in a gaseous state moves in a random motion, making interstitial jumps or following a free volume network, reaches the dispenser surface passing into the external medium; and matrix degradation due to chemical, biological, or physical stresses. The latter method does not, so far, appear promising in the sense of controlled emission. Also, most plastics are environmentally inert, and decomposition under natural influences may require decades. (In fairness though, very little work has been done in the evaluation of the hylobiology and general degradation pathways, intermediates, ultimate products, and the like arising from high polymers exposed to given environmental stresses.)

Leaching as a mechanism of agent release is only tenable in an aquatic situation (water and moist soil). Even so, extraordinary methods may be required to provide long-term, continuous, effective agent emission. Both the degree of porosity and the associated tortuosity of the pore structure are controlling. In general, plastics are not particularly permeable to water, the elastoplastics being exceptions, and even where transmission does occur, the agent particle, which will be considerably larger than the water molecule, may not be accommodated by the pore size developed.

If agent loading is adequately high, then pore growth can be realized through its gradual partition and loss. This method is used with the major class of antifouling paints where cuprous oxide is mechanically bound by an elastomeric or plastic film forming polymer. In this case, as agent molecules on a near interface disperse in the contacting water, pore growth is initiated exposing further agent. Consequently, a gradual development of a porous structure occurs, and efficacious release for up to 1 year or so occurs. However, antifouling paints of this nature must contain over 80% toxicant loading in order to realize continuous pore growth.

In true plastic matrices where no degree of agent solubility is present, the use of a coleachant to enhance porosity generation and growth will likely be essential.

Release through a volatility mechanism can take one of several forms. If the incorporated agent is volatile and the plastic permeable to the vapor phase, then diffusion to the plastic surface will occur with concomitant loss into the surrounding medium. However, experience has shown that this mode of release does not readily lend itself to practical use. Very volatile agents are not incorporable into plastics due to the necessary processing temperatures (230°F or higher). Conversely, agents stable under processing conditions almost always lack sufficient volatility for effective release. Also, plastic materials are not overly permeable to the usually large molecules of interest, such as the organophosphorous and carbamate insecticides. Ambient temperature tends to exert a significant influence on loss rate, and thus maintaining a desired release-rate range may be difficult.

The in-use solution to the noted problems associated with a volatility mechanism lies in the utilization of an additive that serves as a transmission path for the agent molecule. In such systems, the additive is dispersed at relatively high concentration throughout the matrix. The agent, which usually displays some degree of solubility within the additive, moves through it via solution forces, etc. and reaches the dispenser surface where vaporization occurs.

TABLE 1

Controlled Release Monolithic Plastic Dispensers — Commodities in Commerce

Designation	Category	Type	Manufacturer
No-Pest Strip®	Insecticide (volatile)	Plasticized PVC[a]	Shell Chemical Co., San Ramon, Calif.
Roach Tape®	Insecticide (contact)	Laminated plasticized PVC	Health Chem. Corp., New York, N.Y.
Lure 'N Kill™ Fly Tape	Insecticide	Laminated plasticized PVC	Health Chem. Corp., New York, N.Y.
Flea collars ("2-in." etc.)	Insecticide (volatile)	Plasticized PVC	A. H. Robbins Co., and others, Richmond, Va.
Dursban® CR-10	Insecticide (aquatic)	Chlorinated PE[b]	Dow Chemical Co., Midland, Mich.
Ecopro™ 1700	Insecticide	EVA/PE[c]	Environmental Chem. Inc., Barrington, Ill.
Ecopro™ 100	Molluscicide and herbicide	EPM/PE[d]	Environmental Chem. Inc., Barrington, Ill.

[a] PVC, polyvinyl chloride.
[b] PE, polyethylene.
[c] EVA, ethylene vinyl acetate.
[d] EPM, ethylene-propylene copolymers.
PP = polypropylene

IV. CONTROLLED RELEASE PRODUCTS

A. General

A number of controlled release plastic-based dispensers have been made commercially available or are in an advanced experimental stage looking towards near future marketing. The compendium presented below is not exhaustive, but rather depicts those materials known to the authors.

Table 1 excludes the numerous plastic-based antifouling paints. An extensive literature has developed over the years describing the theory, formulation, evaluation, and efficacy of such materials. References 5 through 7 are convenient sources for review of this area.

Formulations, processing technology, and the like for many of the above listed products are trade secrets and consequently not accessible to the authors. Where patents have been issued the universality of claims and pointed lack of data concerning mechanisms, and generalized formulas essentially provide little insight. Consequently, the discussion to follow is primarily based upon studies performed in the author's laboratory.

B. Plasticized Polyvinyl Chloride Dispensers

The first known product based on the monolithic dispersal of an insecticide in polyvinyl chloride (PVC) is the Shell No-Pest® strip. Dichlorvos (dimethyl-2,2-dichlorovinyl phosphate), the active agent in this product, is incorporable in a number of plastics matrices. Release is through a diffusion process, and the emitted vapors are insecticidal.[9] Dichlorvos release is poor from nonplasticized polyvinyl chloride and other plastics. Only PVC compounds containing a plasticizing element have shown long-term controlled emission of dichlorvos at efficacious rates. The Shell product

TABLE 2

Controlled Release Monolithic Plastic Dispensers — Experimental Materials

Category	Type	Developing Agency	Ref.
Herbicide	Plasticized PVC[a]	U.S. Army	8
Insecticide	PE[b], EVA/PE,[c] EPM,[d] EPM/PE	Environmental Chemicals, Inc.	
Molluscicide	PE, EVA, EPM	Environmental Chemicals Inc.	
Trace nutrients	PE, PP,[e] EVA, EPM, EVA/PE, PE/PE	Environmental Chemicals Inc.	
Pheromone	Plasticized PVC	Health Chem. Corp.	5

[a] PVC, polyvinyl chloride.
[b] PE, polyethylene.
[c] EVA, ethylene vinyl acetate.
[d] EPM, ethylene-propylene copolymers.
[e] PP, polypropylene.

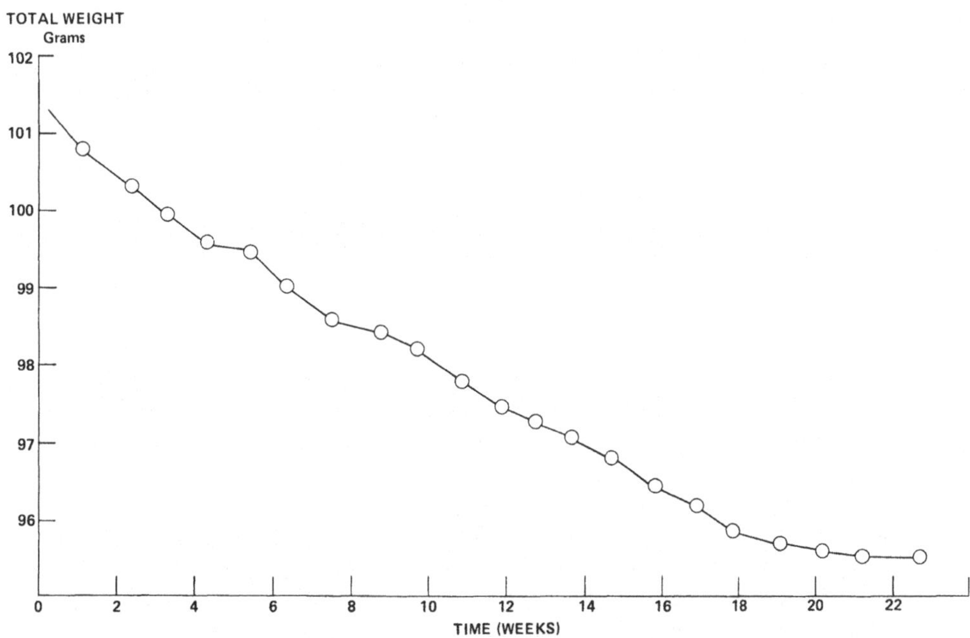

FIGURE 1. Weight loss: NO-PEST® STRIP (April 1969). 20% dichlorvos and related products.

contains 18 to 19% dichlorvos and 1 to 2% related materials by analysis. A plasticizer content of about 30% has been determined, and infrared analysis indicated the material to be dioctylphthalate. (Note: this work was performed in 1968).

In model systems made from plastisols (PVC plus plasticizer), dibutylphthalate provides similar release rates and lifetimes as those measured with the No-Pest® strip. Several phosphate-type plasticizers give longer emission life. Volatile insecticides similar to dichlorvos, such as thionodichlorvos, will probably provide as good results.[10] Figure 1 depicts the weight loss of No-Pest® strip exposed to air. Release beyond the 16th week is too low for adequate insect control under recommended use conditions. Note that total weight loss is about 6%, whereas agent content is 20%. Since both plasticizer and dichlorvos are emitted, respective rates unknown, the decrease in weight is not representative of dichlorvos loss.

In a recent improved material, "Improved Vaporette Insect Strip",* 14.57 g of a 100-g unit was volatilized in 147 days, indicating a 73% volatile loss if only the dichlorvos is emitted.

Temephos *(O,O,O′,O′*-tetramethyl *O,O*-thiodi-*p*-phenylene phosphorothioate) has been incorporated in plasticized PVC and effective release noted over a 12-week period in evaluations against mosquito larva.[11]

It has been noted in unpublished work conducted at the Environmental Management Laboratory of the University of Akron, Ohio, that a number of insecticides can be incorporated in PVC formulations having the following general composition:

PVC	40 to 60%
Plasticizer	20 to 35%
Agent	20 to 40%
Stabilizer	0 to 2%

Pesticides examined include:

Sevin®:	1-napthyl methylcarbamate
Dimethoate:	*O,O*-dimethyl *S*-(*N*-methylcarbamoylmethyl) phosphorodithioate
Ethyl trithion:	*S*-(*p*-chlorophenylthio)methyl) *O,O*-diethyl phosphorodithioate
Methyl trithion:	*S*-((*p*-chlorophenyl thio)methyl) *O,O*-dimethyl phosphorodithioate
Diazinon®:	*O,O*-diethyl-*O*-(2-isopropyl-6-methyl-5-pyrimidinyl) phosphorothioate
Naled®:	1,2-dibromo-2,2-dichloroethyldimethyl phosphate
Malathion®:	*O,O*-dimethyl dithiophosphate of diethylmercaptosuccinate
Temephos:	*O,O,O′,O′*-tetramethyl *O,O*-thiodi-*p*-phenylene
Chlorpyrifos:	*O,O*-diethyl *O*-3,5,6-trichloro-2-pyridyl phosphorothioate

Emission is generally low for nonplasticized formulations, while quite significant for those containing a plasticizer. Table 3 depicts release rates as measured by weight loss in air at 70°F and 70 to 75% relative humidity. Generally, there is a somewhat higher emission during the first few days after preparation, and then a steady state condition is reached. Figure 2 illustrates the difference in emission curves for plasticized and nonplasticized material. Again, it is noted that when present both plasticizer and toxicant are emitted.

Plasticizer emission varies with the type and amount. In control experiments for the formulations listed in Table 3, the emission rate at 35 parts plasticizer loading over an initial 30-day air exposure is 0.08%/day for BBP, 0.03%/day for TOP, and 0.11%/day for DOP. In all cases, the fusion temperature was 300°F for 10 min.

Sevin® and dimethoate do not release from the several PVC formulations unless a plasticizer is present. Dichlorvos and Naled® release, but very slowly; however, methyl and ethyl trithion are not compatible and release very rapidly unless slowed through the use of a plasticizer as shown in Figure 2.

Consequently, the presence of a plasticizer may increase or decrease the loss of the toxic agent. Figure 3 compares the volatility loss from three formulations with the loss of plasticizer alone.

It is believed that any volatile pesticide can be incorporated in PVC, provided that

* Zoecon Co., Palo Alto, Calif.

TABLE 3

Weight Loss of Experimental PVC/Insecticide Strips in Air — Geon 121 base matrix[a]

Insecticide	Insecticide (%)	Plasticizer	Plasticizer (%)	Emission rate volatile less /day
Sevin®	20	BBP[b]	45	0.11%
Dimethoate	20	BBP	45	1.83%
Ethyl trithion	20	BBP	35	0.23%
	20	TOP[c]	30	0.06%
	20	DOP[d]	27	0.16%
	20	TOP	27	0.08%
	40	TOP	20	0.14%
	40	DOP	27	0.19%
Naled®	20	BBP	35	0.22%
	20	TOP	30	0.98%
	20	DOP	27	0.43%

[a] Geon® 121 is a product of the B. F. Goodrich Chemical Co., Cleveland, Ohio.
[b] BBP, butylbenzylphthalate.
[c] TOP, trioctylphosphate.
[d] DOP, dioctylphthalate.

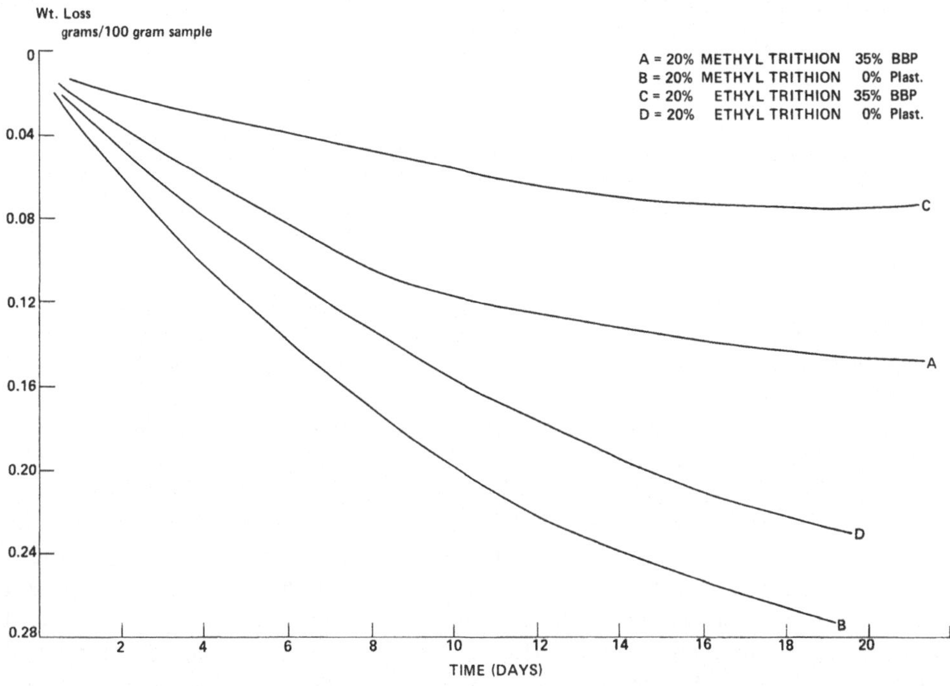

FIGURE 2. Insecticide loss from plasticized and nonplasticized PVC (Geon 121).

the necessary processing temperature is not above the decomposition point. By careful adjustment of the type and amount of plasticizer used, a desirable release rate might be achieved. Processing problems might be expected if the agent is highly volatile, and materials of low volatility may not be efficacious in a given use because delivery to the dispenser surface and/or rate of emission may be too slow.

Presently, the Health Chem Corporation has developed and commercialized a lami-

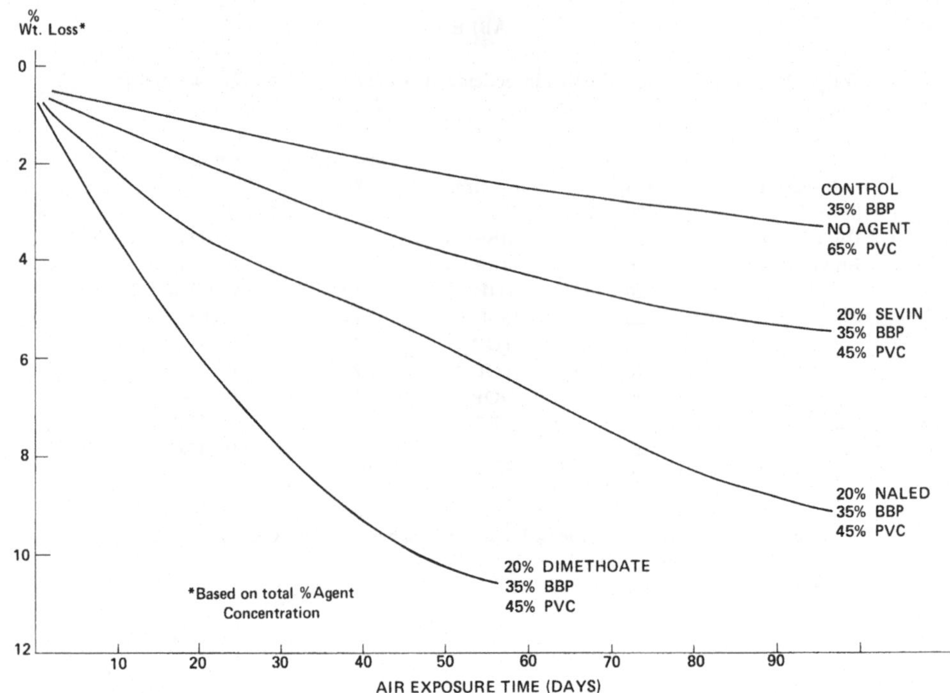

FIGURE 3. Volatile weight loss comparison.

nated PVC insecticidal tape for household insect control. Although the principles underlying movement of the pesticide, Diazinon® or Baygon®, are the same as previously discussed, a unique feature has been used to control surface loss of the toxic element — namely a semipermeable PVC membrane laminated to the surface. This type of dispenser is discussed in Chapter 5.

Experimental work has been performed in the past and presumably in the present using plasticized PVC as the base matrix. Nelson and Whitlaw showed that Diazinon® and Supracide® [*O,O,* dimethyl-*S*-(2-methoxy-1,3,4-thiadiazol-*S*-(4H)-onyl-(4)-methyl-O-dithiophosphate] could be incorporated in a plastisol, and release for an 8-week period achieved.[12] A 1:1 ratio of PVC (GEON® 135)* and dioctylphthalate were used with a 120°C × 30 min cure.

Stokes et al. prepared and evaluated controlled release formulations of Aldicarb® and dimethoate using plasticized PVC and cellulose acetate.[13] Release was rapid, and in most cases, 16% or more of the toxic element was emitted within 24 hr. Variations occurred with the type of cellulose acetate used, the type of plasticizer, and the amount of plasticizer. Cellulose acetate butyrate plasticized with triphenyl phosphate showed the slowest release rate, 8% in 24 hr. Generally, the release rate increased or remained unaffected by the amount of plasticizer used. Triphenylphosphate gave slower release rates, followed by dimethylphthalate, diethyl phthalate, dibutylphthalate, and the most rapid release with dimethyl sebacate. Cellulose acetate with 15% triphenyl phosphate provided the longest dimethoate release life (57% total available toxicant released in 42 days) in moist soil.

C. Nonplasticized Pesticide Emittors

So far as known, only a few products consisting of a pesticide monolithically dis-

* B. F. Goodrich Chemical Co., Cleveland, Ohio.

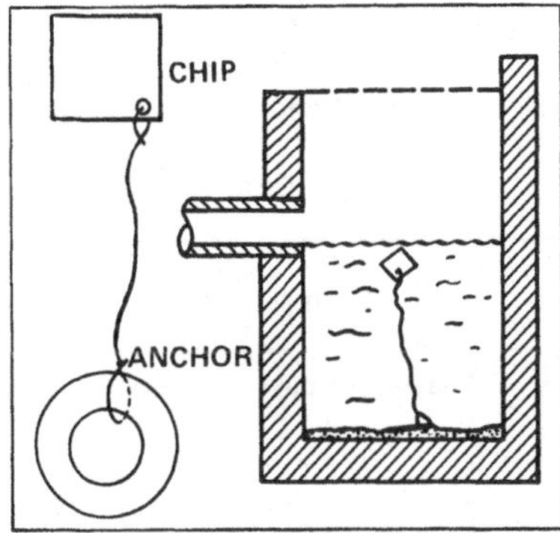

FIGURE 4. Floating larvicide (ECO-PRO 1700), For use in catch basins. One chip is said to be effective for a basin having 30 ft^2 of surface area.

persed in a plastic dispenser have reached the market place. It is presumed that considerable experimental work is in progress. Four materials are presently available in the commercial market. These are Dursban® 10 CR (Dow Chemical Co., Midland, Mich.), Ecopro™ 1700, Ecopro™ 1000, and Ecopro™ 1330 (Environmental Chemicals, Inc., Barrington, Ill.).

The Dursban® 10 CR larvicide, contains 10.6% chlorpyrifos, *O,O*-diethyl-*O*-3,5,6-trichloro-2 pyridyl phosphorothioate, in a chlorinated polyethylene matrix. This material is the culmination of research begun at the U.S. Army Environmental Hygiene Agency.[14] Chlorpyrifos/polyethylene was shown to be effective mosquito larvicides for as long as 18 months with a one time application in artificial pools.[15] Laboratory tests indicated that continuous exposure of mosquito larva to 0.23 ppb chloropyrifos gave 99.7% control as compared to the LC_{90} of 0.9 ppb applied in the conventional manner.[16] This suggested that the same sort of chronic intoxication phenomenon observed with long-term herbicide exposures existed for at least this specific larvicide. Treatment rates of 1.5 ppm-ta* are suggested. In field usage, 12 to 18 weeks of 100% control under various environmental situations have been reported for Dursban® 10 CR.[17] When label recommendations were followed, nontarget biota destruction was not observed. Up to 7 months control has been reported in another field application.[18]

It is suspected that chlorinated polyethylene was selected as the matrix element in order to achieve the requisite specific gravity.

The Ecopro materials marketed by Environmental Chemicals Inc. under various experimental use permits have shown extraordinary long-term control under laboratory conditions. Ecopro™ 1700 contains 7.2% temephos and is currently sold as a sinking granule or a 1½ in. × 1 in. anchored floating chip. (See Figure 4). Whereas the recipe is proprietary, the base matrix consists of an ethylene-propylene-type material with elastoplastic properties modified with a low-density polyethylene. The combination of two polymers provides increased porosity whose degree is dependent upon the disparity of the respective melt indexes. Also, calcium carbonate is used as slow coleachant following the principles enumerated by Cardarelli and Walker.[19]

Table 4 illustrates the long-term temephos release achieved under laboratory conditions with Ecopro™ 1700. At low concentrations, continuous exposure to temephos suppresses normal larva development, and the usual 9-day cycle from egg to pupa may be extended to 16 or more days, depending upon which instar was exposed.

* ta, "Total active ingredient", is used to indicate that the figure given is for the total amount of toxicant added in the dispenser form and not the concentration in the medium at any given time.

TABLE 4

Culex pipiens quinquefasciatus Say Long-term Periodic Bioassay Results with Ecopro™ 1700[20, 21]

Total immersion time (days)	Dosage (ppm-ta)							
	3.6	2.2	1.5	0.55	0.23	0.14	0.06	0.035
0	2 days	2 days	3 days	6 days	6 days	7 days	12 days	16 days
30	2	2	3	2	2	5	5	6
71	2	3	3	3	4	6	9	11
110	2	3	4	3	8	7	13	9
160	3	2	1	6	4	11	10	11

Note: Mortality as LT_{100} (lethal time to 100% larva mortality).

TABLE 5

Culex pipiens quinquefasciatus Say Long-term Periodic Bioassay with Ecopro™ 1330[a,23]

Total immersion time (days)	Dosage (ppm-ta)						
	12.5	8.0	4.1	2.7	2.3	1.4	0.47
30	2 days	3 days	7 days	8 days	10 days	10 days	8 days
110	4	9	13	13	—	13	13
160	4	8	12	—	8	11	12
225	5	7	9	10	12	12	10
310	12	13	13	15	—	—	—

Note: Mortality (LT_{100}) as a function of dosage.

[a] Three replicates of 15, first and second instar larva each, were run using standard rearing and testing techniques.

TABLE 6

Biomphalaria glabrata Long-term Periodic Bioassay with Ecopro™ 1330[22, 24]

Total Immersion Time (days)	Dosage (ppm)					
	12.5	8.2	4.1	2.0	1.6	0.8
0	4 days	6 days	7 days	8 days	9 days	12 days
60	7	8	9	9	17	19
120	9	10	14	17	26	30
240	8	11	14	23	—	—

Note: Mortality (LT_{100}) as a function of dosage.

Ecopro™ 1330, containing 30% tributyltin flouride in a polyethylene modified ethylene vinylacetate copolymer, was developed as a molluscicide. However, envionment impact assessment tests indicated that long-term biocidal effect on mosquito larva were present.[22] Table 5 depicts the results of long-term immersion testing with this material against mosquito larva, and Table 6 provides results against *Biomphalaria glabrata* snail hosts of *Schistosoma mansoni*, one of the causative agents in human schistosomiasis. These are laboratory results, and while field testing is in progress, data has yet to be accumulated and assessed.

It was noted that, unlike temephos, tributyltin fluoride is active against pupae. At higher dosages, such as 30 ppm, a LT_{100} of 7 days was observed after 481 days of continuous pellet immersions.[25]

Other insecticidal systems are under development that utilize a plastic matrix element. Bogaard reports the use of a polyethylene-paraffin mixture containing a metallic salt hydrate as an additive and 15% dichlorvos which provides long-term effective vapor release.[26] Harris has incorporated Fenac® 2,3,6-trichlorophenylacetic acid in polyethylene and demonstrated active release over a 160-day period.[27] Porosity, the water solubility of the agent, and the agent loading determined the rate of migration and release through leaching.

D. Controlled Release Trace Nutrients

The recent development of controlled release plant and animal trace nutrients, though still experimental, offers considerable promise to agriculture.[28] It is well recognized that life forms require small amounts of various elements in their diet for proper growth and resistance to disease. Lack of these nutrients in the foodstuffs consumed by man and his domesticated livestock results in difficiency disease. Elements universally essential to nutrition in most higher plants and animals (and classified as trace nutrients) are sulfur, calcium, boron, copper, iron, manganese, molybdenum, zinc, cobalt, magnesium, selenium, silicon, fluorine, and iodine. Deficient soils are typically treated with zinc, manganese, boron, sulfur, copper, molybdenum, cobalt, and iron. In some cases, application of the appropriate inorganic salt or oxide of the element is made directly, such as the spraying of citrus trees with zinc solution.

The problems associated with the conventional application of the nutrient to the target plant or animal are similar to those experienced with the use of pesticides. A relatively short use life is observed with the need for periodic retreatment at frequent intervals. Higher than needed dosages are used to insure a sufficient residual over the growing season. As with pesticides, conventional treatment is wasteful of chemicals, higher in cost, and lower in efficacy than could theoretically be achieved with the controlled release methodology.

In a program sponsored by Environmental Chemicals Inc. nine elements were monolithically incorporated in several plastic matrices. Table 7 lists the materials selected for evaluation and indicates major agricultural usages. Solubility data is given, since the solution rate in ground waters is a major determinant of dispenser release rate.

The basic thrust of this work was to formulate a given chemical in a plastic matrix in such a way as to provide prolonged release in moist soil at a rate commensurate with efficient nutrition of a selected crop. Materials of commercial interest were developed, but perhaps as important, considerable insight was gained into the creation of leaching-type systems.

The key to continuous long-term emission of inorganic ions from a plastic matrix lies in either the use of a porosity enhancing coleachant or "porosigen" or in the use of a mix of plastic materials. If the agent loading is sufficiently high and the agent is highly water soluble, then the requisite pore structure develops very rapidly, and a retardant ("antiporosigen") is necessary.

A generalized formulation is shown below:

Polyolefin matrix element	45 to 75%
Dispersant	0.5 to 2%
Porosigen	0 to 15%
Retardant	0 to 10%
Agent	20 to 45%

TABLE 7

Trace Element Selection and Use[28]

Element	Compounds evaluated	Cold water solubility (g/100 G)	Major crop usages
Zinc	$ZnSO_4 \cdot H_2O$	100 +	Nuts, citrus, fruits, beans,
	$ZnCl_2$	432	potatoes, coffee, flax,
	ZnO	2×10^{-4}	wheat, forage corn
	$ZnCO_3$	1×10^{-3}	
Copper	$CuSO_4 \cdot H_2O$	32	Oats, potatoes, legumes,
	$CuCO_3$	$<1 \times 10^{-4}$	wheat, pine, grapes, ap-
	Cu_2O	$<1 \times 10^{-4}$	ples, forage, citrus
	$Cu_2(OH)_3Cl$	10	
Iron	$FeCl_3 \cdot 6H_2O$	92	Citrus, pome fruits, cu-
	$FeSO_4 \cdot 7H_2O$	16	curbits, ornamentals
	Fe_2O_3	$<1 \times 10^{-4}$	
Manganese	MnO_2	$<1 \times 10^{-4}$	Barley, oats, wheat, pas-
	$MnSO_4 \cdot H_2O$	98	ture grasses
	$MnCl_2 \cdot 4H_2O$	151	
Selenium	Na_2SeO_4	84	Sheep and cattle forage
Boron	H_3BO_3	6	Legumes, beets, soy,
	$Na_2B_4O_7$	1	beans, grapes, turnips, to-
			matoes
Cobalt	$CoSO_4 \cdot H_2O$	40 +	Pasture legumes
Molybdenum	MoO_3	44	Clover, wheat, cucurbits
	Na_2MoO_4		
Magnesium	$Mg(C_2H_3O_2)_2$	200 +	
	$MgCO_3$	0.01	
	$MgSO_4$	26	

The porosigen is of course lost to the ingressing moisture. A highly soluble material is emitted at a faster rate than one of low water solubility. Ammonium sulfate and sodium bicarbonate are considered to be "fast" porosigens, while calcium carbonate and silica function as "slow" porosity-enhancing materials.

It has been discovered that the use of two plastic components can dramatically effect porosity. For instance, considering low-density polyethylene (LDPE) only, if the two plastics are close in respective melt indexes, porosity is greater (and unexpected), whereas a wide disparity in the indexes leads to a decrease in porosity.

One particular formulation shown below has demonstrated long-term release of zinc ion. (Recipe — Formulation EC-1800)

Ethylene vinyl acetate (M.I. = 9.0)	26.7%
Low-density polyethylene (M.I. = 8.5)	26.7%
Zinc stearate (dispersant)	1.1%
Ammonium sulfate (porosigen)	1.7%
Zinc sulfate	43.8%

NH_4^+ is completely emitted after about 300 hr water immersion which develops porosity to the level where Zn^{++} emission is continuous, reaching a plateau at 400 hr or so (about 40% total loss.

Figure 5 depicts Zn^{++} emission as a function of matrix type and additives used.

Release of Zn^{++} in moist soil is at a much lower rate than that observed in water. Using soy beans as the test plant grown in zinc-poor soil enriched with controlled release zinc emittors, the following data (Table 8) were reported.[28] 1800 D contains 43% $ZnSO_4$ in an ethylene vinyl acetate/(EVA/LDPE) base while 1801 H has 44% $ZnCO_3$ in an EVA base.

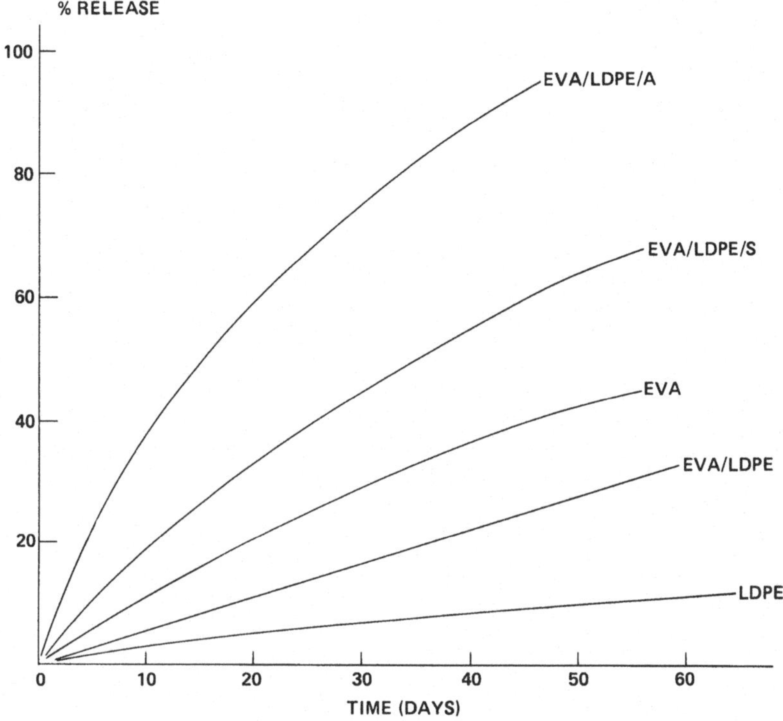

EVA = ethylene vinyl acetate copolymer, M.I. = 9.0
LDPE = Low Density Ployethlene, M.I. = 8.5
A = Ammonium sulfate
S = Sodium Bicarbonate

FIGURE 5. Zn^{++} emission from ZnSO (43.8%) as a function of matrix in water.

TABLE 8

Growth Rate of Soy Beans in Zinc-poor Soil Treated with 1800 D, a Fast Zn^{++} Dispenser, and 1801 H, a Slow Zn^{++} Dispenser

CMPD	Pot Dosage (g)	Zn^{++} Leachate Average per day (ppb)	Zn^{++} Soil Content Average per day (ppb)	Daily growth (height) soy bean (cm)	Average Zn^{++} Content (56 days) Root (ppm)	Stem (ppm)	Leaf (ppm)
1800 D	1	1.5	—	2.75	—		
	0.5	1.1	59	1.94	29	2.3	3.3
	0.2	0.7	25	1.40	26	0.7	1.3
	0.1	0.2	6	1.14	7.1	0.4	1.6
	0.0	0.08	~0.3	1.09	0.8	0.5	3.9
1801 H	1	0.3	—	3.13	6.3	1.9	5.1
	0.5	0.4	9	2.20	2.4	0.6	2.2
	0.2	0.2	4	2.14	2.7	1.9	2.8
	0.1	0.08	1	1.95	2.0	1.3	2.5
	0.0	0.08	0.4	1.30	1.1	2.9	4.6

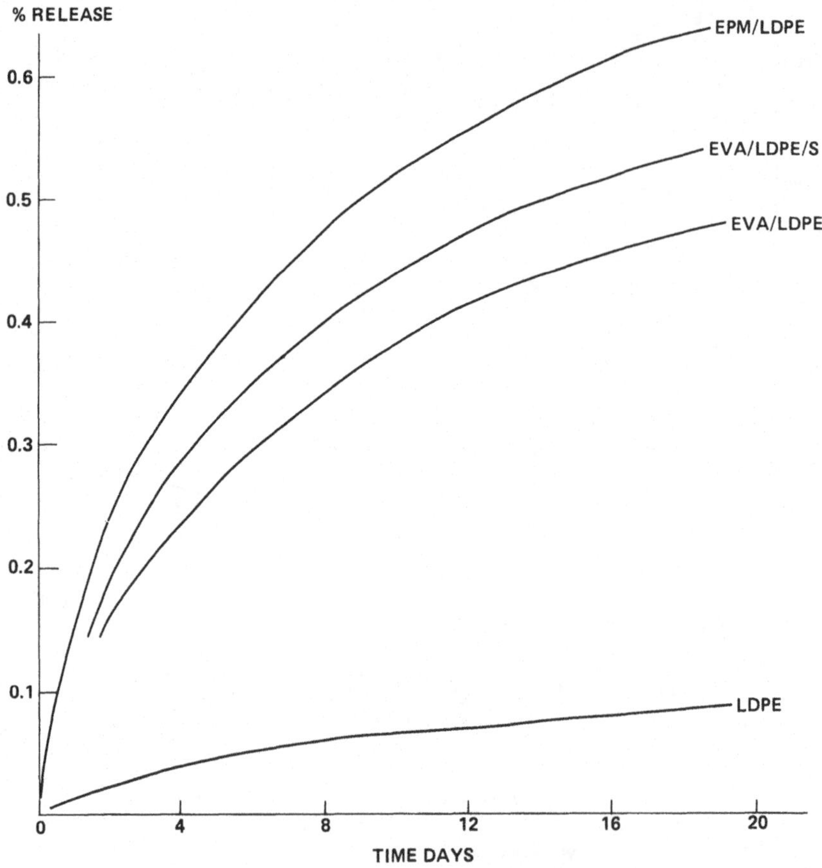

FIGURE 6. Controlled release selenium matrices.

Ethylene-propylene copolymers (EPM) show similar release characteristics, i.e., modification with LDPE of similar melt index increases the porosity and hence the loss rate as shown in Figure 6. In the group of four sodium selenate-loaded materials, EPM/LDPE shows faster release than EVA/LDPE, even when $NaHCO_3$ is used as a porosigen (EVA/LDPE/S). The pattern is the same for manganese emittors (Figure 7).

Ecopro^{TM} 100, containing 43% copper sulfate, was developed as a molluscicide and shows about 4 months Cu^{++} release at biocidal levels. However, it can also be utilized as a trace nutrient dispenser, since release rates in moist soil are considerably slower, and several years emission can be expected. It is of note that copper emission and presumably the loss rate for zinc and other nutrients are temperature dependent. Figure 8 depicts Ecopro^{TM} 1000 immersion in water maintained at 40°F and 90°F, respectively. This factor may be advantageous in agricultural practices.

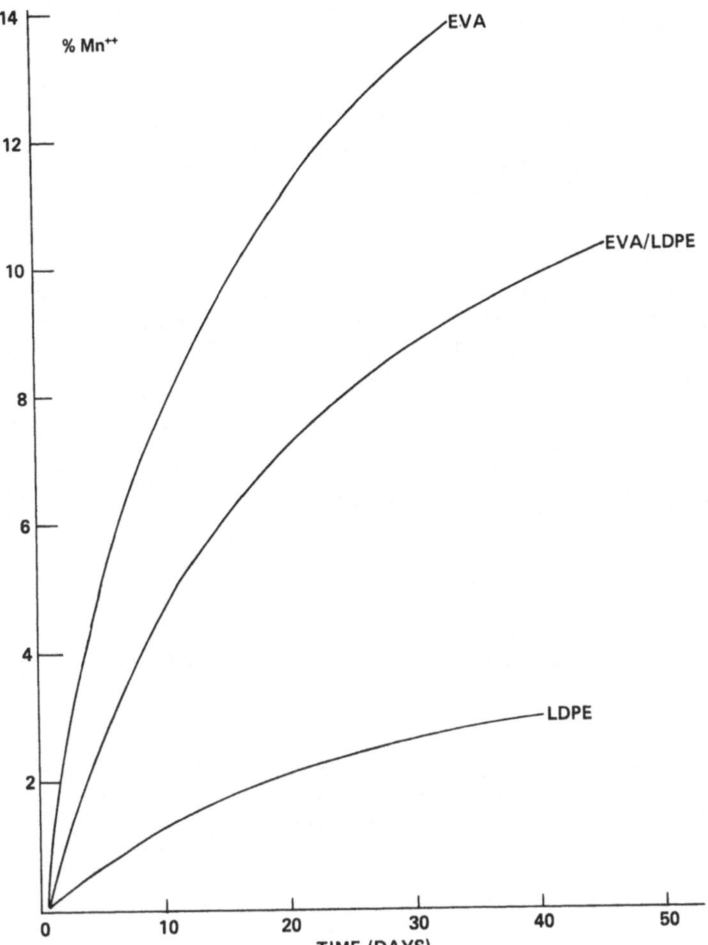

FIGURE 7. Mn^{++} emission from $MnCl_2$-loaded plastic matrices immersed in water.

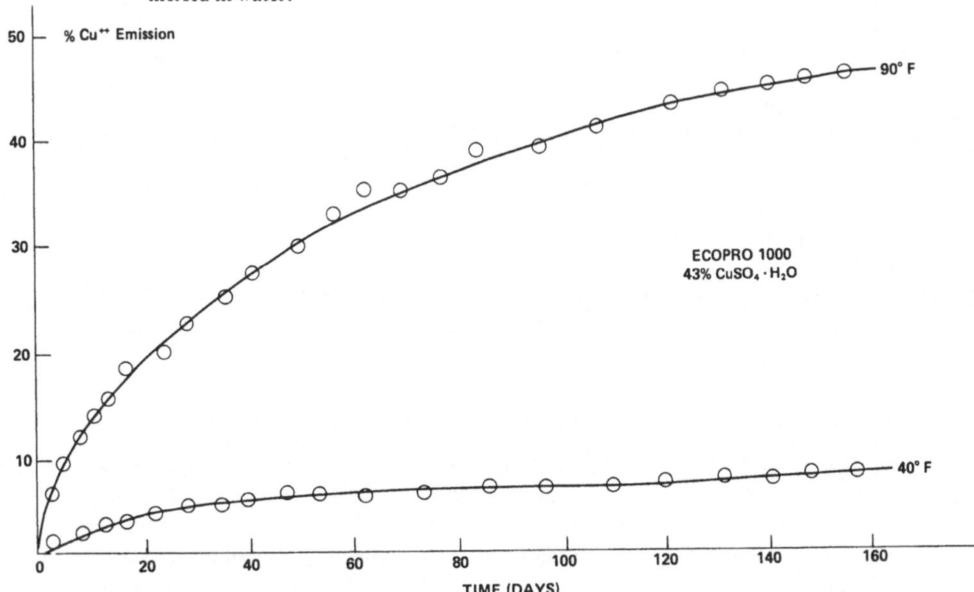

FIGURE 8. Copper loss as a function of water temperature.

The work with nonplasticized polyolefin materials is new, and our understanding of release mechanism is vague indeed. However, the low materials and processing costs should be economically attractive. Since feasibility has now been demonstrated, it is expected that the future will see the incorporation of agricultural pesticides, especially herbicides, in such matrices and possibly bulk fertilizing materials.

REFERENCES

1. **Philibert, J.**, Diffusion in solids, *Sci. Technol. (N.Y.)*, 47, Aug. 1968.
2. **Ferry, J. D.**, *Viscoelastic Properties of Polymers*, John Wiley & Sons, New York, 1961.
3. **Rogers, C. E.**, Structural and Chemical Factors Controlling the Permeability of Organic Molecules through a polymer, pp. 17-28, in *Controlled Release Pesticides*, Scher, H. B., Ed., ACS Symposium Series 53, American Chemical Society, Washington, D.C., 1977.
4. **Crank, S.**, *The Mathematics of Diffusion*, Clarendon Press, Oxford, 1975.
5. **Cardarelli, N. F.**, *Controlled Release Pesticide Formulations*, CRC Press, Cleveland, 1976.
6. Anon., *Marine Fouling and Its Prevention*, United States Naval Institute, Annapolis, Md., 1952.
7. **Morgans, W. M.**, *Outlines of Paint Technology*, CRC Press, Cleveland, 1969.
8. **Nelson, J. H., Evans, E. S., Pennington, N. E., and Young, W. W.**, The U.S. Army Environmental Hygiene Agency activities in the area of slow release insecticides, *Proc. Int. Controlled Release Pesticide Symp.*, Cardarelli, N. F., Ed., University of Akron, Ohio, 1974, 30.1.
9. **Folckemer, F. B., Hansen, R. E., and Miller, A.**, Resin Compositions Comprising Organophosphorus Pesticides, U.S. Patent 3,318,769, 1967.
10. **Soloway, S. B. and Morales, J. G.**, Volatile Insecticide, U.S. Patent 3,740,428, 1973.
11. **Taylor, R. T. et al.**, Controlled release formulations for use against *Aedes aegypti*, *Proc. 56th Ann. Mtng. New Jersey Mosquito Extermination Assoc.*, pp 80-86, Atlantic City, NJ, Mar. 19-21, 1969 (CDC Reprint No. 460).
12. **Nelson, L. L. and Whitlaw, J. T.**, Laboratory Release Rate Studies of Diazinon® and Supracide® from Polyvinyl Chloride Formulations, Entomol. Special Study 31-019-71/72, Rep. (AD-736422), U.S. Army Environmental Hygiene Agency, Edgewood Arsenal, Md., March, 1971.
13. **Stokes, R. A., Coppedge, J. R., Bull, D. L., and Ridgway, R. L.**, Use of Selected Plastics in controlled release granular formulations of Aldicarb and Dimethoate, *Agric. Food Chem.*, 21(1), 103, 1973.
14. **Nelson, J. H., Evans, E. S., Pennington, N. E., and Young, W. W.**, The U.S. Army Environmental Hygiene Agency activities in the Area of Slow Release Insecticides, Rep. No. 30, *Proc. Int. Controlled Release Pesticides Symp.*, Cardarelli, N. F., Ed., University of Akron, Ohio, 1974.
15. **Miller, T. A., Nelson, L. L., and Young, W. W.**, Polymer formulations of mosquito larvicides. IV. Larvicidal effectiveness of polyethylene and polyvinyl-chloride formulations of chlorpyrifos during an 18 month field rest, *Mosq. News*, 33, 172, 1973.
16. **Nelson, Y. Y., Miller, T. A., and Young, W. W.**, Polymer formulations of mosquito larvicide. V. Effect of continuous low-level chloripyrifos residues on the development of *Culex pipiens quinquefasciatus* Say populations in the laboratory, *Mosq. News*, 33, 396, 1973.
17. **Keenan, C. M.**, Use of a controlled release larvicide in Southern Maryland, *Mosq. News*, 38, 203, 1978.
18. **Yates, M.**, Distribution Devices for the Application of Dursban 10 CR Insecticide, *Down Earth*, 34(3), 32, 1978.
19. **Walker, K. E. and Cardarelli, N. F.**, Slow Release Copper Toxicant Compositions, U.S. Patent, 4012221, 1977.
20. **Cardarelli, N. F.**, Evaluation of Environmental Chemicals Inc. Controlled Release Mosquito Larvicides, Interim Rep. Environmental Chemicals, Barrington, Ill., May 20, 1977.
21. **Walker, K. E. and Cardarelli, N. F.**, Laboratory Evaluation of Controlled Release Abate® as a Mosquito Larvicide, Rep. No. 87, 1977 Joint Meeting of AMCA and the Louisiana Mosquito Control Assoc., New Orleans, March 27-30.
22. **Cardarelli, N. F., Gingo, P. J., and Walker, K. E.**, Laboratory and Field Evaluations of Controlled Release Molluscicides and Schistolarvicides, Ann. Rep., Edna McConnell Clark Foundation, New York, N.Y., July 1, 1977, 1.
23. **Cardarelli, N. F.**, Controlled release organotins as mosquito larvicides, *Mosq. News*, 38, 328, 1978.
24. **Cardarelli, N. F.**, Controlled release organotin molluscicides, *Bull. WHO*, submitted for publication.
25. **Cardarelli, N. F.**, A Method and Composition for the Long Term Controlled Release of a Nonpersistent Organotin Pesticide from an Inert Monolithic Thermoplastic Dispenser, Patent approved, 1979.

26. **Bogaard, T. D.**, Controlled release of vapor forming compounds from a new matrix, *Proc. Int. Controlled Release Pesticide Symp.*, Harris, F. W., Ed., Wright State University, Dayton, Ohio, 1975, 132.
27. **Harris, F. W.**, Controlled-release herbicide-polymer formulations for aquatic weed control, *Hyacinth Control J.*, 11, 61, 1973.
28. **Cardarelli, N. F.**, Controlled release trace nutrients, *Proc. Controlled Release of Bioactive Materials Symp.*, National Bureau of Standards, Gaithersburg, Md., 1978, p. 9.1.

Chapter 3

MONOLITHIC ELASTOMERIC MATERIALS

Nate F. Cardarelli

TABLE OF CONTENTS

I. INTRODUCTION

An elastomer is essentially a polymeric material that will stretch under tension and, when the applied force is released, will return to its original dimensions. This "elastic" condition arises from weak interchain forces. In contrast, plastic materials having moderate interchain forces will stretch upon application of a tensile force, but they then retain the new dimensions even though the applied force is released. All elastomeric materials are liquids at room temperature and more or less follow the well-known principles regarding the kinetics and thermodynamics of liquids.

Elastomers in their original state generally lack useable properties. In order to utilize them in practical applications, it is necessary to crosslink the smaller molecules into macromolecules and, usually, to develop environmental resistance, color, etc., through the addition of various agents. This process is termed "compounding."

The initial step in processing the raw elastomer, often referred to as a "gum" or "gum stock," consists of softening the polymer through mechanical working and, usually, elevated temperatures on a two-roll rubber mill or within several types of internal mixing devices. Once in the softened state, additives can be homogeneously blended into the elastomer — much as a housewife blends together the ingredients of a cake. The materials so added serve distinct purposes and create desired properties in the stock. Carbon black or finely powdered silicates are used to impart strength, clays may be added to extend volume, antioxidants offer protection against oxidation, various oils or fatty acids assist in the dispersion of other additives or increase processibility through what is essentially a lubrication process, and vulcanizing agents provide crosslinking during later processing; literally thousands of chemicals are used in various proportions to impart useful properties to the finished stock. This compounding step is vital to the process and, though much studied over the last 80 years, remains essentially more of an art than a science.

After compounding, the material, now designated as a "batch stock," is subjected to elevated temperatures, almost always in the 260 to 320°F range, for a determined period of time in order to establish crosslinking and the creation of macromolecules. This so-called "curing" step can be performed through pressing, molding, and extrusion at elevated temperatures, or by exposure to steam heat or hot air within an oven or autoclave. A few room-temperature elastomer-curing systems are known, but tend to be expensive, require chemicals of a hazardous nature such as lead peroxide, and generally provide a material having inferior physical properties.

Table 1 depicts the function of various additives, giving one or more examples of each. The selection of the proper additive and the amount used is usually an empirical determination based upon past experience guided by scientific insight.

It has long been known that pesticidal substances can be added to elastomer compounds where they are usually retained through mechanical bonding. Fungicides and mildewcides of various types are added to impart resistance to microorganisms. Indeed, a number of the sulfur-bearing accelerators are fungicidal. The release of an incorporated pesticide into the exterior environment is a distinctly different question. One historical problem facing the rubber-goods industry has been the tendency of selected additives to slowly migrate through the polymer matrix and accumulate at the surface — a phenomenon termed "blooming". The unsightly discoloration of white sidewalls on tires was one such specific problem. Through the years, the rubber compounder has worked diligently in trying to avoid this problem, and not to make use of it!

It is not the intent of this chapter to educate the reader in the intricacies of rubber compounding. Many good texts are readily available. Rather, it is noted that the com-

TABLE 1

Typical Compounding Ingredients

Category	Examples
Vulcanizing agents	Sulfur, zinc oxide, various peroxides
Accelerators (increase cure rate)	Benzothiazole disulfide, tetramethyl thiuram disulfide
Activators (enhance acceleration)	Stearic acid, zinc oxide
Antioxidants	Various secondary amines
Non-reinforcing fillers	Kaolinite, chalk
Reinforcing fillers	Carbon black, precipitated silica, ultrafine carbonates
Softening agents	Paraffin oils and waxes, pine oil, fatty acids

pounding process is complex and essential to developing favorable properties in processed elastomers. Specific, and hopefully useful, compounding information will be provided throughout the chapter in regards to the retention and release of pesticidal materials.

This section of the text is limited to the description of monolithic controlled-release pesticidal elastomers. A definition is in order. "Monolithic" means the agent, in this case a pesticide, is dispersed throughout the elastomeric matrix, retaining its chemical nature. In comparison, the coacervation process, which results in a more or less spherical envelope surrounding a quantity of the agent, may use an elastomeric material in the retaining membrane, but the resulting microcapsule is not a monolithic device. Theoretically, the phases are discrete. No agent is present in the envelope, and molecules of the enveloping material are not present within the interior pesticide aggregate. The recent development of polymerized pesticides wherein the agent is chemically appended to a polymeric backbone so that a single phase material results will be discussed elsewhere. Figure 1 illustrates the distinctions between the three types.

II. HISTORICAL DEVELOPMENT

In 1964, the author was faced with a serious fouling problem involving rubber-encased sonar domes. Normally, objects immersed in marine environments are temporarily protected against the attachment and growth of sessile organisms, such as barnacles, through the use of toxic paints. However, antifouling paints are rapidly eroded from sonar domes due to cavitation arising from the passage of high-density sound waves, and the foul-free duration may be extremely short. One approach to prolonging the antifouling life was to add the toxic element directly to the elastomer used and tailor the compound in such a way as to (1) induce the slow movement of the agent to the rubber surface at a rate commensurate with fouling protection while (2) retaining the necessary acoustical and strength properties. Known antifoulants were incorporated at various concentrations in the chloroprene formulations used. They were tested for foul-free life through immersion and periodic inspection in heavily fouled coastal waters. This approach was successful and later a sheet-rubber version was commercialized for use on ship hulls, buoys and other marine objects. Such antifouling coverings have been described in detail elsewhere.[1-3]

The development of a pesticidal rubber and the surprisingly long-term fouling protection noted — years compared to months for conventional antifouling systems — prompted the search for other areas where a controlled release pesticidal elastomeric

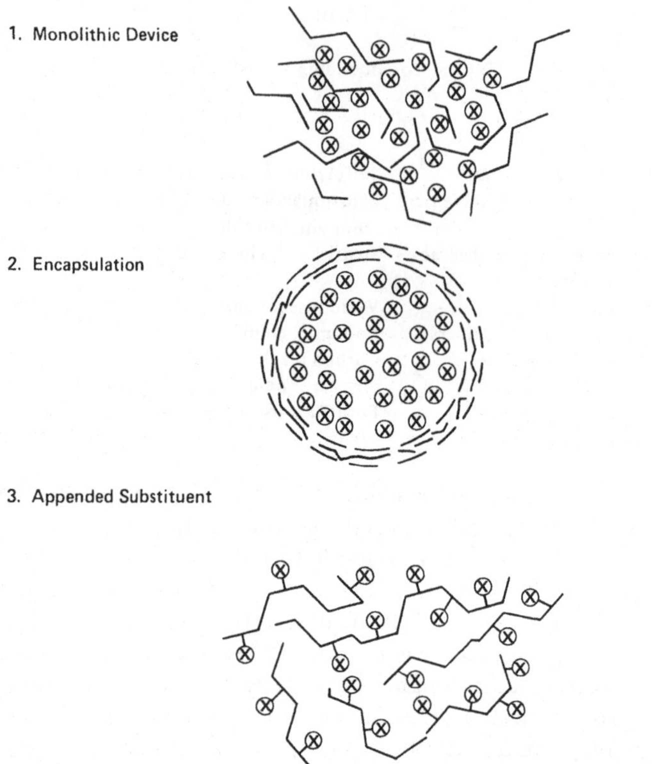

FIGURE 1. Three types of elastomer/pesticide combinations.

material might be useful. It was noted in 1965 that the antifouling preparations showed considerable toxicity to insect larva and species of snails implicated in disease transmission. By 1966, effort — which continues today — was initiated towards the development of a long-lasting sustained-release molluscicide.

Agents found effective in the antifouling formulations were the specific trialkyl and triaryl organotins listed in Table 2. Conventional toxicants such as cuprous oxide and salts of mercury and lead can be incorporated in elastomers, but will not release upon immersion so that no effective antifouling property is imparted. The rationale is discussed in the next section. Initial work with the chloroprenes was extended to other elastomers — natural rubber, styrene-butadiene copolymers, ethylene-propylene-diene terpolymers, polyacrylonitriles, polyisobutylene, polyurethanes, and others. Organotin agents could be incorporated in plastic materials such as polyvinyl chloride (where they function as stabilizers), but could not be made to release at rates sufficient for imparting long-term antifouling or larvicidal properties.

Organotin-loaded elastomers were found effective against mosquito larva at relatively low concentrations, 1 ppm or less, for prolonged periods under laboratory conditions.[4,5] Two major difficulties arose in that rather limited field tests indicated that TBTO was substantially deactivated in the presence of organic matter,[6] and the organotins not only lacked registration for use as aquatic larvicides, but elementary questions regarding toxicity and environmental impact were unanswered. As a result, various in-use pesticidal materials (organophosphates, halogenated hydrocarbons and carbamates) were formulated in elastomers and evaluated. Although sustained release was achieved in a number of instances, a temephos/ethylene-propylene-diene com-

TABLE 2

Elastomer-Soluble Organotin Agents[3,15-17]

Material	Formula	Room temperature state	M.P./B.P. (°C)	Soluble H_2O/25°C (ppm)	Soluble elastomers (wt. %)	LD_{50} oral, rat (mg/kg)
TBTO[a]	$(C_4H_9)_6Sn_2O$	Liquid	−45/305	2.7	4—16	148—234
TBTS[b]	$(C_4H_9)_6Sn_2S$	Liquid	/393	<6	3—12	
TBTA[c]	$(C_4H_9)_3SnOCOCH_3$	Solid	80/	<30	3—?	99
TBTF[d]	$(C_4H_9)_3SnF$	Solid	218/	4.8	4—?	200
TPTA[e]	$(C_6H_5)_3SnOCOCH_3$	Solid	122/	<30	6—12	136—490
TPTC[f]	$(C_6H_5)_3SnCl$	Solid	106/240	<30	2—8	

[a] *bis*(tri-*n*-butyltin) oxide.
[b] *bis*(tri-*n*-butyltin) sulfide.
[c] tributyltin acetate.
[d] tributyltin fluoride.
[e] triphenyltin acetate.
[f] triphenyltin chloride.

pound showed the greatest promise in both laboratory and field situations. Schultz and Webb observed 100% *Culex pipiens quinquefasciatus* Say mortality over 26 three-day immersion and test cycles.[5] Such compounds were patented.[7]

The degree of success achieved with organotin-loaded elastomer-based molluscicides has been far greater. The extreme hydrophobicity of these agents results in a very short molecular lifetime in a watery medium. It is believed that the organotin molecules are rapidly absorbed by organic matter and soils, and snail contact occurs while browsing the soil for nutrients. Conventional molluscicides, with the exception of copper sulfate, have yet to be successfully incorporated in elastomers. Ethanolamine niclosamide systems have shown some small promise and perhaps with further study useful formulations may develop.[1] Trifenmorph has never been successfully incorporated in elastomers, and detoxification through chemical rearrangement is believed to occur.

A small amount of work against water-borne bacteria and fungi has indicated that organotin/elastomers may be functional in water-purification systems.[8]

In 1969, effort was initiated towards the development of a practical sustained controlled release herbicide based upon an elastomeric binding element. A natural-rubber formulation utilizing around 20% w/w of the butoxyethanolamine ester of 2,4-dichlorophenoxyacetic acid was developed and found efficacious against various aquatic weeds in both laboratory and field studies.[9-11] During the course of this work, it was discovered that water-weed mortality is a function of both time of exposure and concentration when treatment is at ultralow dosages, with the duration of exposure the dominant factor.[12,13] Herbicidal materials other than various 2,4-D chemicals have been incorporated in natural rubber, styrene-butadiene copolymers, etc. and found to provide efficacious activity under laboratory conditions.[14]

III. RELEASE MECHANISMS

During the development of antifouling rubber systems, it was observed that agents, such as the organotins, which are soluble in the elastomeric matrix would slowly devolve from that matrix when immersed in water. In contrast, nonelastomer-soluble chemicals such as cuprous oxide would not release unless the loadings were so high as to seriously affect the physical integrity of the sheet-rubber stock. Organotin chemicals and their solubility limits are listed in Table 3. Solubility limits with solid organotins are difficult to determine in that incorporation may include both a soluble phase and a mechanically held, insoluble phase. For instance, TBTF is soluble to 6% or so in natural rubber, but up to 30% by weight can be added without drastic change in physical properties. Liquids such as TBTO and TBTS display sharp solubility limits. Exceeding such limits results in reversion during processing and/or dramatic loss in physical properties.

A material soluble within an elastomeric matrix will disperse evenly through the batch stock during processing so that a uniform condition of solution equilibrium exists. Upon immersion in water, agent molecules on or very near the elastomer surface will pass slowly into the interface through dissolution processes. As the surface layer depletes, a localized solution disequilibrium results and internal solute molecules, driven by solution pressure, diffuse outward towards the depleting zone. A continuous process is thus activated. The mechanism is appropriately termed "diffusion-dissolution". Either first or second order kinetics are followed, depending upon whether diffusion or dissolution is rate controlling.

When solubility within the elastomeric matrix is lacking, reliance must be placed upon other processes. In antifouling paints, matrix-insoluble inorganic or organic chemicals are usually compounded with a plastic material. Release is achieved through

TABLE 3

Pesticidal Materials Showing 1% or Better Solubility in Elastomers[3,7,10,15,31,32]

bis(Tri-*n*-butyltin) oxide	Chlordane[a]
bis(Tri-*n*-butyltin) sulfide	Heptachlor[b]
Tributyltin acetate	Hexachlorocyclohexane
Tributyltin fluoride	Diazinon®[c]
Tributyltin adipate	Fenthion[d]
Tributyltin chloride	Baytex®[e]
Tributyltin resinate	Dichlorvos[f]
Triphenyltin acetate	Rabon[g]
Triphenyltin chloride	Butoxyethanolester of 2,4-D
Niclosamide	2,4-D
Ethanolamine niclosamide	Fenac®[h]
Malathion[i]	Fenuron®[j]
Dasanit[k]	Diquat[l]
Chloropyrifos	Silvex®[m]
Temephos	Carbaryl[n]
Tributyllead acetate	Dibrom[o]
Triphenyllead acetate	Ethoprop[p]

[a] Octachloro-4,7-methano-tetrahydroindane.
[b] Heptachloro-4,7-methanotetrahydroindane.
[c] *0,0*-diethyl-*0*-(2-isopropyl-6-methyl-5-pyrimidinyl) phosphorothioate.
[d] 2-(1-methylethoxy) phenol methylcarbamate.
[e] *0,0*-dimethyl-*0*-[3-methyl-4-(methylthio)phenyl] phosphorothioate.
[f] Dimethyl-2,2-dichlorovinyl phosphate.
[g] 2-chloro-1-(2,4,5-trichlorophenyl)vinyl dimethylphosphate.
[h] 2,3,6-trichlorophenylacetic acid.
[i] *0,0*-Dimethyl phosphorothioate.
[j] 3-phenyl-1,1-dimethylurea.
[k] Sulfinyl phosphorothioate.
[l] 6,7-dihydrodipyrido (1,2-*a*:2′,1′-*c*) pyrazinediium dibromide.
[m] 2-(2,4,5-trichlorophenoxy) propionic acid.
[n] 1-naphthyl methylcarbamate.
[o] Dimethyl-1,2-dibromo-2,2-dichloroethyl phosphate.
[p] *0*,ethyl-*s,s*-dipropyl phosphorodithioate.

either a leaching process or exfoliation. In the former case, the rate of development of porosity and its tortuosity are rate controlling along with water solubility. To achieve the requisite porosity, relatively high agent loadings, 40% or greater on a dry-film basis, are necessary. Cuprous-oxide-bearing antifouling paints are not operable below about 85% loadings. In an exfoliating type surface, the polymeric matrix either undergoes a slow solvation or degrades through chemical processes, releasing the mechanically held agent into the interfacial zone. Leaching-type monolithic elastomeric systems are difficult to design, and with one exception, have not displayed desirable long-term efficacy in antifouling or molluscicidal studies.

Diffusion-dissolution phenomena where dissolution is rate controlling can be mathematically described with considerable precision. If release is from one side of a flat surface then:

$$\frac{dQ}{dt} = -KC = -K\frac{Q}{a} \tag{1}$$

where C = agent concentration in the matrix, K = proportionality constant, Q = amount of agent in the matrix per unit area, and a = thickness of the sheet, pellet, etc.

integrating

$$Q = Q_0 \, e - \frac{Kt}{a} \tag{2}$$

where Q_o = initial agent concentration, and t = elapsed time.The amount of agent released, Q_R, is given by:

$$Q_R = Q_0 \left(1 - e - \frac{Kt}{a}\right) \tag{3}$$

In the case of antifouling sheet rubber, the foul-free life, T, is the time required for the initial loading C_o to drop to C_t, the concentration at which fouling attachment occurs.[1]

$$T = \frac{a}{K} \ln \frac{C_o}{C_t} \tag{4}$$

The K factor varies from elastomer to elastomer, the degree of crosslinking, and the presence or absence of additive materials that will increase or retard diffusion.

Equation 4 relates to sheet thickness, a vital consideration for antifouling materials. However, insecticidal and molluscicidal formulations would normally be applied as pellets or granules to the use environment. Efficacious release life can be related to the ratio of the surface area to the volume of the dispensor. Increasing this ratio increases the loss rate. Thus, pellet geometry becomes important to any consideration of effective lifetime for a given dosage.

In diffusion-dissolution systems where dissolution processes are more rapid than diffusion, i.e., where water solubility is relatively high such as with the 2,4-D herbicidal materials, diffusion becomes the rate-controlling step, and loss is proportional to $t^{1/2}$. Simple diffusion of a solute in a polymer has been dealt with at length by various authors and will not be repeated here.[18] The governing rate equations are based upon Fick's law of diffusion:[19,20]

Zero order kinetics. (K/d is large):

$$\frac{dQ}{dt} = \frac{C_o D}{hk} \tag{5}$$

(K/d is small):

$$\frac{dQ}{dt} = C_o \left(\frac{D}{\pi t}\right)^{1/2} \tag{6}$$

where c_o = initial agent concentration, D = diffusion coefficient, K = partition coefficient, and h = thickness of boundary layer.

Investigation of the complex diffusion profile involved in multiphase elastomeric systems has provided some insight into the relationship of compound parameters and diffusion rates under varying environmental conditions. The following equations developed by Kanakkanatt[21,22] were examined empirically and found to reasonably describe observed sorption and desorption phenomena:[23]

$$D_f = \left(\frac{2}{3} - V_r\right) \left\{D_o + \gamma \left[(T_2 - Tg_2) - (T_1 - Tg_1)\right] - \left[\alpha - \delta (Vs_2 - Vs_1)\right] \left(dc_2 - dc_1\right)\right\} \quad (7)$$

$$D_f = D_o + \gamma \left[(T_2 - Tg_2) - (T_1 - Tg_1)\right] - \left[\alpha + \delta (Vs_2 - Vs)\right] (dc_2 - dc_1) - \beta (bc_2 - bc_1) \quad (8)$$

where D_f = final diffusion coefficient, D_o = original diffusion coefficient, V_r = volume fraction of rubber, Vs_1 and Vs_2 = two different volume fractions of the pesticide-polymer composite, T_1 and T_2 = two different ambient temperatures, Tg_1 and Tg_2 = glass-transition temperature of the two different elastomers, α = cross-link-density coefficient, β = carbon-black-content coefficient, γ = temperature coefficient, δ = slope of the curve α vs. Vs, dc = cross-link density (= 10^4/Mc, where Mc = molecular weight between crosslinks), bc = carbon-black content.

If the agent is not soluble in the polymer, the following general equation provides a first approximation of release from a planar porous matrix:[24]

$$\frac{dM}{dt} = \frac{D\epsilon (2C_o - \epsilon C_f) C_f}{4\tau t} \quad (9)$$

where dM/dt = flow of permeant at time t, D = diffusion coefficient of agent in permeant, ε = matrix porosity, τ = tortuosity factor, C_o = initial concentration of permeant in the matrix, and C_f = solubility of the agent in the permeant.

Laboratory determination of the factors involved can be extremely difficult. Desai, Simonelli and Higuchi describe a method for measuring tortuosity,[25] and Harris[26] provides a method of determining porosity. Standard means of determining diffusion rates of various materials in polymers are well known.[27]

IV. FORMULATION

A. Polymer

Candidate controlled release formulations of the monolithic elastomeric type may rely either upon a diffusion-dissolution- or a leaching-release mechanism. In work performed to date, it is readily apparent that the diffusion-dissolution system will provide extreme long release duration, whereas leaching systems are much shorter lived. Basically, three factors will govern the type of system desired, (1) the solubility limit of the agent in the matrix, (2) dispensor geometry, i.e., S/V ratios, and (3) desired usage. Generally, inorganic substances are insoluble in elastomers, and no choice exists save to create a leaching-type of system. The envisioned usage pattern will often dictate the desired biological life. For instance, fouling protection for a number of years would certainly be expected, whereas a controlled release, preplant, agricultural herbicide may be useful if the life is in the three-to-four-month range.

Since establishment of a diffusion-dissolution mechanism is dependent upon solubility parameters, the choice of base polymers is limited to those displaying solvation power for a given agent. Solubility can be measured using several simple techniques. If the agent is a liquid, then the gum rubber or a nonfilled, cured rubber can be immersed in the agent and periodically weighed after wiping excess from the surface. The elastomer will absorb the agent until the solute concentration reaches the solution limit. This is a rough, though usually adequate, method. For exacting determination, the ability of an elastomer sheet to mechanically retain solvent within whatever pore structure exists must be determined and this value used as a correction factor.

TABLE 4

Elastomers Exhibiting Significant Solubilizing Power Towards Pesticides

Natural rubber
Synthetic natural rubber (Ameripol® SN 600)[a]
Styrene-butadiene copolymers (Ameripol® 1007[a], 1001[a], and 4616[a]) NBS 387[b]
Ethylene-propylene-diene terpolymers (Nordel® 1070[c], Epcar 5465[a])
Polisobutylene (Polysar 301[d], NBS 388[b])
Polyester urethane (Estane 5701[a], Estane VC[a])
Nitrile butadiene (Hycar Series 1001, 1013, 1014[a])
Chloroprene (Neoprene® WRT, AD, GN[b])
Cis-polybutadiene (Ameripol® CB220)[a]

[a] B. F. Goodrich Chemical Company, Cleveland, Ohio.

[b] National Bureau of Standards.

[c] E. I. DuPont de Nemours and Company, Wilmington, Delaware.

[d] Polymer Corporation, Sarnia, Canada.

Determination of the solubility of solids in an elastomer requires considerably greater effort. One technique used by the author consisted of preparing two smooth, thin panels, one of the masterbatch formulation and the other of the masterbatch with a known concentration of the agent. Whether the solid-agent concentration in the test panel exceeds the solubility limit is not critical, although concentrations greater than 20% by weight are not advisable. Both panels are cured, weighed, and pressed firmly together. Several hundred psi pressure is essential for integral contact. If the agent is soluble in the loaded panel, there will be a constant migration under solution pressure into the nonloaded panel. This process continues until both panels contain the same concentration of agent (assumes that the loaded-panel concentration is at or below the solubility limit), or the blank panel reaches an equilibrium weight which represents the solubility limit. Allow several months for this condition to occur. Thin panels of 0.02 to 0.03 in. thickness are preferable. The technique is likewise suitable for use with liquid agents where migration is generally much more rapid.

Another method of determining solubility is through the observation of certain physical properties of loaded panels. Usually such factors as tensile strength and modulus show little change with agent loading until the solubility limit is reached. At that point, dramatic variation may, though not always, be noted. For instance at 9.6% bis (tri-*n*-butyltin) oxide in chloroprene rubber (DuPont's Neoprene WRT®), tensile strength was 2750 psi, elongation at break 725%, and the 300% modulus, 1400 psi. At 9.8% TBTO, tensile dropped to 375 psi, elongation was 200%, and the 300% modulus was, perforce, zero.[28] Results are not as dramatic with solids as a rule. The tensile strength of TBTF/Neoprene WRT® was 2675 psi at 10.5% loading, dropping to 1450 psi at 10.8%.

The presence or absence of solubility can be determined by microtoming the sheet stock into very thin sections and examining each section by SEM techniques[29] or X-ray fluorescence.[30]

An exhaustive study of solution limits of pesticides in elastomers has never been attempted. Table 3 depicts those pesticides found to be appreciably soluble (over 1%) in elastomers. The degree of solubility is dependent upon the respective chemical structures. Halogenated materials usually display a higher solubility in halogenated elastomers, such as the chloroprenes and brominated polyisobutylene than in natural rubber, for instance. Elastomers examined are depicted in Table 4.

TABLE 5

TBTO Solubility and Antifouling Lifetime of Butadiene Acrylonitrile Copolymers of Varying Acrylic Content

Acrylic content (Hycar series)[a]	TBTO Solubility[1,3] (21°C)	Foul-free life (av.) (Miami Beach, Fla. emmersion)
18%	15.8	6 months
22%	12.3	15 months
26%	8.7	44 months
33%	7.9	35 months
40%	4.2	23 months
52%	0.1	0 months

[a] Hycar is a trade name of the B. F. Goodrich Chemical Company of Cleveland, Ohio.

The degree of solubility is dependent upon the chemical and physical structure of the polymer. In general, high molecular weight and/or high cross-link-density materials show lower agent solubilities than low molecular weight and/or lower cross-link densities of the same general monomeric structure. It is, therefore, possible to polymerize to an end point commensurate with the desired solubility. In copolymer systems, the monomer ratio can be crucial (see Table 5).

A high degree of agent solubility does not produce a long biological life in itself. Usually, the higher the loading the greater the loss rate. In contrast, conditions of very low solubility have always been nonefficacious. Table 5 depicts the antifouling life of several butadiene acrylonitrile copolymers.

In leaching-type systems where the agent loading must of necessity be high, polymer selection is usually based upon the ability of the given matrix to mechanically hold large toxicant concentrations without an unacceptable loss in physical properties. Ethylene-propylene-diene terpolymers, natural rubber, *cis*-polybutadiene and moderate-molecular-weight styrene-butadiene copolymers have been found by the author to be capable of retaining 70% or better by weight of fillers such as copper sulfate monohydrate. Epcar 5465*, for example, will hold 87% by weight of both the monohydrate and pentahydrate of copper sulfate.[33]

B. Effects of Fillers

Highly structured carbon blacks, petroleum waxes, and various finely powdered substances such as kaolinite and montmorillonite clays, calcium carbonate, etc. when added to a given elastomer/pesticide system will influence, sometimes profoundly so, the agent loss rate. Both the amount of carbon black used and its structural type must be considered. The greater the absorptive surface area present, the greater the effective diffusion path length in diffusion-dissolution-type systems, and thus, the slower the delivery of the agent to the depleting surface. Since a minimal amount of the toxic element must undergo dissolution in order to maintain the required biological dosage, the agent delivery rate is critical to both efficacy and lifetime. Considering first the effect of carbon-black structure, Table 6 illustrates antifouling performance of 0.72% TBTO/neoprene types. Loss is too rapid from the poorly structured blacks, and fouling commences within a few months as agent evolution decreases below the fouling threshold.

* B. F. Goodrich Chemical Co., Cleveland, Ohio.

TABLE 6

Effect of Carbon-black Type on Antifouling Performance[3]

Carbon-black type	Particle size (nm^2)	Oil absorptivity (gals/100 ls approximate)	5-Month fouling performance in tropical marine waters
SAF	11—19	16.5	No fouling
ISAF	20—25	15.0	No fouling
EPC	26—30	11.0	No fouling
FF	31—39	9.5	No fouling
FEF	40—48	8.2	Fouling, month 4
HMF	49—60	7.0	Fouling, month 3
SRF	61—100	6.3	Fouling, month 2
FT	101—200	5.2	Fouling, month 2
MT	201—500	4.4	Fouling, month 2

Note: Determination of the diffusion coefficient for TBTO in carbon-black (FEF)-loaded natural rubber indicates decreasing values with increasing black concentration:[34]

0.0% FEF black	3.0×10^{-6} cm^2/sec
3.8% FEF black	2.20×10^{-6} cm^2/sec
7.3% FEF black	1.91×10^{-6} cm^2/sec
12.2% FEF black	1.82×10^{-6} cm^2/sec

Consequently, the amount of carbon black can dramatically affect lifetime. Using 0.18% TBTO in a chloroprene base at 7.2% ISAF carbon black, no fouling is evident for 6 months, whereas doubling the black content results in only 2 months of foul free life. A similar pattern is seen for 2,4-D BEE (butoxyethanol ester of 2,4-dichlorophenoxy acetic acid) with varying carbon blacks. Absorptivity is dependent upon structure as well as particle size (Table 7). Petroleum waxes will also serve as diffusion rate regulants.[36]

In leaching-type systems, carbon-black loading, regardless of black type, affects loss rate. The accumulative 1 month loss of copper ion from a compound with varying black content was reported as 40% (10 part loading), 9% (20 part loading), 8% (30 part loading), and 7% (50 part loading).[33]

C. Agent Loading

The release lifetime of a given diffusion-dissolution formulation is basically dependent upon (1) the polymer used, (2) the type and amount of regulant used, such as carbon black, (3) the vulcanization conditions, (4) the product geometry, and (5) the agent loading. Figure 2 depicts the results observed when agent loadings are varied, the other factors held constant, with antifouling elastomers. It is observed that excellent fouling protection is achieved with organotin concentrations of 6% or less in antifouling sheet rubber, compared to the necessary 40% or greater organotin and 85% or greater cuprous oxide loadings necessary in paint preparations.[1] This contrast is startling and illustrative of one favorable aspect of diffusion-dissolution systems as compared with leaching- or exfoliating-release mechanisms.

Agent loadings in excess of the solubility limit can be achieved at least in certain select systems through the utilization of special additives to create a third phase. For instance, TBTO can be added to an inert carrier material such as carbon or phenolic hollow microspheres. The mix is then added to the elastomer batch.[37] Microballoons*

* BJO-0930®, Union Carbide Corporation, Niagara Falls, New York.

TABLE 7[35]

2,4-D BEE Absorption by Carbon Black

Black type	ASTM code	Particle size (Av, nm)	BEE Absorption (g BEE/kg black)
SAF	N110	19	820
SCF	N294	22	640
CF	N293	27	620
ISAF	N231	29	600
FEF	N550	79	600
ISAF	N220	29	580
ISAF	N219	30	520
HAF	N326	40	480
HAF	N330	45	330
SRF	N774	160	330
SRF	N770	128	320
MT	N990	462	280
FT	N880	256	250

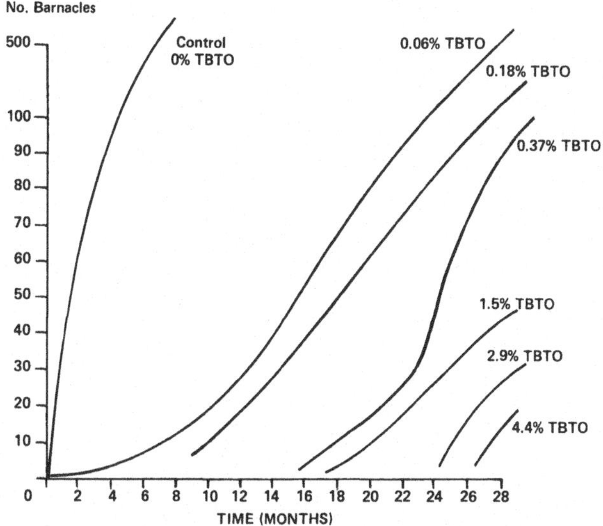

FIGURE 2. Barnacle fouling pattern on 120 in.^{2} TBTO-loaded Neoprene® formulation immersed in tropical waters.

having a bulk density of 0.105 g/cm^3 were slurried with TBTO for 24 hr or so. The agent penetrates the phenolic wall and fills the interior void. This composite is then added to an elastomer batch during mill mixing. The phenolic spheres are insoluble and mechanically held by natural rubber, chloroprene, and other elastomers. During use, the microballoons disperse TBTO to the surrounding matrix as that matrix depletes below the solubility limit. Organotin concentrations in excess of 20% by weight can be developed without loss in processibility or pertinent physical properties. Similarly, ethanolamine niclosamide, which is soluble to 11 g/100 g of neoprene, can be added up to 35 g/100 g using 40-μm-diameter carbon microspheres. Commercially available microspheres range from 10 to 60 μm in diameter with wall thickness varying over 0.1 to 3 μm.

D. Curing Systems

Vulcanization time and temperature affect the release rate of an agent in a diffusion-dissolution-type formulation. The particular cure system used, if it is appropriate to the elastomer in question, does not seem to greatly influence loss rates. Using 0.18% TBTO in a 10% carbon black (ISAF) compound, panels were press cured for varying times at a constant 149°C. After 6 months, tropical immersion samples cured at 20 and 30 min showed no barnacle fouling, at 45 and 60 min levels several barnacles were present, and after 90 min cure, considerable fouling was observed. Cross-link density increases with cure duration, thus the diffusion coefficient decreases.

In leaching-type systems, the type of vulcanizing agents used appear to play a minor role in loss rate. However, cure temperatures can be critical. In a typical recipe, the 9-day copper-ion loss from panels cured at 290°F for 30 min was 22% of the total available. At 320°F, this loss increased to 81%.[33]

It should be noted that various agents will sometimes react with compounding ingredients, especially during the elevated pressure and temperature state experienced during cure. Bis-(tri-*n*-butyltin) oxide, for instance, has been noted as reacting with sulfur-bearing curatives, fatty acids, amines, and even with the chloroprene matrix.[1] Consequently, agent(s) emitted by the product may not be the same, chemically, as the agent added during the mixing operation.

E. Typical Controlled Release Formulations

Over the years, a number of elastomeric compounds have been found to be especially proficient in retaining pesticidal agents and releasing these *in water* at a rate appropriate to pest control. Several typical ones are provided here along with known incorporable agents.

1.[3]	Neoprene WRT®	100 parts
	FEF black	14 parts
	Phenyl-β-naphthylamine*	2 parts
	Zinc oxide	5 parts
	Lauric acid	3 parts
	Magnesium oxide	4 parts
	Ethylene thiourea	0.75 part
	Benzothiozyldisulfide	1 part

Agent: up to 10 parts TBTO or TBTS; 30 parts TBTF; 3 to 5 parts malathion, temephos, carbaryl; 40 parts ethanolamine niclosamide.

Cure: 20 to 45 min at 290° to 310°F.

2.[3]	Natural rubber	100 parts
	HAF black	40 parts
	PBNA	1 part
	Zinc oxide	5 parts
	Stearic acid	3 parts
	(Mercaptobenzothiozole disulfide)	0.6 part
	Sulfur	2.5 parts

Agent: will retain copper sulfate 50%; TBTO and TBTS 8%; TBTA 12%; TBTF 14%; Dibrom®, Fenac®, 2,4-D BEE, Silvex®, Diquat to 20% or greater.

Cure: 30 min at 290°F.

* "PBNA" oxidant was recently removed from commercial use due to suspected carcinogenic activity. Any phenylamine-type antioxidant should suffice.

3.[3]	Butadiene acrylonitrile copolymer	100 parts
SRF black	40 parts
Zinc oxide	3.6 parts
Lauric acid	3.0 parts
Tetramethylthiuram disulfide	0.4 part
Sulfur	1.5 parts

Agent: using materials of 22 to 40% bound-acrylic content, TBTO and TBTS can be successfully released.
Cure: 20 min at 300°F.

4.[3]	Polyisobutylene (Polysar 301®)	100 parts
Zinc oxide	5 parts
Stearic acid	3 parts
Mercaptobenzothiozole disulfide	0.5 parts
Tetramethylthiuram disulfide	1.0 part
EPC black	50.0 parts
Sulfur	2.0 parts

Agent: will release TBTO and TBTS, about 12% maximum loading.
Cure: 307°F for 25 min.

5.[36]	Styrene-butadiene (NBS 387)	100 parts
Zinc oxide	5 parts
Sulfur	2 parts
Benzothiazyl disulfide	1.75 parts
Channel black (EPC)	40 parts

Agent: Ameripol® 1007 can be substituted for NBS 387. Will release TBTO, 2,4-D BEE, Fenuron®, Fenac®, Diquat, triphenyllead acetate, TBTS.
Cure: 290°F for 25 to 35 min.

6.[1]	Natural rubber	100 parts
Zinc oxide	1 part
Sulfur	1 part
Stearic acid	1 part
PBNA	2 parts
Silica	25 parts
Mercaptobenzothiazyl	1 part
Methyl zimate	0.25 part
Hexamethylenetetramine	1 part
Sulfasan®	1 part
Ammonium acetate	1 part

Agent: will retain and release up to 8 parts TBTO. Transparent material.
Cure: 280°F to 300°F for 30 min.

7.[1]	*cis*-Polybutadiene (Ameripol CB®)	100 parts
ISAF black	30 parts
Zinc oxide	3.6 parts
Lauric acid	3.0 parts
Santocure®	1.7 parts
Sulfur	2.5 parts

Agent: up to 10 parts TBTS can be added without drastic alteration in physical properties.
Cure: 290°F for 35 min.

8.[1]	EPDM (Nordel 1070® ethylene-propylene polymer)	100 parts
	ISAF black	85 parts
	Zinc oxide	5 parts
	Stearic acid	0.5 part
	Mercaptobenzothiazole disulfide	0.5 part
	Tetramethylthiuram disulfide	1.5 parts

Agent: will hold and release up to 9 parts TBTO or 7 parts TBTS.
Cure: 307°F for 60 min.

9.[1]	Epcar 5465 (ethylene-propylene-diene terpolymer)	100 parts
	Sulfur	1 part
	Mercaptobenzothiazole	0.2 part
	Zinc oxide	3 parts
	HAF black	50 parts
	Tetramethylthiuram disulfide	1.0 part

Agent: will hold up to 87% copper sulfate.
Cure: flat cure over 285°F to 320°F for 18 to 40 min.

10.[1]	Natural rubber	100 parts
	Sulfur	0.5 part
	HAF black	20 parts
	Mercaptobenzothiazyl disulfide	2 parts
	Zinc oxide	2 parts
	Stearic acid	0.5 part
	Tetramethylthiuram disulfide	1.0 part

Agent: will retain and release 20 or more parts of 2,4-D BEE.
Cure: 290°F for 30 min.

V. ANTIFOULING ELASTOMERS

Antifouling rubber, now commercialized by the B. F. Goodrich Company (Akron, Ohio) as Nofoul®, was the initial controlled release elastomer invention. Unlike antifouling paints, this material is (1) a solid-sheet formulation attached through an adhesive system to the object to be protected, and (2) operates through a diffusion-dissolution mechanism. Conventional antifouling paints suffer several disadvantages that are overcome through the use of this material. These include (1) a relatively short biological life under practical application and use conditions, especially in heavily fouled tropical waters, (2) poor film integrity due mainly to the very high agent loadings necessary to sustain a developing porosity-release mechanism, (3) poor cavitation resistance, (4) the propensity to substantially contaminate harbors and other water bodies, and (5) the human hazards, especially to the eyes, associated with the use of cuprous oxide materials. The lifetime of a conventional paint, ignoring manufacturers' claims, is well recognized by the U.S. Navy and other users as 8 to 12 months at best in tropical waters. Figures 3 and 4 illustrate barnacle fouling. On areas such as sonar domes, rudders, and propellers where erosion is great, life may be only a few weeks. In contrast, Nofoul® covered objects have been foul free for as long as 9 years (U.S. Coast Guard buoys). In no known case, when properly applied, has biological failure occurred in less than 5 years. This material, containing only around 6% active organotin, remains a specialty item due to the high costs, associated mainly with application.

FIGURE 3. Barnacle encrusted panel after 2-months exposure at Miami Beach. (Photo by Miami Marine Test Station. With permission.)

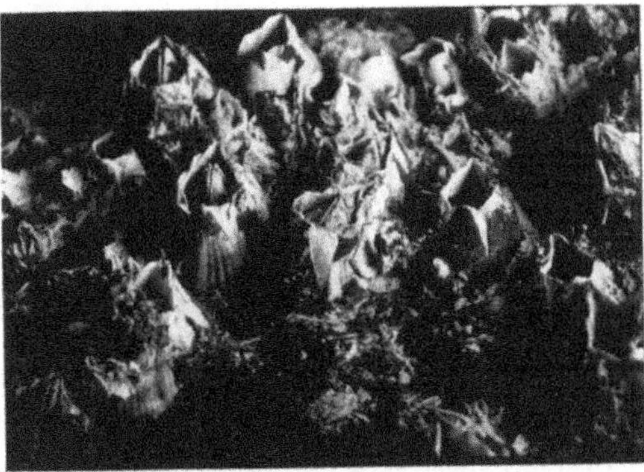

FIGURE 4. Barnacles on depleted conventional antifouling paint. Note the lifting effect as the shell slices under the paint to reach (and corrode) the metal substrate. (Photo by Miami Marine Test Station. With permission.)

Nofoul® is used in the rubber covering of sonar domes. A sheet material, usually buffed on both sides to enhance adhesion, is sold for use on ship hulls and other objects subject to a marine environment. The usually 0.080-in. thick by 36-in. wide sheets are applied over one of several primers using a company-developed adhesive system. Although considerable improvement in application procedures has been made over the years, the process is akin to the application of wallpaper, and there are various problems associated with complete adhesion, seaming, etc. It is time consuming compared to spray painting. Nofoul® sheet and the adhesive system used result in a strong protective covering able to withstand tensile, cavitation, sheer, and tear forces far greater than that found with paint coatings.

Antifouling rubber protects not alone against the attachment of the larval forms of sessile foulants, but also prevents damage by Teredinidae borers, "ship worm", and *Limnoria* (see Figure 5), the "gribble" or so-called "marine termite".

Typical antifouling paints are sometimes elastomer based, but commonly use a non-elastomer, soluble cuprous-oxide agent, thus relying on a leaching-type system or, in a few instances, on a composite leaching-exfoliation system. High agent loadings, 90 to 92% on a dry-film basis, are normal and necessary for the proper development of porosity with depletion. Consequently, such films have little physical strength and are easily damaged. Also, paints releasing ionic substances, such as the copper ion, require a barrier subcoat to prevent electrolytic attack on both metal and *fiberglass* substrates.

The trialkyl and triaryl organotins used in more modern antifouling paints are non-ionic in nature. Thus, a corrosion barrier may not be needed. Unfortunately, paint concerns have selected bases for organotins that lack solvent properties, epoxies, vinyls, alkyds, etc. Therefore, even presently marketed materials rely upon the organotin leaching from the matrix. This requires 40% or greater agent loadings and generally poor film strength is noted. The organotin paints evaluated by this author and others show no appreciable advantage over the less expensive cuprous-oxide materials save the aforementioned lack of galvanic action and the ability to make a wide range of attractive colors. Cuprous-oxide paints vary from dark red through brown to black. Other colors require great expense in preparation.

Partially depleted cuprous-oxide-paint films have been sectioned by microtome and examined for copper content using X-ray spectrophotometry. It is evident that the nature of the loss mechanism is dramatically different from that observed with No-foul® rubber. In the former case, those sections nearest the water/film interface are completely devoid of copper, whereas those sections nearest the substrate have a copper content equivalent to that of a nonexposed film. This is as would be expected with a leaching system. In contrast, partially depleted organotin containing elastomers show an identical tin atom content in each section, although the total tin concentration is less than that prior to immersion. This would, of course, be the case with a diffusion-dissolution mechanism.

Fouling of a depleting conventional paint characteristically shows a pattern such that in one month little if any attachment is evident and the next month a dramatic fouling coverage occurs. Antifouling elastomers as they deplete show a very gradual fouling-community growth. First algae appears, then within a few months, barnacle accumulation initiates. This is followed first by encrusting bryozoans, then oysters, *Bugula*, and tubeworms. It may take 24 months or longer for a test panel to completely foul from the time of first appearance. Fouling threshold values have been established for TBTO based upon dissolution rate as 5.5 $\mu g/cm^2$-day (algae), 0.80 $\mu g/cm^2$-day (barnacles), 0.55 $\mu g/cm^2$-day (bryozoans), 0.25 $\mu g/cm^2$-day (*Bugula*), 0.20 $\mu g/cm^2$-day (tubeworms) and 0.20 $\mu g/cm^2$-day for tunicates.[1] It is believed that TBTO repels the larval forms of sessile organisms, especially barnacles and tubeworms, and causes an accumulative poisoning of algae. Blue and blue-green algae will attach, grow for a short period of time, and then succumb to organotin intoxication. In vivo tests of TBTO against fresh water algae show relatively little toxic effect. Either marine algae are more susceptible, or the secondary organotins evolved from TBTO-loaded elastomers, such as dibutyltin dilaurate, tributyltin sulfide, etc., are superior algicides.

Several organolead materials have been found soluble in a number of elastomers. These release in water at some unknown rate showing appreciable long-term sustained-release antifouling.[1,38] Organolead elastomers show a number of simularities to the organotins both as antifouling agents and molluscicides. Organolead antifouling paints show a diffusion-dissolution-type release mechanism, long-term release measured in

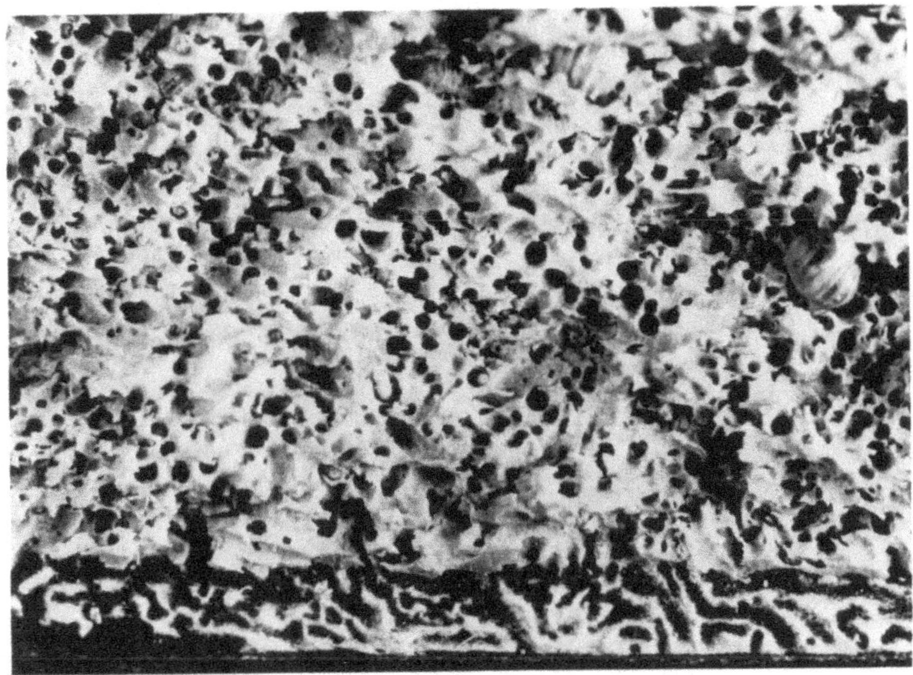

FIGURE 5A.

FIGURE 5B.

FIGURE 5. (A) *Limnoria* destroying wood. (B) Creosoted wood piling nearly destroyed by gribble attack after 5 years in Miami waters. (Photos by the author.)

FIGURE 6. Immersion test rack containing eight rubber panels. (Photograph by the author.)

years, and the evolved products tend to be chemical reactants of the added organolead with base components.[39] Triphenyllead acetate "TPLA" appears to be the agent of choice. Carr and Kronstein evaluated polyisoprene, chlorosulfonated polyethylene, a styrene-butadiene copolymer, and others. TPLA combines with the polymer fraction to form water-soluble metallic soaps or complexes and such are released into the polymer/water interface.[39,40] As with the organotins, organolead content as low as 2.5% provides long-term fouling resistance.[40] An antifouling life of over 5 years for such paints has been observed on test panels under tropical conditions. In a recent advance of the art, Carr and Kronstein have found that the reactivity and performance of TPLA paints can be enhanced by prereacting this agent with the selected polymer prior to compounding.[41] This action permits the use of lower TPLA loadings and better fouling resistance. Organoleads, when added in small amounts to conventional cuprous-oxide paints, provide a synergistic effect substantially prolonging the antifouling life.[1] Insecticides, such as DDT, have long been used as additives to antifouling paints in order to enhance the antifouling potency of the various metal salts and oxides used as agents.

Nonmetallic organic pesticides have been formulated in different elastomeric bases and long-term antifouling potency discovered. These include polychlorinated hydrocarbons possessing the methanobenzene structure;[42] niclosamide, nicotinamide,[43] and other molluscicidal materials, and 2,4,5,6-tetrachloroisophthallonitrile.[1]

Antifouling elastomeric formulations are evaluated by curing panels of 6 × 6, 10 × 12, or 12 × 12 in. size and immersing in marine waters. Such panels are adhered or bolted to a substrate (wood, aluminum, steel, fiberglass, etc.), placed in a compartmentalized rack (see Figure 6), and slung by Teflon®-coated wire cables from a dock into a standard water depth. Panels are retrieved monthly and examined for the amount and type of fouling organisms present. Major U.S. test sites are at Miami Beach and Fort Lauderdale, Fla., and Honolulu, Hawaii. In various programs, tem-

FIGURE 7. Panel-loaded rotating-concentric-drum device used to test antifouling rubber at various velocities. (Photo by Miami Marine Test Station. With permission.)

porary sites have been established at San Diego, Calif., Seattle, Wash., and various seaside locations in Massachusetts, Maryland, and Virginia. Relatively little investment in manpower or equipment is necessary to establish and maintain a test station. The maintenance expense is considerably higher in tropical and subtropical areas than in the temperate latitudes.

The effect of water velocity on the agent loss rate can be determined by use of a motorized, concentric-drum device shown in Figure 7. Rotation at 2, 4, and 8 knots was achieved with this apparatus at the Miami Marine Test Station. At the given velocities, and in accordance with theory, the evolution rate of the organotins from elastomers is unaffected.

Smaller panels, commonly 4 × 4 in., have been attached to a larger plastic substrate panel and immersed at varying depths to below 7000 feet by attachment at selected intervals on an anchored cable suspended from a submerged buoy. Figure 8 depicts one such panel group. Periodic analysis of panels for residual organotin indicated that water pressure does not accelerate agent loss rate from formulations releasing through diffusion-dissolution.

VI. CONTROLLED RELEASE MOLLUSCICIDES

Both diffusion-dissolution- and leaching-type controlled release molluscicides have been developed. The antifouling organotin formulations were initially adapted to molluscicidal usages. Then, somewhat later, compounds were prepared especially for this use. Commercial organotin containing molluscicides are listed below:[44]

FIGURE 8. Panel composite for deep-sea immersion. (Photo by the author.)

- BioMet-SRM,^{TM} 6% TBTO in natural rubber (M&T Chemicals Inc., Rahway, N. J.).
- CBL-9B, 20% TBTF in natural rubber (Creative Biology Laboratory, Barberton, Ohio).
- Ecopro^{TM}-1330,® 30% TBTF in ethylene-propylene copolymer (Environmental Chemicals Inc., Barrington, Ill.).
- Ecopro^{TM}-1043, 43% $CuSO_4 \cdot HOH$ in an ethylene-propylene copolymer (Environmental Chemicals, Inc., Barrington, Ill.).

BioMet-SRM^{TM} and CBL-9B function through a diffusion-dissolution mechanism. They exhibit in-field biologically effective lives in excess of 2 years, and by extrapolation, beyond 7 years. Ecopro^{TM}-1330 appears to be a leaching-type system, yet displays both highly potent molluscicidal and insecticidal life in excess of 480 days![45]

Leaching-type systems have been developed using copper sulfate pentahydrate and monohydrate in ethylene-propylene-diene elastomers, ethylene-propylene plastics, and others. Commercially available materials include:

- Incracide E-51,® 50% $CuSO_4 \cdot HOH$ in an ethylene-propylene-diene elastomer (International Copper Research Assoc., New York, N. Y.).
- Incracide C-10,® 50% $CuSO_4 \cdot HOH$, in an unknown thermosetting material (same source).

A. Monolithic Elastomeric Controlled Release Organotin Molluscicides

It has been known since 1962 that various organotins were potent molluscicides.[15] Cost, difficulty in formulation, and a general lack of environmental data retarded development until the advent of controlled release materials. Conventional application of organotins to snail habitats was difficult. The extremely low water solubilities and hydrophobicity render the use of solutions impractical, and emulsions break down rapidly upon contact with water. Results obtained in the few field experiences reported were about equivalent to those seen with niclosamide and copper sulfate, with less damage to fish and phytoplankton.[46] Appreciable downstream carriage was noted.[47]

The development of long-term, continuous, controlled release organotin molluscicides was initiated by the author and various colleagues in 1966. This work culminated in the development of the BioMet SRM^{TM} material and CBL-9B.[3,28,37] Recipes for various materials are given in the pertinent patents.[3,37,48] The CBL-9B formulation is depicted below:

CBL-9B[49,a]	Parts by weight
Natural rubber	100
HAF carbon black	10
Zinc oxide	2
Stearic acid	0.2
PBNA[b]	1.0
Sulfur	2.5
Altax[c]	2.0
TBTF	30.0

[a] Cure conditions are 290°F for 30 min.
[b] Phenyl-β-naphthylamine (antioxidant).
[c] Benzothiazyl disulfide (accelerator).

Although a large number of elastomers will accept a sizeable TBTF or TBTO loading, natural rubber is preferred from the standpoint of cost and biodegradability. The critical elements as regards rate of release are the cure conditions and the amount and type of carbon black utilized. HAF, ISAF, SRF, and FEF blacks appear to give the best results. The Altax-sulfur-zinc oxide or Captax®-sulfur-zinc oxide cure systems may be somewhat superior. Peroxide and isocyanate cures seem somewhat inferior. With TBTF, a black loading of 8 to 20 parts in natural rubber, CB rubber, and the chloroprenes provides the longest effective release life. In TBTO systems, 15 parts or less of the carbon black is preferred.

Evaluation of TBTO and TBTF controlled release molluscicides in laboratory microenvironmental bioassay has shown that proper dosage control will permit snail destruction without concomitant significant effect upon select aquatic plants or small fish.[28]

A number of small-scale field evaluations have been conducted at sites around the world.

1. Tanzania

A pelletized chloroprene formulation containing 3% TBTO and 8% niclosamide was applied to a 42,000 ℓ pit and a 28,000 ℓ pool fed by water seepage in 1967.[50] Results were favorable, though inconclusive. Sites were monitored for only a few months.

2. St. Lucia (West Indies).

Upatham treated 28 m^2 of marsh at 5 g/m of BioMet SRM® pellets. A pretreatment snail count of 846 dropped to 6 by posttreatment day 137, and then a gradual increase in snail population was noted.[51] Other pellets were bagged to prevent being washed away and suspended in a stream flowing through a ravine. A snail-population drop from 290 to 77 by day 54 was noted. At this time, the bagged pellets were removed, and the snail population declined further to 55 per count by day 109. Then population resurgence was initiated. On day 123, pellets were reapplied, and by day 151, only 3 living snails per survey were found.[51]

In recent tests, Christie has applied BioMet SRM® pellets to several swamps at 10 and 20 g/m^2.[52] Preliminary results indicate a drastic reduction in snail population.

3. Iran

Snail-infested ponds were treated at various dosages with BioMet SRM®.[53] After several months, snail populations dropped to as low as 1% of the pretreatment level.

4. Philippines

Six kg of BioMet SRM® were added to an irrigation canal with caged snails placed 50, 100, and 150 m downstream from the point of application. At the 8-week check, mortality was 100, 97, and 97%, respectively.[54] However, pellets buried 1.5 cm under the mud bottom in a stream provided only a 25% mortality after 60 days. (TBTO movement through mud in laboratory soil-burial bioassays has been observed, but mortality seldom exceeds 50% at practical dosage levels.[55])

5. Brazil

da Souza and Paulini applied various controlled release organotins to ditch and pond sites on a small, irrigated truck farm near Belo Horizonte, Brazil. Snail control was achieved over a 4-month test period using pelletized 5.8% TBTO/butyl rubber.[56] While it required 30 days or longer exposure to substantially reduce adult snail populations, the absence of young snails and eggs was noted within a relatively few days. A superior treatment method seemed to be to affix sheet rubber in a vertical position (5.8% TBTO/neoprene) to the bottom pool surface using two or three wooden stakes, thus avoiding silting. In this arrangement, adult snails were destroyed within a few days.

In a later program, BioMet SRM® pellets were applied to pits, ponds, canals, and a well at dosages varying from 3 ppm active to 12 ppm active.[57] Complete reduction of snail populations occurred within a week in stationary-water bodies. Residual activity after initial snail kill varied from 6 to over 9 months in most sites.[58] Long-term control appeared to be related to water quality rather than to the dosage used. Satisfactory control was not achieved using pellets in flowing-water systems.

Tin analysis was performed on pellets recovered from four test sites after 9 months immersion. TBTO loss varied from 7% to 22%, indicating the loss of 3 ppb to 12 ppb/day and an estimated longevity in excess of 3 years.[59,60]

Gilbert et al. incorporated TBTO into a rubber/clay composition and extruded a rod containing 11% active agent.[61] Various sites were treated including a 3.5 km canal and irrigation system. Thirty such sites were reported free of snails at the latest observation (12 to 18 months posttreatment). Where treatment failed, the water was noted to be of significantly higher alkalinity. It was concluded that focal treatment (i.e., applying the agent at the site where human-snail contact occurs) is superior to area-wide treatment.

6. Rhodesia

A number of test sites in Rhodesia were treated with various organotin-containing elastomers. At Dam 4, Farm 30 (Hippo Valley) a pretreatment snail count of 1,090 dropped to 0 viable snails within 2 months at 30 g/m^2 of BioMet SRM® applied in a 300 m^2 quadrant. However, a neighboring untreated quadrant also dropped to 0 within 3 months, indicating toxicant transport beyond the treated areas.[62] Shiff and others have noted the dramatic drop in egg-laying activities which appears prior to adult mortality.[62,63] Ill effects on nontarget biota do not appear to be significant. An early suppression in some phytoplankters is observed, but populations rapidly resurge. Fish loss is not noted.

Ninety meters of the shoreline of Coldstream farm reservoir having a 950 m perimeter were treated using Nofoul® rubber strips. Snail populations fell rapidly after treatment to 0 within a few months.[64] About 9 m^2 of sheet stock cut into strips was used. Contiguous nontreated areas also exhibited dramatic drops in snail populations. No damage to fish or frogs was observed.

CBL-9B pellets were applied to a 480 m^2 sector of a large pond (Rothbury farm) at 20 g/m^2. Snail populations and bottom fauna were monitored in the test area and contiguous area. Sampling dropped from 481 living snails to 3 living snails in the treated section within 3 months.[65] An adjacent nontreated area showed about a 60% population decrease over the same time frame. Ostracods, hydrachnellid mites, oligochaetes, chironomids and other insecta appeared to be relatively unaffected.

After 7 months immersion, BioMet SRM® pellets were recovered from field sites along with biotrope materials in the vicinity. These were examined for residual tin content with the results shown below:[66]

1. Mud scraped from pellets, 0.24 to 0.58 ppm Sn.
2. Mud taken from below pellets, 0.09 to 5.07 ppm Sn.
3. Surface-water samples, no Sn detected.
4. Bottom-water samples, no Sn detected.
5. Mud-contaminated bottom waters, 0.18 ppm Sn.
6. Pellets, 1.28 % Sn (compared to 2.15% Sn in unimmersed controls).

B. Environmental Impact, Toxicity, and Chemodynamics of Organotin/Elastomer Formulations

Evidence from field evaluations has indicated a minimal effect of TBTO and TBTF on nontarget biota. However, it is well recognized that both materials are potent biocides to various animal phyla and, at least, to the lower plants.[15] Vascular plants seem to be unaffected in field evaluations, whereas significant toxicity has been observed in the laboratory.[67,68] Until recently, environmental-impact assessment and toxicological measurements have dealt with generally nonformulated organotins, or those in highly concentrated emulsion or wettable-powder systems. In contrast, controlled release organotin/elastomer materials release their agents at a very low daily rate (ppb or ppt range) into the treated waters. Such concentrations at molluscicidal dosages are usually below the phytotoxic limit and *apparently* nontoxic to animal life *excepting* certain insecta and snails. It is inconceivable that TBTO and TBTF would only be toxic to snails! It is believed that the intolerance demonstrated by various snail species arises from the chemodynamics of organotin/water/soil systems and the kill mechanism involved. The Environmental Management Laboratory, among others, has conducted a number of experimental programs keyed to elucidating the effects of controlled release organotins on nontarget biota and the effect of environmental parameters upon the agent release rate, transport mollusc contact, and ingestion by the snail.

It is well recognized that a state of chronic intoxication is induced in the snail through continuous exposure to ultralow organotin concentrations. Unlike conventional usage of molluscicides where the intoxication syndrome is acute and moribundity occurs within the first few hours (or minutes!) of exposure, followed by death within 6 to 24 hr, controlled release molluscicides require 1 to 4 weeks to destroy the snail. There is no symptomology manifest for 3 to 5 days posttreatment. The question arises as to the effect on nontarget organisms similarly exposed over extended periods to the organotins. There is no question about the long-term efficacy of organotin molluscicides. The present tasks are to assess the degree of hazard associated with usage and to develop a field methodology for various types of infested habitats.

TBTO and TBTF kill mechanisms have not been elucidated. Examination of snail tissue after exposure to TBTO (6% active in chloroprene) indicates massive cell damage occurs in connective tissue and various membranes. Within 15 min of exposure, cell-wall destruction can be found gradually increasing to the point where function is lost. Muscle and nerve tissue do not appear to be affected. Proteolysis is evident and selective. Death ensues from the destruction of membraneous structures allowing infusion of hemolymph with massive internal hemorrhage.[15] In vitro examination of TBTO and TBTF in the presence of amino acids has to date indicated rapid reaction with glutamic acid and alanine. Present investigations of the reactivity of TBTO and TBTF with 24 amino acids and various peptides will be published in due course.

TABLE 8

Tolerance Studies: *Biomphalaria glabrata*

	Parent Generation		Offspring	
	Control	Test	S1	S2
Maturation period	5 weeks	7 weeks	16 weeks	24 + weeks
Prehatch ova mortality	7%	55%	74%	90 + %
Infant (0—3 day post-hatch) mortality	12%	84%	96%	?
LT_{100} (1 ppm-ta)	—	9 days	2 days	0.4 days

TABLE 9

Biomphalaria glabrata Ova Bioassay With BioMet SRM®

Conc. (ppm)	No. eggs exposed	Ova LT_{100} (days)	No. hatch	Juvenile LT_{100} (days)
10	130	23	0	—
1	87	27	2	2
0.1	161	32	11	8
0.01	118	—	94	∿18
0.00	103	—	101	—

If the above interpretation of histological evidence is correct, the snail target species ought not to be able to develop resistance. In order to further evaluate this hypothesis, two groups of 100 snails each were exposed to the LT_{50} for CBL-9B and BioMet SRM®. Survivors were removed to nontoxic aquaria, eggs were collected, and the S1 generation raised. The following results (Table 8) were noted for BioMet SRM®.[44]

The same pattern was observed with CBL-9B. Experiments were repeated with essentially the same results. The effect of parent exposure to TBTO and TBTF results in a dramatic weakening of the offspring even to the third generation. Although outside confirmatory studies are needed, it is hypothesized at this time that tolerance decreases in succeeding generations.

Organotins are not only ovacidal, but snails that do hatch are dramatically less able to survive. Table 9 depicts *Biomphalaria glabrata* ova bioassay with eggs exposed to various BioMet SRM® levels.

Egg laying is suppressed. Snails exposed in groups of 50 to BioMet SRM® at 0.6 ppm for 0 (control), 24, 48, and 72 hr and removed to nontoxic water laid, in a 30-day period, the following number of egg masses, 434, 247, 251, and 169, respectively.

Water quality affects the time span between initial exposure and eventual mortality of snails treated with controlled release organotin. Essentially, the final result is the same, but the LT_{100} varies. The effect of inorganic ions (NO_3^-, $HPO_4^=$, Na^+, Ca^{++}, $CO_3^=$, HCO_3^-, NH_4^+, NO_2^-, $SO_3^=$, Mg^{++}, Fe^{++}, Fe^{+++}, Sr^{++}, I^-, F^-, Cl^-, $SiO_3^=$, and K^+) on the LT_{100} at maximum expected concentration in natural waters was determined.[49] In general, such ions under acidic water conditions have little effect. They lengthen the LT_{100} under alkaline conditions by several days. The effect of pH alone is depicted in Table 10.

The presence of suspended or soluble organic matter (yeast, humic acid, fulvic acid) increases the LT_{100} with both TBTO- and TBTF-releasing materials. The degree of effect is greater with TBTO.[15] Inorganic clays in suspension seem to be of little consequence in acidic waters, but significantly lengthen the LT_{100} under alkaline conditions.

TABLE 10

LT_{100} Values for Controlled Release Molluscicides Against Adult *Biomphalaria glabrata*

	LT_{100} (days)					
	6.0 pH	6.5 pH	7.0 pH	7.5 pH	8.0 pH	8.5 pH
BioMet SRM®	6	6	5	5	3	3
CBL-9B	3	4	6	2	9	6

FIGURE 9. Compartmentalized aquaria for toxicant transport studies.

Laboratory and field work by Shiff indicated that snail contact and ingestion of organotins arises through browsing activities on the bottom soils.[69] Implications were that dosage requirements could be based upon the bottom surface area of a given water course and not upon total volume of water. In order to verify the Shiff hypothesis, a number of tests were performed. In the initial tests, BioMet SRM® and CBL-9B pellets were placed at the soil surface (50% potting soil, 50% sand mixture) in aquaria containing various water depths. Bioassay results showed a positive correlation with concentration/cm^2 of surface area and a negative correlation with concentration/ cm^3water volume.[49,70] Analysis of soil and water tin content from one such test showed that the inorganic tin concentration in the soil was 0.29 ppm at 10 days posttreatment. It increased to 1.08 ppm by the 18th day. Water analysis showed 0.041 ppm (10th day) and 0.073 ppm (18th day).[49] Isolation studies showed that freely moving snails browsing the bottom soil succumbed to organotin intoxication at a dramatically greater rate than snails in the same environment isolated by netting from such contact.[71] Water removed from treated aquaria wherein an LT_{100} was developed was found to be nontoxic to snails, even though snails continued to succumb to organotin intoxication within the treated aquaria. Soil samples removed from treated aquaria and placed in bioassay against target snails was found to rapidly lose toxicity. Christie estimates that such soil detoxifies in 2 to 4 weeks.[49] Removal of the toxic pellets from the test aquaria and continuous introduction of adult snails likewise indicated rapid detoxification.

CBL-9B pellets were placed upon the soil surface of one series of compartmentalized aquaria (Figure 9) and floated on the water surface of another series. Bioassays were

TABLE 11[67]

Toxicant Movement Study in Compartmentalized Aquaria by Snail Bioassay, CBL-9B[a]

Av. pellet-target distance	LT_{50} (days)		LT_{100} (days)	
	Soil browsing	Isolated	Soil browsing	Isolated
12 cm	6	8	9	12
24 cm	7	7	10	13
36 cm	8	10	13	14
48 cm	10	16	13	20
60 cm	12	18	14	20

[a] CBL-9B = 2.03 ppm active.

TABLE 12

Analysis of Soil and Water For Residual Tin From Several Aquaria After Exposure for 28 Days to 1 CBL-9B Pellet[a]

Pellet location	Water depth	Dosage (w/w)	Residual Tin	
			Soil	Water
Soil surface	10.5 cm	90.3 ppm	0.236 ppm	0.009 pm
	21.1 cm	45.2 ppm	0.136 ppm	0.012 ppm
	24.4 cm	39.0 ppm	0.154 ppm	0.013 ppm
Water surface	10.5 cm	90.3 ppm	0.181 ppm	0.014 ppm
	21.1 cm	45.2 ppm	0.342 ppm	0.011 ppm
	24.4 cm	39.0 ppm	0.436 ppm	0.014 ppm

[a] A soil bottom of 2 in.

performed along with soil and water analysis. Compartments were situated at 12, 24, 36, 48, and 60 cm from the toxic pellet. In each compartment, 20 snails were free to browse the bottom soil and an additional 10 test snails were isolated. Table 11 illustrates one set of results for CBL-9B. From this and other experiments, it is concluded that TBTF and TBTO transport is mainly through a soil path, and snail contact is mainly by ingestion of soil, detritis, and organic absorbents of the organotins.

Whether the controlled release organotin pellet was placed on the subsoil surface or suspended just below the water surface made little difference in kill rate. Analysis of water and soil from this type of experiment using the Trachman et al. technique[72] provided results shown in Table 12.

When either TBTO or TBTF are released in a water-soil system, the majority of the organotin content is absorbed by the soil. In the several studies performed to date, an ∿15:1 soil content to water content ratio is observed.

BioMet SRM® pellets containing carbon-14-labeled TBTO where suspended in a water-soil system, and water aliquots were examined qualitatively for beta emission. After 24 hr immersion, the water count was 2.75 times that of the background (pellet surface count was 240 times background). After removal of the pellet, the water count dropped to the background level in about 4 days, indicating that the soil had purged the organotin from the water.[67]

Work to date indicates that the absorption of organotins from a given water course creates an equilibrium condition between release from the pellet surface and absorption so that actual water concentration seldom exceeds 12 ppb at relatively high experimental dosages. At recommended field dosages, it is doubtful if organotin concentrations in field waters would ever exceed 2 ppb. Removal of the toxicant emission source leads to rapid natural purging of the residual toxicant, i.e., TBTO and TBTF are biologically nonpersistent in the classical sense. Apparently, ligands are formed within the soil, and the organotin moiety is tightly bound. Repeated washing of organotin-bearing soils with known solvents leads to very little recovery.

Downward and vertical organotin movement in soils is under investigation. Bioassay through an intervening soil surface and thermodynamic considerations based on hydrophobicity lead one to hypothesize that such downward movement is restricted to the first few vertical centimeters. Pickup by growing plants of organotin from a surface-soil treatment has not been noted, and very heavy applications of TBTO and TBTF are necessary to induce toxic symptoms in terrestrial vascular plants.[15,73]

Laboratory bioassay of mice populations drinking TBTO- and TBTF-laden water at pellet dosages of 100, 10, 1, and 0.1 ppm through the parent and three subsequent generations have shown no mutagenetic or teratogenetic effects.[74] Some minor toxic involvement of the hepatosplenic system is noted at the highest dosages. No loss of fecundity was observed. Chronic intoxication is observed in crayfish ensuing after 70-days exposure to 100 ppm BioMet SRM®. No symptomalogy is observed at 10 ppm in these animals.

Although the degradation mode for TBTO and TBTF is not empirically known, several schemes have been postulated. In following the Sheldon breakdown pattern,[75] the expected degradants have been examined in terms of environmental impact emphasizing toxicity to fish, mammals, and aquatic vascular plants. It is noted, as Monaghan has observed,[76] that the Sheldon hypothesis depends upon a TBTO → dibutyltin oxide transition which may not predominate under conditions of extremely low water concentrations.

Degradation of TBTO exposed to UV radiation arises from homolytic cleavage and dealkylation.[75] However, pellets exposed for weeks to tropical sunlight and then immersed show no loss of activity. Table 13 presents a compendium of 30-day bioassay of controlled release organotins and possible degradants against a number of species. Concentrations bioassayed are far higher than expected under natural conditions.

The general scheme proposed is that TBTF degrades to tri-butyltin carbonate, then to TBTO, rates being dependent on the pH, the CO_2 content, etc. of the water medium. TBTO in turn dealkylates to dibutyltin oxide then to dibutyltin dihydroxide, butyltin trihydroxide, butylstannoic acid, and finally, to stannic oxide, the ultimate product.

C. Controlled Release Organolead Molluscicides

It has long been recognized that specific organolead compounds are molluscicidal.[15] Triphenyllead acetate, "TPLA", and Tributyllead acetate, "TBLA", have been formulated in a number of elastomeric compounds at various concentration levels. True solubility is present in several elastomers. However, it is possible to incorporate TPLA and TBLA to levels considerably in excess of that limit without inducing unacceptable problems in the processing or physical characteristics of the cured materials. Organolead loadings in TPLA and TBLA are listed in Table 14.

The standard recipes used in this study and also with copper sulfate, trifenmorph, and ethanolamine niclosamide are shown in Table 15.

In periodic bioassay against *Biomphalaria glabrata*, it was evident that (1) release is slow and continuous, (2) a chronic intoxication syndrome is manifested, and (3) longevity of release exceeded 9 months, with biological efficacy through 6 months or better with the superior candidates. It is hypothesized that a diffusion-dissolution mechanism existed. Increasing cure time resulted in decreased agent-release rate.[78]

TABLE 13[49,74]

Compendium of 30-Day Toxicity Bioassay of Controlled Release Molluscicidal Materials and Degradants Against Various Species

Material	Total active conc (ppm)	Species										
		E[a]	V[b]	L[c]	H[d]	GA[e]	MF[f]	ML[g]	M[h]	G[i]	S[j]	SL[k]
BioMet SRM™	10.0	+[l]	+	+	−[m]	−	+	+	−	+	+	+
	1.0	+	+	−	−	−	−	+	−	−	+	+
	0.1	−	−	−	−	−	−	+	−	−	+	+
	0.01	−	−	−	−	−	−	+	−	−	−	+
CBL-9B	10.0	+	+	+	+	−	+	+	−	+	+	+
	1.0	+	+	+	−	−	−	+	−	+	+	+
	0.1	−	+	−	−	−	−	+	−	−	+	+
0.01	0.01	−	−	−	−	−	−	+	−	−	−	+
Ecopro-1330™	10.0	+	+	+	−	−	−	+	−	+	+	+
	1.0	+	+	−	−	−	−	+	−	−	+	+
	0.1	−	+	−	−	−	−	+	−	−	+	+
	0.01	−	−	−	−	−	−	+	−	−	+	+
TBT-CO_3[n]	100.0	+						+	+	+	+	+
	10.0	+						+	−	−	+	+
	1.0	+						+	−	−	+	+
DBTO[o]	10.0	+							+			
	1.0	−							−			
	0.1	−							−			
	0.01	−							−			
DBTD[p]	10	+										
	5	+										
	1	+										
	0.1	−										
	0.01	−										
BTTH[q]	10.0	+										
	5.0	+										
	1.0	−										
	0.1	−										
	0.01	−										

SnO_2	100.0	–	–	–	–	–	–	–	–	–
	10.0	–	–	–	–	–	–	–	–	–
	1.0	–	–	–	–	–	–	–	–	–

[a] *Elodea canadensis,* vascular, aquatic, rooted plant.
[b] *Vallisneria americana,* vascular, aquatic, rooted plant.
[c] *Lemna minor,* "duckweed", vascular, floating, aquatic plant.
[d] *Eichornia crassipes,* "water hyacinth", vascular, floating, aquatic plant.
[e] Green algae.
[f] *Myriophyllum spicatum,* "Eurasian watermilfoil", vascular, rooted plant.
[g] *Culex pipiens quinquefaciatus,* mosquito larva, 1st and 2nd instar.
[h] Laboratory mouse, COBS strain, albino.
[i] *Lebistes reticulatus,* "guppy".
[j] *Biomphalaria glabrata,* snail, Puerto Rican albino strain.
[k] *Schistosoma mansoni,* larval trematode, cercarian form.
[l] + = toxic response.
[m] – = no toxic response.
[n] TBTO-CO_3 = tributyltin carbonate.
[o] DBTO = dibutyltin oxide.
[p] DBTD = dibutyltin dihydroxide.
[q] BTTH = butyltin trihydroxide.

TABLE 14[77,78]

Organolead/Elastomer Loadings

	TPLA		TBLA	
Matrix element	Estimated solubility limit %	Max loading %	Estimated solubility limit %	Max loadings %
SBR 1001	42	78	8	36
SBR 4616	40	73	12	38
Natural rubber	47	67	9	42
EPT 5465	43	59	6	58
CB 220	45	71	7	36

TABLE 15

Standard Base Recipes

	Code				
Ingredient	SBR 1001	SBR 4616	Natural rubber	EPT 5465	CB 220
Ameripol® 1001[a]	100	—	—	—	—
Ameripol® 4616[b]	—	100	—	—	—
Natural rubber[c]	—	—	100	—	—
Epcar 5465[d]	—	—	—	100	—
Ameripol® CB 220[e]	—	—	—	—	100
HAF black	—	—	10	—	15
FEF black	15	15	—	—	—
SRF black	—	—	—	10	—
Zinc oxide	1	1	2	—	3.0
Stearic acid	—	—	0.2	0.5	2.0
Sulfur	1.5	1.5	2.5	1.25	1.4
TMTDS[f]	—	—	—	1.0	—
NOBS[g]	—	—	—	—	1.2
Captax®[h]	1	1	—	0.75	—
Sulfads[i]	—	—	—	0.75	—
PBNA	—	—	1	—	—

Note: The standard recipes used in this study and also with copper sulfate, trifenmorph, and ethanolamine niclosamide.

[a] Styrene-butadiene copolymer (B. F. Goodrich Chemical Co., Cleveland, Ohio).
[b] Styrene-butadiene copolymer (B. F. Goodrich Chemical Co., Cleveland, Ohio).
[c] Ribbed, smoked sheet RSS # 3 (Natural Rubber Bureau, Hudson, Ohio).
[d] Oil-extended ethylene-propylene-diene terpolymer (B. F. Goodrich Chemical Co., Cleveland, Ohio).
[e] *cis*-Polybutadiene (B. F. Goodrich Chemical Co., Cleveland, Ohio).
[f] Tetramethylthiuram disulfide (E. I. DuPont de Nemours & Co., Wilmington, Delaware).
[g] *N*-Oxydiethylene benzothiazole-2-sulfenamide (American Cyanamid Co., Bound Brook, N. J.).
[h] Mercaptobenzothiazole (R. T. Vanderbilt Co., New York, N.Y.).
[i] Dipentamethylene thiuram hexasulfide (R. T. Vanderbilt Co., New York, N.Y.).

TBLA/SRR 1001, TPLA/CB 220, and TPLA/Nat. Rubber were biologically superior.

Serious questions regarding environmental impact and toxicity of the organoleads must be addressed prior to the use of such materials. In a few cursory studies performed to date, a degree of phytotoxicity has been observed.

D. Controlled Release Ethanolamine Niclosamide

Salts of ethanolamine niclosamide have been used as molluscicides since about 1958,

TABLE 16[15]

Niclosamide/Neoprene WRT® Release Data

Compound code	Ethanolamine niclosamide loading (phr)	Ethanolamine niclosamide (%)	Loss (μg/in.²-hr)	Diffusion coefficient (cm²/sec × 10^{-9})	Extrapolated half-life (years)
1121B	12	7.1	3.48	2.7	2.2
1121D	20	13.4	6.65	2.1	2.4
1121E	30	18.9	10.30	1.9	2.9
1121F	50	28.0	10.50	1.7	3.8
1121G	70	35.0	37.5	1.2	1.5
1121H	100	43.5	51.9	1.0	1.6

even though such materials are highly piscicidal and phytotoxic.[15] One method of ameliorating the unfavorable environmental impact would be to formulate with elastomers and create controlled release materials. Bayluscide® (sodium salt of ethanolamine niclosamide) was incorporated as the 70% wettable powder in neoprene elastomers in 1966.[3,48] Continuous release from pellets cut from cured sheet stock was observed over a 3274 hr period.[15] Release rates showed no correlation with agent loadings, indicating that a diffusion-dissolution mechanism was operable.

Work at the Creative Biology Laboratory (Barberton, Ohio) demonstrated that ethanolamine niclosamide could be incorporated in natural rubber, ethylene-propylene-diene terpolymers, chloroprene, acrylonitrile polymers, and styrene-butadiene copolymers.[79,80]

Several materials provided five or more months efficacious release (100% snail mortality) at dosage levels that would permit economic field application. Superior recipes are shown below.

Ingredient	Additives (parts by weight)	
	XM-19	XM-23
Ameripol® 1510[a]	100	—
Ameripol® CB 220	—	100
FEF black	15	15
Zinc oxide	1	1
Sulfur	1.5	1.5
Captax®	1.0	1.0
Ethanolamine niclosamide (70 w.p.)	15.0	30.0

[a] Styrene-butadiene copolymer, B. F. Goodrich Chemical Co., Cleveland, Ohio.

In general, an increase in the niclosamide (ethanolamine salt) content results in an increase in the release rate. However, the rate is not proportional to the loading. Table 16 is a niclosamide/Neoprene WRT® series depicting loss-rate-analysis data. All materials shown provided 6 months, 100% snail mortality at 10 ppm (active) dosage levels.

Solubility of the agent in Neoprene WRT® appears to be around 30 pph (18.9%). Below this limit, loss is due to a diffusion-dissolution mechanism. When the limit is exceeded, both diffusion-dissolution and leaching occur simultaneously, leaching being predominant.

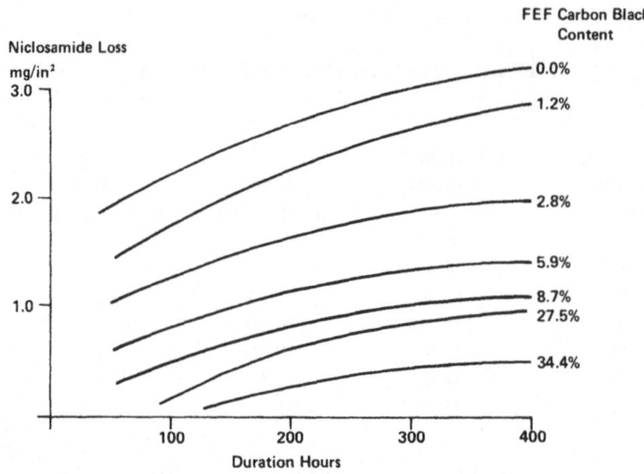

FIGURE 10. Effect of carbon black content on ethanolamine niclosamide release from natural rubber.

By increasing the carbon-black-elastomer ratio and/or increasing the carbon-black structure (fineness), agent-loss rate is decreased as shown in Figure 10.

Unfortunately, incorporation of niclosamide in an elastomeric matrix does not mitigate its effect on nontarget organisms. At dosage levels necessary to destroy snails, fish are likewise killed and considerable phytotoxicity noted.

E. Controlled Release Copper Sulfate

Copper sulfate pentahydrate has been used as a molluscicide since the 1920s, and possibly as early as 1916. Advantages include low cost, low toxicity towards many nontarget organisms, stability to UV, light, and biodegradants, high water solubility, and general availability. Chemical reactivity with dissolved minerals is a major disadvantage. The resulting low persistency in natural waters requires relatively massive dosages, and downstream carriage is poor. While copper sulfate is not piscicidal at dosages required to kill snails under *laboratory* conditions, the field situation often requires concentrations 20 times as great or greater, so that some degree of fish loss occurs.

Copper ion is the toxic element involved, and other copper salts of some degree of water solubility are molluscicidal. Once ionized, copper rapidly combines with a number of negative ions forming insoluble materials, the carbonate, bicarbonate, and hydroxide appear to predominate. Such materials precipitate to the water-course bottom where, apparently, browsing snails avoid ingestion, or the oral mode of entry does not lead to intoxication. Also, organic solubles and suspendents, and inorganic suspendents, adsorb or absorb copper ion, thus desolubilizing it and likely destroying its toxicity.[15]

Probably the major disadvantage affecting the use of soluble copper salts has been low persistency. Consequently, a controlled release material should overcome the detoxification processes found in natural waterways and permit the use of less material while extending the between-treatment interval. A leaching- or exfoliating-type system would be necessary in that copper salts are not rubber soluble. A development program sponsored by the International Copper Research Organization (New York) ensued.

The pentahydrated copper sulfate normally used in conventional snail-control strategies was found to be incompatible with elastomers. During the vulcanization process, temperatures of 275°F or higher are necessary to initiate crosslinking. At 220°F, water

TABLE 17

Controlled Release Copper Sulfate Materials[81]

	Parts by weight			
	1	2	3	4
Epcar 5465	100	100	—	—
Ameripol® 1510	—	—	100	100
Carbon black (HAF)	10	50	15	15
Stearic acid	—	—	1	1
Sulfur	1.0	1.0	1.5	1.5
Mercaptobenzothiazol	0.2	0.2	1.0	1.0
Tetramethylthiuram disulfide	1.0	1.0	—	—
Zinc oxide	3	3	3	3
$CuSO_4 \cdot H_2O$	118.2	157.2	182	—
$CuSO_4$	—	—	—	182
Ammonium sulfate	2—3	3—5	0—3	0—3

of hydration is released as steam, and the pentahydrate degrades to the monohydrate. In a press cure, the steam cannot escape and reacts with the monohydrate, possibly as follows:

$$2CuSO_4 \cdot HOH + H_2O \xrightarrow{290^\circ F} 2H_2SO_4 + 2Cu + O_2\uparrow$$

Both the liberated oxygen and the sulfuric acid degrade the rubber hydrocarbons, leaving a weak nonprocessable gum. Metallic copper plates to the surface! A copper patina is clearly visible, while analysis discloses the presence of sulfuric acid. The reaction may be unusual and was certainly not expected, perhaps clearly illustrating the complex nature of the largely unknown chemical cauldron that exists when temperature and pressure are applied to highly reactive compounding elements.

The solution, of course, was to use the monohydrate form which is stable to temperatures as high as 340°F. $CuSO_4 \cdot HOH$ was found to be compatible with all styrene-butadiene copolymers examined, polyisobutylene, at least several chloroprenes, natural rubber, and *cis*-polybutadiene.[33] Using the twin criteria of high loading and slow release into a watery environment, an ethylene-propylene-diene terpolymer provided superior performance. In the now commercially available Incracide E-51® (International Copper Research Association, New York), a B. F. Goodrich Chemical Co. product, Epcar 5465®, was used as the binding matrix. Copper sulfate loadings as high as 87% were possible, and a release life of 6 to 9 months observed. High loadings, however, do not indicate longer life, but rather, faster release. The optimum $CuSO_4 \cdot HOH$ content is around 50% by weight. Higher loadings lead to more rapid development of porosity and a shorter release life.

Copper sulfate release from elastomers was found to be short lived in alkaline water conditions due to the rapid formation of an impermeable copper carbonate/copper hydroxide film on the pellet surface. This problem was overcome by adding ammonium sulfate as a coleachant, thus altering the pH at the pellet/water interface and preventing film formation.[81] Table 17 depicts several controlled release copper sulfate materials that will release for 6 months or longer. Such compounds are mixed on the mill or in internal mixers without difficulty. Processing operations are those normal to the rubber industry. Since adequate dispersion of the agent requires that a fine-dust form be used, workers need filter masks to prevent inhalation of the dust. Although copper is a vital trace nutrient to humans, prolonged breathing of copper sulfate dust can lead to a reversible intoxication mainly involving the liver.

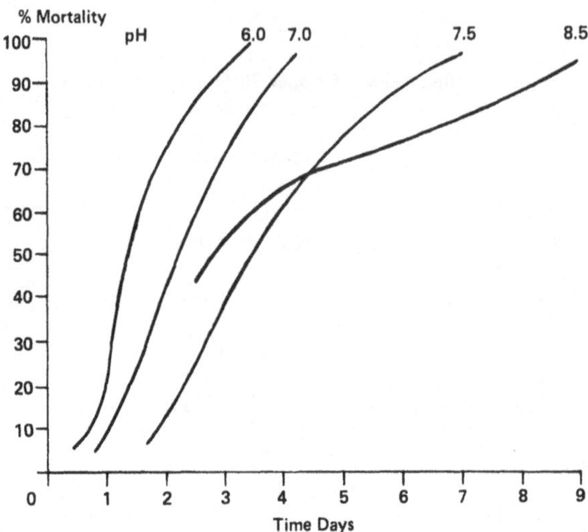

FIGURE 11. Incracide E-51®-caused mortality at varying pH levels. Average of 10 replicates of 10 *Biomphalaria glabrata* each, 17.5 ppm-ta.

Unlike diffusion-dissolution systems, the copper release from elastomers tends to be independent of carbon-black content, at least over a 5 parts to 50 parts range, and cure conditions. In fact, uncured gum rubber will release at a rate only slightly higher than stocks cured for 30 min over a 280°F to 320°F range.

Copper release is influenced by pH (Figure 11) and temperature. Molluscicidal efficiency is dependent upon water quality to a far greater extent than that seen with the nonionic organotins. Table 18 depicts the LT_{100} for Incracide E-51® against adult *Biomphalaria glabrata* under varying water quality conditions.

The increase in LT_{100} noted with increasing alkalinity may only be due to the lowering of copper-ion solubility.

Incracide E-51® has been examined in various laboratory microenvironments previously described.[82,83] It was demonstrated that a copper-ion release between 0.065 and 0.13 ppm/day will destroy *B. glabrata* over a 30- to 90-day period with no gross effect on most aquatic weeds examined, or on *Lebistes reticulatus* ("guppy"), the sentinel fish used. Copper ion released at 0.13 ppm/day reached a total maximum concentration of 0.36 ppm where detoxification processes, mainly precipitation, were equal to copper-ion loss from the pellet material. At 0.2 ppm/day release rate, the *B. glabrata* LT_{100} was 5 days, and no significant *L. reticulatus* kill was noted until 34 days, and then only 3%.

An LD_5 (lethal dose for 5% of the population) for *L. reticulatus* is estimated at 0.038 ppm/day (60 days), while the LD_{99} for *B. glabrata* is around 0.025 to 0.030 ppm/day (60 days).

Controlled release copper sulfate pellets buried at a depth of 2 in. in the bottom soil of microenvironmental bioassay test containers showed some effect on browsing snails, but were inadequate for control purposes. It is thus suspected that silting over will destroy the molluscicidal property of sinking pellets under actual field conditions.

Copper loss from Incracide E-51® pellets varies with the S/V (surface area to volume ratio) and, as with copper ion from antifouling paints, water velocity. Kemper and Albright note that small granules of E-51 lose 52.2% of their copper content in 16 weeks, while larger granules show only 19.2% loss in that time.[84] Shiff et al. have reported that 0.4 g E-51 discs provide 0.21 ppm/day of copper ion for at least 205 days.[85]

TABLE 18[15]

Potency of Incracide E-51® Under Various Water Conditions Against *Biomphalaria glabrata* (Dosage 17.5 ppm-ta[1])

Condition		pH 6		pH 7		pH 8	
Ion	Conc[a] (ppm)	LT_{100} (days)	S V[b]	LT_{100} (days)	S V	LT_{100} (days)	S V
None		3	0[c]	4	0	7	0
K^+	30	4	0	6	0	7	0
Na^+	300	3	0	2	0	9	0
Ca^{++}	600	2	0	5	0	4	0
Mg^{++}	125	8	–[d]	6	–	10+[e]	–
Sr^{++}	30	8	–	4	0	10+	–
Fe^{++}	10	8	–	5	0	N/A	N/A
Cl^-	150	7	+	6	–	9	0
$SO_4^=$	570	7	0	6	0	9	0
$SO_3^=$	10	3	0	5	0	4	+
$CO_3^=$	300	8	–	2	0	7	0
HCO_3^-	350	7	–	6	0	5	0
NO_3^-	300	3	0	5	0	4	+
NO_2^-	0.5	3	0	5	0	5	0
$PO_4^=$	100	7	–	6	0	4	+
$SiO_3^=$	30	4	+	5	0	4	0
F^-	7	8	–	5	0	5	0
Yeast suspension	100	6	–	5	0	6	0
Kaolin suspension	100	2	0	2	0	2	+
Bentolite clay suspension (0.8 meq/g)	100	2	0	2	0	2	+
Montmorillonite clay suspension A (1 meq/g)	100	5	0	6	0	4	+
Montmorillonite clay suspension B (1 meq/g)	100	5	0	5	0	5	+
Montmorillonite clay suspension (0.9 meq/g)	100	6	–	5	0	4	+
Humic acid	100					7	0

[a] The inorganic ion concentrations used herein are the maximum likely in natural waters.
[b] SV indicates significant variation in mortality curve.
[c] A "0" rating indicates no significant variation in mortality.
[d] A "–" indicates slower kill (the mortality curve being shifted to the left.)
[e] A " + " depicts a significantly more rapid kill.

Only a few small field trials with E-51 have been attempted. A trial in St. Lucia showed fast destruction of the original snail colony, but rapid repopulation followed, and no long-term effect was noted.[86] Arfaa has discussed the treatment of 14 ponds ranging in volume from 710 to 960 *ℓ*. *Bulinus, Lymnae,* and *Physa* spp. were destroyed within 3 weeks at 100 ppm/pellet (17.5 ppm-ta). At 80 ppm/pellet (14 ppm-ta) and 90 ppm/pellet (15.75 ppm-ta), 100% mortality was reached in 4 weeks.[87] Reintroduction of *Bulinus* again led to 100% mortality in 2 weeks. Appleton and Stiles treated five experimental sections of a farm storage dam of approximately 1750 m^2 surface area. Treatment rates were at 50, 100, 200, and 400 0.43 g pellets per 10 m^2. Elimination of live snails appeared to have been achieved in 46 days in all treated areas, as well as in a contiguous control area.[88] Species apparently controlled were *B. globosus, L. natalensis,* and *M. tuberculata.*

The presence of a chronic effect of copper ions on snails exposed to low concentrations was first noted by Paulini.[89] The standard formula

$$LD = T \times LC$$

where LD is the lethal dose, LC the lethal concentration, and T time, does not apparently fit bioassay results with copper.[89-91] Cheng and Sullivan have reported that high $CuSO_4$ concentrations for short-term exposures and low $CuSO_4$ concentrations for long-term exposures both provide rapid and irreversible decline in the respiratory rate of *B. glabrata.*[92] Since the required LD_{90} for *B. glabrata* in laboratory microenvironments is 1 ppm-hr or greater, the observation that mortality occurs at much lower concentrations, e.g., 0.025 to 0.030 ppm continuous exposure, strongly indicates that a "chronic-intoxication syndrome" is present.[93]

F. Other Molluscicides

Several nonmetallic organic chemicals uses as fungicides have been found soluble in natural rubber and other elastomeric materials, will release over an extended period at a continuous rate, and are effective against target snails at relatively low concentrations. 2,4,5,6-tetrachloroisophthalonitrile and 3,3,4,4,-tetrachlorotetrahydrothiophene-1,1-dioxide demonstrated considerable merit in this respect.[15] The latter is particularly attractive being effective at 0.1 ppm, relatively inexpensive, nonpersistent, has a mammalian (rat) LD_{50} of 112 mg/kg, and is metabolized by fish and mammals.

VII. BAITS

A "pesticidal bait" requires three basic elements: a toxic agent, an attractant (which may or may not be toxic), and a binding material. In addition, the chemoreceptors of the target animal must be sensitive to the presence of an attractant, and chemotaxis must exist. Although baits are not usually considered as controlled release materials, it is obvious that they possess similar properties. In both, a biological agent must be slowly released from a binding element. A bait, of course, has two agents. The *toxicant* is usually bound because release is not desired, and the *attractant* is emitted through dissolution, exfoliation, leaching, vaporization, or other processes.

If an attractant is present, snails will ingest natural rubber and nitrile polymers, providing that the bait is in a finely granulated form. Carboxylated polyacrylics, which possess both elastomeric and plastic properties, were found to be superior binders for the various attractants used.[15] Various inexpensive attractants such as casein, fish flour, and wheat paste are bound by various elastomers and release through a leaching mechanism. A generalized recipe can be given as follows:

Binding elastomer	100
Attractant	5—30
Toxicant	1—5
Curative	1—2
Accelerator	0.1—0.2
Fungicide	0.05—0.1

A fungicide is essential to prevent rapid biodegradation by varied microorganisms abundantly present in the tropics. The Carboset®* group of carboxylated polyacrylates are especially good binding materials.

* B. F. Goodrich Chemical Co., Cleveland, Ohio.

TABLE 19[95,97]

St. Lucia Field Evaluation: Superior Baits

Ingredient	Recipe (%)		
Carboset 526	69.8	—	64.9
Carboset 525	—	64.9	—
Biorell®/Tetramin®	24.9	—	—
Wheat germ	—	25.0	25.0
Trifenmorph	5.3	5.3	5.3
Carboset 515	—	4.7	4.7
Tetrachloroisophthallonitrile	—	0.1	0.1
Dosage/10 ft. (g)	22	30	30
Mortality (%)	72.8	86.4	67.8
Exposure time (hr)	188	72	120
Dosage/10 ft. (g)	100		
Mortality (%)	72.2		
Exposure (hr)	120		

The molluscicidal fungicides mentioned in the preceding section were used to discourage microbial attack. Neither are suitable as the molluscicidal agent, however, in that concentrations much above 0.1% or so discourage snail ingestion. Similarly, the organotins, chlorinated phenols, copper-bearing materials, and niclosamide were found repellent. The snail will approach such bait chips, rasp off a bit, but rapidly regurgitate it, demonstrating distress. In a sense, he "tastes" an insufficient quantity for lethality and will not take the second bite. Only trifenmorph was found to be readily ingested by the eighteen species of aquatic snails tested, with usually fatal results.[15,94] In contrast, the amphibious snail *Oncomelania formosana* would readily accept niclosamide-laced baits. The superior *Oncomelania*-bait recipe is shown below:

Carboset 525®	66%	Binder
Wheat germ	25%	Attractant
Niclosamide	5%	Toxicant
Tetrachloroisophthalonitrile	1%	Fungistat
Calcium chloride	3%	Curative

This formulation can be mill mixed (preferable) or cast dried from an alcohol slurry. The dried material is pulverized to a 200+ mesh powder to facilitate snail ingestion.

Walker et al. evaluated a number of baits against *B. glabrata* in isolated banana drains on the island of St. Lucia. Snail mortalities as high as 86.4% were achieved within 72 hr at a dosage of 30 g/10ft of ditch.[95,96] Several superior formulations are shown in Table 19.

Baits were not attractive to fish or crustaceans, and no particular environmental disturbance was noted. Tadpoles, water insects, various worms, and similar water life did not appear to suffer any ill effects. In fact, one nematode species bored into the bait chips, and a number were found dwelling therein. However, trifenmorph is toxic to many of the animal habitats of the ditches. This was dramatically illustrated in the St. Lucian study when one experimental binder solubilized under the alkaline pH conditions and released about 5 g of trifenmorph. This agent essentially destroyed all piscine and other animal life for approximately 20 ft downstream.

Carp were fed bait chips at 1 g/gal of water under laboratory conditions and no ill effects noted,[96] but when the same amount of trifenmorph was used as an emulsion (ca. 12 ppm), carp mortality was near 100% within a few hours.

VIII. ELASTOMER-BASED SCHISTOLARVICIDES

The transmission cycle of snail-borne parasites can be interrupted at a number of points. Conventionally, attack has been centered upon the destruction either of the snail host and snail eggs through the use of molluscicides, or of the schistosome within the infected human through chemotherapy. However, other methods of intervention can be conceptualized. The cercaria could be rendered noninfectious prior to human contact. The miracidium could be destroyed prior to snail contact. The sporocyst could be interfered with so as to prevent the development of infectious cercariae.

Conventional tactics used in snail control are not transferable to schistolarva control even though molluscicides may be schistolarvicidal in nature. Destruction of the larval forms without concomitant snail control is meaningless. Intoxication of an infested water course through the application of conventional molluscicides or molluscicides used as schistolarvicides lasts but a few hours to a few days. However, infested snails shed cercariae daily. Consequently, a schistolarvicide must be continuously present unless the snail host is destroyed. The same is true of miracidia. In the normal situation, schistosome egg input into the water course tends to occur at frequent intervals. Miracidial destruction as a means of intervention would be useless unless the human source were removed or the infested waters were continuously rendered miracidiacidal.

It is believed that schistolarvae can be destroyed or made noninfectious at toxicant levels far lower than those necessary for molluscicidal action. In theory therefore, use of a schistolarvicide should have significantly less environmental impact than typical chemical methods of snail control have. Also, since much lower dosages are used, a degree of economy should be realized. Another potential advantage is that schistosome larvae have a limited ability to react to the presence of toxicants. Unlike snails, they cannot leave the treated water course, burrow into the mud, or engage in other avoidance behavior. A poisoned snail may recover, whereas an intoxicated schistolarva lacks time in which to effect recovery. Sublethal dosages of cercariacide or miracidiacide dramatically affect the ability to infect, and this phenomenon is basic to schistolarval control. In this situation, mortality is not the critical criterion in the interruption of transmission.

Laboratory investigations have demonstrated that TBTO, TBTF, and TBTA at water concentrations in the 10^{-6} to 10^{-7} range cause cercarial mortality within 30 min. Ritchie et al. evaluated TBTO against *Schistosoma mansoni* cercariae and miracidia and noted a change in mobility and infectivity.[91] They have reported that both cercariae and miracidia were immobilized within 5 min after exposure to TBTO in the 0.1 ppm to 0.01 ppm range. Near complete suppression of infectivity for both larval forms occurred at 0.1 ppm within 30 or 40 min.

Although data is incomplete, it has been hypothesized that TBTO and TBTF cercaria interactions follow this pattern for 1 hr exposures:[52]

- 0.01 ppm. Complete mortality.
- 0.001 ppm. A significant portion of the test population is immobilized.
- 0.0001 ppm to 0.001 ppm. The suppression of infectivity. Cercariae penetrate the host animal, but do not develop into adult trematodes.
- 0.0001 ppm. There is interference with sporocyst and schistosomule development.

Infected snails exposed to organotin levels at 1/100 the LC_{90} no longer shed cercariae.

Copper sulfate and niclosamide are not particularly toxic to cercariae below 0.01 ppm, a practical concentration for long-term controlled release treatment as shown in Figures 12 and 13.

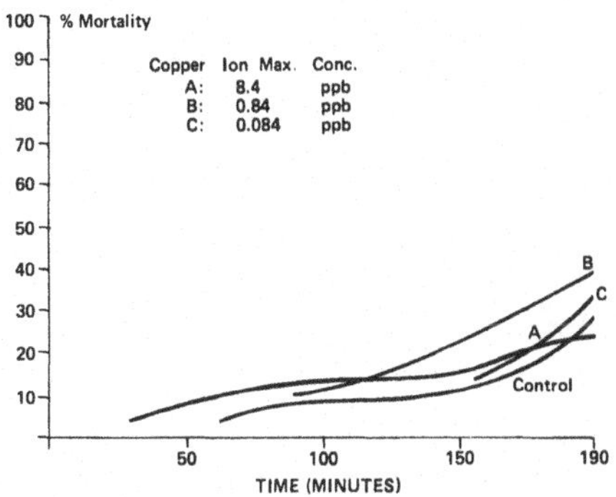

FIGURE 12. Incracide E-51® effectiveness in killing *Schistosoma mansoni* cercaria.[52]

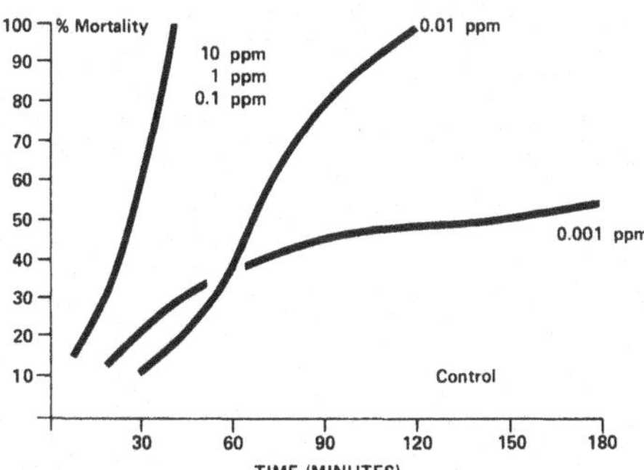

FIGURE 13. Cercaria mortality due to various concentrations of niclosamide.[52]

A number of controlled release materials have been evaluated in the laboratory as cercariacides in the low ppb concentration range. Due to analytical limitations, exact concentrations cannot be determined. Although triphenyllead is cercariacidal, the exposure time necessary to induce mortality may be too great (Figure 14). Tributyllead is superior as shown in Figure 15. Niclosamide from a neoprene matrix (Figure 16) is not effectual. Both CBL-9B (20% TBTF) and BioMet SRM® (6% TBTO) show rapid kill in the low ppb range, while Ecopro^{TM}-1330* (30% TBTF) is considerably slower, as shown in Figure 17. CBL-9B appears to be the superior controlled release cercariacide, as shown by the mortality curves in Figure 18.

Since reduction of infectivity is the paramount criterion, a number of experiments were performed wherein cercaria-laden water was treated with controlled release materials, or their breakdown products, at a given concentration over a time span. Mice were then exposed to the water for a time, 30 min being common, removed, maintained, and sacrificed at the end of a 7-week incubation period. Livers were perfused, and the worm burden determined. The exposure apparatus is depicted in Figure 19.

* Ecopro-^{TM}1330 is a leaching-type system wherein the binding matrix is an ethylene-propylene thermoplastic. The commercial product is manufactured by Environmental Chemicals Inc. of Barrington, Ill.

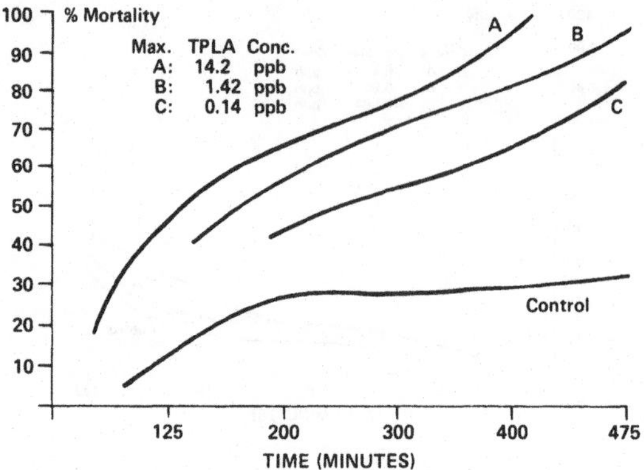

FIGURE 14. CB/TPLA effectiveness against *Schistosoma mansoni* cercaria.[52]

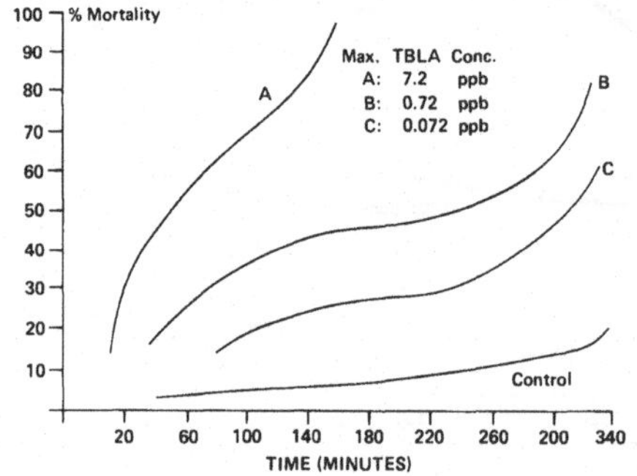

FIGURE 15. SBR/TBLA effectiveness against *Schistosoma mansoni* cercaria.[52]

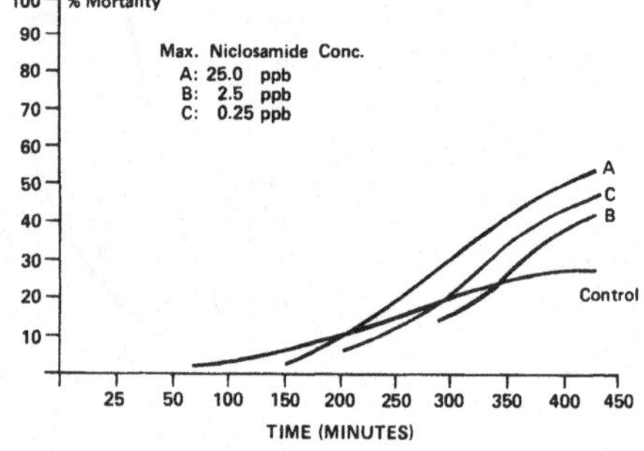

FIGURE 16. XM-167 effectiveness against *Schistosoma mansoni* cercaria.[52]

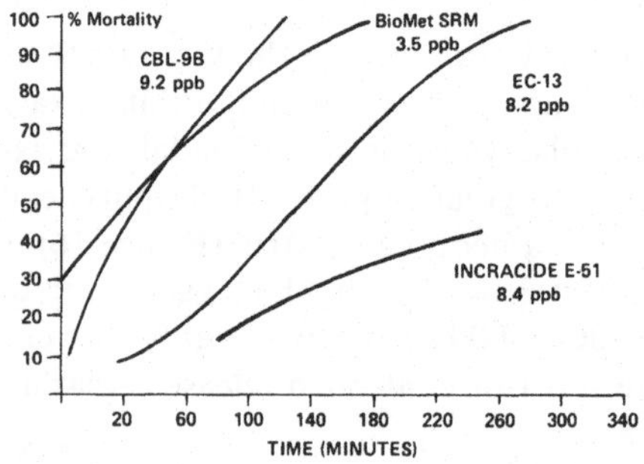

FIGURE 17. Comparison of controlled-release molluscicides evaluated as cercariacides.[52]

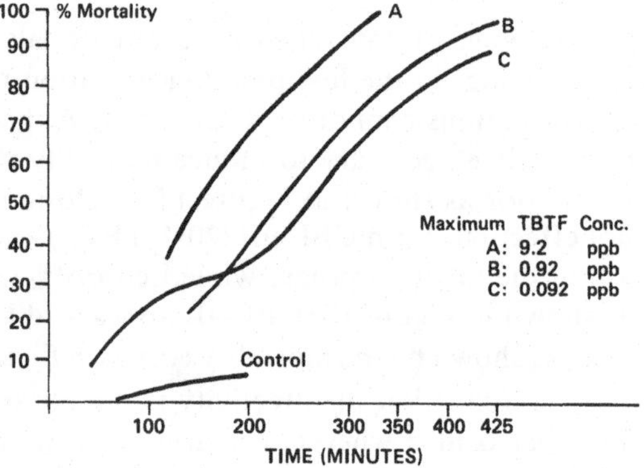

FIGURE 18. CBL-9B effectiveness against *Schistosoma mansoni* cercaria.[52]

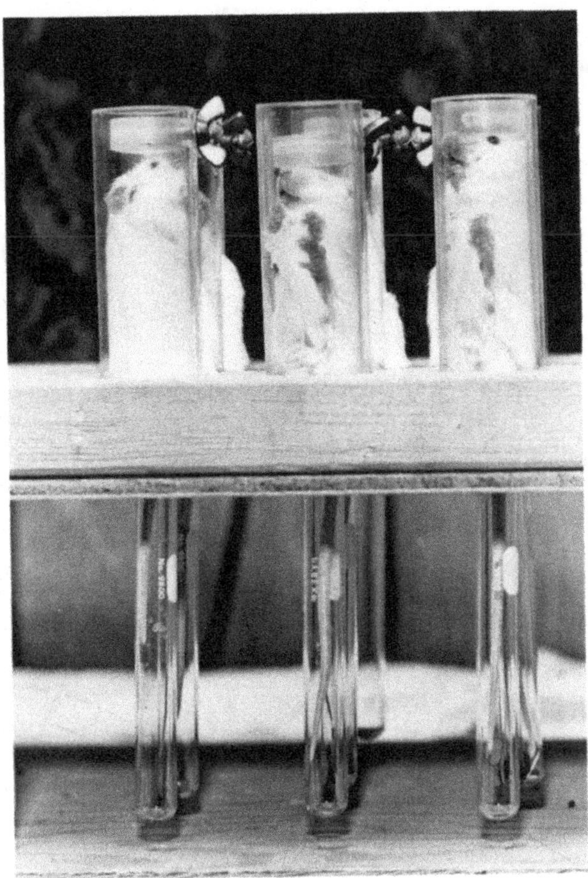

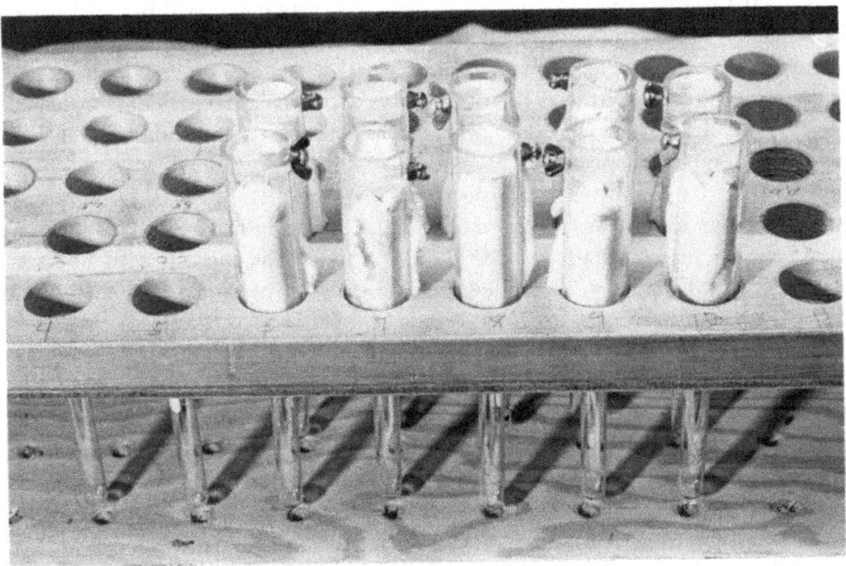

FIGURE 19. Mouse tail-dip apparatus. Mice are caged in plexiglass cylinders mounted on a wooden frame. Tails are suspended in test tubes containing *Schistosoma mansoni* cercariae exposed to a given cercariacide concentration. (Photographs from the Environmental Management Laboratory, University of Akron, Akron, Ohio. With permission.)

TABLE 20[52,98,99]

Schistosome Burden in Mice Exposed to Molluscicides and Controlled Release Molluscicides

Material	Agent conc. (ppm)	Cercariae exposure time (min)	No. mice exposed	No. mice infested	Average Schistosome burden
Cu^{++}	25.4	30	10	0	0.0
	12.7	30	10	5	2.7
	6.3	30	10	8	7.2
	0.0	—	10	10	9.7
CBL-9B	0.1	60	9	8	3.0
	0.01	60	10	9	3.3
	0.001	60	10	3	1.4
	0.00	—	10	9	12.9
BioMet SRM®	0.1	60	10	6	0.6
	0.01	60	10	3	1.1
	0.001	60	10	5	1.7
	0.0	—	10	8	3.3
	0.1	90	10	4	1.5
	0.01	90	10	7	3.5
	0.001	90	10	5	2.3
	0.0	—	10	10	7.6
Niclosamide 50 W.P.	0.1	90	10	0	0
	0.01	90	10	0	0
	0.001	90	10	2	3.5
	0.0	—	10	10	5.5
Trifenmorph 96% technical	0.2	60	10	3	3.3
	0.0	—	10	10	8.8

Table 20 depicts representative infectivity values for the various determinations. In each case, 50 freshly shed cercaria were used as the single-mouse exposure dose.

The dramatic reduction in the cercarial infectivity when exposed to controlled release copper has been observed in studies with *S. mansoni* cercariae performed by Sornmani in Thailand.[100]

Ritchie et al. have shown that the organotins are miracidiacides at ultralow concentrations.[91] Only one study, to date, has been performed in evaluating a controlled release material in this respect.[101]

After 90 days of preimmersion, BioMet SRM® pellets at 1 and 0.1 ppm-ta were placed in bioassay against *S. mansoni* miracidia. Although the total TBTO concentration is unknown, it would not exceed 0.0002 ppm at the most (based upon a 4-year release life). 10 miracidia × 10 replicates were used at each dosage. All animals succumbed within 30 min of exposure. No control mortality was noted.

IX. CONTROLLED RELEASE HERBICIDES

The development and evaluation of controlled release herbicidal materials was a natural outgrowth of the antifouling and molluscicidal effort. Initial work was based on the use of butoxyethanol ester of 2,4-dichlorophenoxyacetic acid, "2,4-D BEE", and the dimethylamine, "2,4-D DMA", in monolithic elastomeric dispensors. The thrust of this early effort was to develop a practical, long-term, controlled release material that could be processed in various forms to allow selective phytozone treatment. Since aquatic weeds occupy specific ecological niches, why not tailor the compound to release only in the target habitat rather than treat the entire body of water? Floating pellets would conceivably be useful against floating vascular plants, rooted weeds would be attacked through the use of a high-density pellet or granule that would sink in the bottom mud, and so on.

Elastomers can be processed in a multitude of forms, pellets, granules, powders, sheets, tapes, discs, tubes, rods, etc. Conventional herbicide carriers, however, are severely limited in geometric form. Density can be adjusted using inexpensive fillers. Elastomeric strips can even be created that will "hover" in the phytozone of attack, say from the surface to 8 in. for the water hyacinth.

Several concepts relevant to phytozone treatment were investigated; sinking, floating, suspending, and timed-release suspending strips. Each was successfully created and tested on a laboratory scale in 1969.[10]

2,4-D BEE was found highly soluble in natural rubber, ethylene-propylene elastomers, and several styrene-butadiene copolymers. The DMA form showed 2% or less elastomeric solubility and was excluded from development work. Carbon black was used as the release regulant with both the loading and the structure of the black being critical. Particle size is the main determinant, but structure plays an important role. SAF black showed greatest effectiveness. However, natural-rubber systems when properly compounded and vulcanized require little or no black, but appear to release at propitious rates without the need of a regulant.

2,4-D BEE release from natural rubber and other elastomers in static-water systems was found to "layer out". That is, a floating pellet released near the water surface, and the toxic-agent concentration was found to be high on and near the surface. There was very little downward movement. Release in the center of a 6-ft, water-filled tube (laboratory) showed that toxicant transport is generally horizontal over a zone extending about 6 in. above and below the pellet location, with very low vertical transport. In dynamic systems, considerable mixing is present, and confinement of 2,4-D BEE to a given zone, save, possibly, the bottom soil, is not possible.[1]

2,4-D BEE recipes have been described elsewhere.[1,10] The major formulations that have passed into a field-test situation are shown below (Table 21). Compound 14ACE-B is used as the weight end for the 14ACE-A or the 14ACE-B1 compound (Figure 20).

Time-release materials utilized a clay-acrylic binder that slowly disintegrated in water, releasing floating pellets or suspenders that rose to the surface.[10]

2,4-D BEE release rates were determined for various formulations. The rate-controlling factor is diffusion within the elastomeric matrix and not dissolution. Consequently, release rate is dependent upon three factors which can be predetermined, (1) the diffusion coefficient of the agent within the binding matrix, (2) the agent loading, and (3) the pellet geometry, mainly the surface area to volume ratio. The diffusion path length can be altered through the use of regulants such as carbon black, finely structured clay, and probably, a number of absorptive materials. However, so far as known, the actual tailoring of a matrix to provide a desired release condition has not been attempted.

Small-scale field tests conducted against *Eichornia crassipes* (water hyacinth) and *Myriophyllum spicatum* (Eurasian watermilfoil) using controlled release 2,4-D BEE were successful.[102] Figures 21 and 22 show a test site before and after treatment. Suspending formulations were found superior against floating plants such as the water hyacinth.

Although most of the known effort with herbicide/elastomer systems has concentrated on 2,4-D, several other agents have been examined. 2,4-D acid (2,4-dichlorophenoxy-acetic acid), Diquat (6,7-dihydrodipyrido (1,2:a:2,1-*c*) pyrazidiium dibromide), Fenac® (sodium salt of 2,3,6-trichlorophenylacetic acid), Fenuron® (3-phenyl-1,1-dimethylurea), endothall [7-oxibicyclo (2·2·1) heptane-2,3-dicarboxyllic acid], and

TABLE 21

2,4-D BEE Formulations

Ingredient	Code (by parts) 14-ACE-B1	14-ACE-B2	11-ACE-D	14-ACE-A	14-ACE-B
RSS # 4	—	—	100	—	—
PA-80	75	75	—	75	75
SMR-5	25	25	—	25	25
2,4-D BEE	30	32	30	30	30
Zinc oxide	1.0	1.0	—	1.0	60
Sulfur	0.5	0.5	2.5	0.5	0.5
Altax®	—	—	1.0	—	—
PBNA	—	—	1.0	—	—
MBTS	—	—	2.0	—	—
HAF	—	—	—	1.0	—
CBTS	2.0	2.0	—		2.0
Stearic acid	0.5	0.5	—	0.5	0.5
TMTDS	1.0	1.0	—	1.0	1.0
Iron oxide	—	10	—	—	—
Type	Floater	Sinker	Floater	Floater	Sinker
% Active	20.4	21.8	21.8	22.2	22.2

Note: Ingredient Code

RSS # 4 = natural rubber, ribbed, smoked sheet.
PA 80 = pale crepe natural rubber.
SMR-5 = refined, ribbed, smoked sheet containing an added lubricant.
Altax® = benzothiazole disulfide.
PBNA = phenyl-β-naphthylamine.
MBTS = mercaptobenzothiazole disulfide.
CBTS = *N*-cyclohexyl-2-benzothiazyl sulfenamide.
HAF = carbon black.
TMTDS = tetramethylthiuram disulfide.

Silvex® [2-(2,4,5-trichlorophenoxy) propionic acid] have been incorporated in a number of elastomeric formulations. Master recipes are noted in Table 22, and recipes for several efficacious formulations are given in Table 23.

Processibility varies for each material. 2,4-D Acid, Silvex,® and Fenac® present no particular difficulty in mixing or curing. Diquat, technical grade, is easily processed in the mechanical aspect, but presents a health hazard. The dust generated, if inhaled, gives rise to nasal hemorrhage. Positive mill, press, and extruder ventilation, as well as a tight-fitting face mask and shield, are essential. Fenuron® melts out of all materials during press or oven cure, and a 10 to 20% agent loss as flashing is observed. Also, Fenuron® vapors affect the nose and throat, resulting in a mild allergic reaction. A dust mask is insufficient for personnel protection. Endothall processes without mechanical difficulty, but the vapors affect not only the nose and throat, but also the eye. Vision becomes blurred, and the condition, as the author can attest, lasts for several hours. No sequalae were present. While the above compounds, except nonformulated Diquat, are generally considered innocuous and simple precautions, such as gloves, are all that is needed for save handling and use, this is not the case when processing in rubber at elevated temperatures. In any scale manufacturing, closed systems would be needed along with complete protection of the human operator from dusts and vapors. Endothall and Diquat present the greater concern. 2,4-D Acid and Silvex® present little, if any, health hazard.

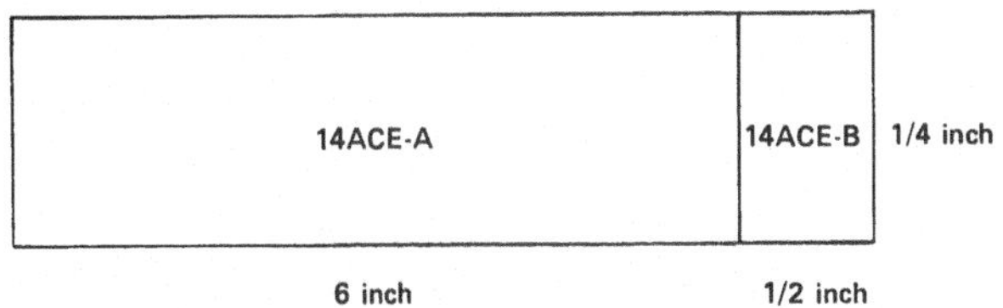

FIGURE 20. Suspender design.

FIGURE 21. Ditch covered by floating water-hyacinth plants prior to controlled-release 2,4-D BEE (suspender) treatment. (Photograph courtesy of U.S. Army Corp of Engineers, New Orleans District.)

Loadings at the maximum holding level are of little value in terms of longevity. As a rule, 50% or so of the maximum provides the longest release life. 2,4-D Acid and Silvex® appear to be soluble in elastomers, and release curves are indicative of a diffusion-dissolution mechanism. Fenac® is definitely not soluble, and the observed loss is due to a leaching process. The others remain unknown.

Loss is affected both by the agent concentration and temperature as shown in Figure 23.

Chronic dosages for 2,4-D acid, Silvex,® and Diquat against several aquatic plants are shown in Table 24. It is noted that dosages above 0.05 ppm/day are probably not practical under field conditions. LdD_{90}* values are obtained at considerably lower dosages.[1]

* Lethal daily dose for 90% mortality in a given population.

FIGURE 22. Same water hyacinth test area 7 months after controlled-release treatment. (Photograph courtesy of the U.S. Army Corps of Engineers, New Orleans District.)

TABLE 22[31]

Controlled Release Herbicide Master Recipes

	Master code (recipe by parts)					
Ingredient	1001	4616	220	600	NRX	EP5465
Ameripol® 1001[a]	100	—	—	—	—	—
Ameripol® 4616[b]	—	100	—	—	—	—
Ameripol® CB-220[c]	— =	—	100	—	—	—
Synthetic NR[d]	—	—	—	100	—	—
Natural rubber	—	—	—	—	100	—
Epcar 5465[e]	—	—	—	—	—	100
ISAF Black[f]	15.0	15.0	—	—	—	—
HAF Black[g]	—	—	15.0	15.0	15.0	—
SRF Black[h]	—	—	—	—	—	10.0
Zinc oxide	3.0	3.0	3.0	3.0	1.0	—
Sulfur	0.0	2.0	1.4	2.0	—	1.25
Altax®[i]	1.75	2.5	—	—	—	—
Stearic acid	—	1.5	2.0	0.5	0.5	0.5
NOBS #1[j]	—	—	1.2	—	—	—
CBTS[k]	—	—	—	2.0	2.0	—
TMTDS[l]	—	—	—	—	1.0	1.0
Captax®[m]	—	—	—	—	—	0.75
Sulfads[n]	—	—	—	—	—	0.75

[a] Hot-polymerized styrene-butadiene copolymer (B. F. Goodrich Chemical Co., Cleveland, Ohio).

[b] Cold-polymerized styrene-butadiene copolymer (source as above).
[c] *cis*-Polybutadiene, Ameripol® CB-220 (source as above).
[d] Synthetic natural rubber, Ameripol® SN-600, (sources as above).
[e] Epcar 5465, oil extended ethylene-propylene-diene terpolymer (source as above).
[f] ISAF black, Vulcan® 6, particle size average 30 nm (Cabot Corp., Akron, Ohio).
[g] HAF black, Vulcan® 3, particle size average 45 nm (Cabot Corp., Akron, Ohio).
[h] SRF black, Sterling® S, particle size average 160 nm (Cabot Corp., Akron, Ohio).
[i] Benzothiazyl disulfide (R. T. Vanderbilt Co., New York).
[j] *N*-oxydiethylene benzothiazole-2-sulfenamide (American Cyanamid Co., Bound Brook, N.J.)
[k] *N*-Cyclohexyl-2-benzothiazolesulfenamide (Monsanto Chemical Co., Akron, Ohio).
[l] Tetramethylthiuram disulfide (E. I. duPont, Wilmington, Del.).
[m] Mercaptobenzothiazole (R. T. Vanderbilt, New York).
[n] Dipentamethylenethiuram tetrasulfide (R. T. Vanderbilt, New York).

TABLE 23[31]

Maximum Loadings of Several Herbicides in Controlled Release Matrices

	Loading (%)					
Agent	1001	4616	220	600	NRX	EP-5465
2,4-D Acid	37	80	82	57	61	53
Diquat	54	69	54	56	66	73
Silvex®	80	75	66	47	48	50
Fenac®	33	67	52	58	44	68
Endothall	49	47	55	73	46	50
Fenuron®	54	58	51	50	57	68

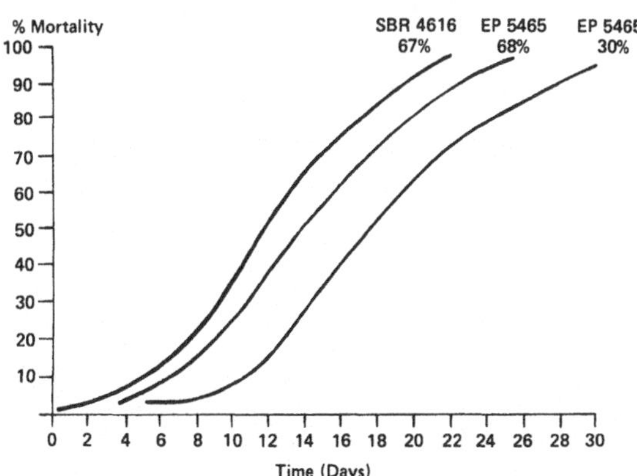

FIGURE 23. Controlled release Fenac® bioassay against Eurasian watermilfoil. Dosage, 10% ta.[31]

The effect of agent loading is shown in Figure 24, and the effect of water temperature on release is depicted in Figure 25.

Extrapolation of release curves and analysis of mortality data permits the estimate of field life for the various controlled release aquatic herbicides. These values vary from 12 months to over 50 months depending upon the particular species attacked, phytotoxicant used, and dosage.[1,10,31]

Little work has been done in the application of controlled release herbicides to soil as a means of controlling weeds of agricultural importance. The rate of release of the

TABLE 24[31]

Chronic Dosages For 30- and 60-day Mortality

Agent	Time (days)	*Cabomba caroliniana* (ppm/day)	*Vallisneria americana* (ppm/day)	*Myriophyllum spicatum* (ppm/day)	*Elodea canadensis* (ppm/day)
2,4-D Acid	30	0.1	0.2	0.008	0.03
	60	0.1	0.2	0.001	0.001
Silvex®	30	0.2	1.1	0.08	0.07
	60	0.2	0.1	0.005	0.01
Diquat	30	1.3	0.01	0.005	0.06
	60	1.3	0.001	0.005	0.01

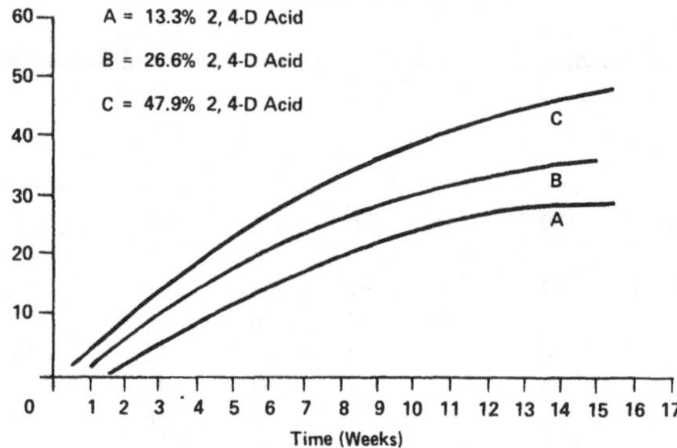

FIGURE 24. 2,4-D Acid loss as a function of loading.[31]

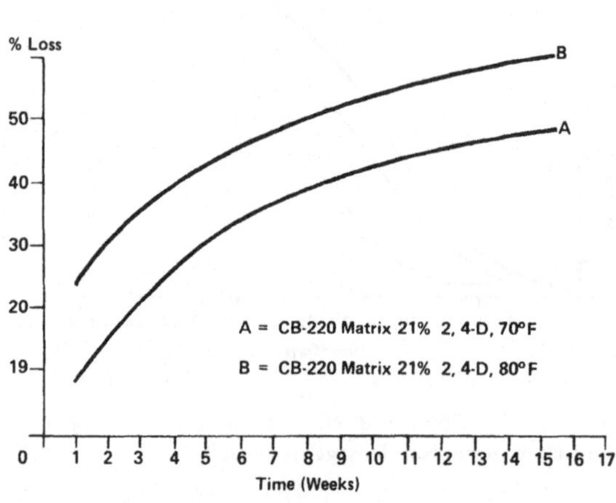

FIGURE 25. Effect of temperature on 2,4-D acid loss from elastomeric matrices.[31]

toxic element from the binding matrix will almost certainly depend upon the amount of moisture present, as is true of conventional materials.[103] Soil properties, especially pH and adsorptive properties, will also be important.[104] In the only known soil-application experiment performed to date, formulation 14ACE-B, 20% 2,4-D BEE, was found phytotoxic to *Brassica* species when tested in the laboratory.[105]

It has been repeatedly observed during the evaluation of controlled release herbicides that the presumptive Ct (concentration times time) relationship does not hold when aquatic weeds are exposed to ultralow biocide concentrations for prolonged durations.[9,13] For instance, Diquat herbicide conventionally used at 1 or 2 ppm against *Elodea* provides total mortality within a day. Continuous exposure at 0.001 ppm for 43 days likewise results in 100% mortality.[13] Thus, a total dosage of 0.043 ppm provides the same results as 1 ppm. This phenomenon, termed the "chronicity effect," has been noted with 14 herbicides against 8 plant species. Observation regarding long-term mosquito larva and snail exposure to organotins suggests, as will be elucidated, that a similar effect exists. The underlying theory remains unknown, although, based upon the stochastic interpretation of Dinman[106] and various threshold studies as described by Kraybill,[107] one mechanism has been suggested.[108] The essential element in this hypothesis is that the target organism is unable to detect very minute quantities of the toxicant, and thus, the normal protective mechanisms triggered by the presence of larger concentrations of the biocidal agent do not become operative.

REFERENCES

1. **Cardarelli, N. F.,** *Controlled Release Pesticide Formulations,* CRC Press, Cleveland, Ohio, 1976.
2. **Cardarelli, N. F. and Neff, H. F.,** Compositions de preservation les incrustacions sous-marines, French Patent 1,506,704, 1966.
3. **Cardarelli, N. F. and Neff, H. F.,** Biocidal Elastomeric Compositions, U.S. Patent 3,639,583, 1972.
4. **Cardarelli, N. F. and Jackson, D. L.,** Slow-release organotin larvicides, report of the Entomol. Soc. Am. Symp., Dallas, Texas, Dec. 4, 1968.
5. **Schultz, H. A. and Webb, A. B.,** Laboratory bio-assay of pesticide-impregnated rubber as a mosquito larvicide, *Mosq. News,* 29, 38, 1969.
6. **Boike, A. H. and Rathburn, C. B.,** An evaluation of several toxic rubber compounds as mosquito larvicides, *Mosq. News,* 33, 501, 1973.
7. **Cardarelli, N. F.,** Floating Larvicide, U. S. Patent 3,590,119, 1971.
8. **Cardarelli, N. F. and Major, C.,** Biocidal rubber for water reclamation systems, Air Force Systems Command Rep. AMRL-TR-69-17, Wright-Patterson Air Force Base, Dayton, Ohio, June 1969.
9. **Cardarelli, N. F.,** Slow release herbicides, *Weeds Trees Turf,* 11(5), 16, 1972.
10. **Bille, S. M., Mansdorf, S. Z., and Cardarelli, N. F.,** Development of Slow Release Herbicide Materials for Controlling Aquatic Plants, Final Rep., DACW 73-70-C-0030, Office of the Chief Engineer, Department of the Army, Washington, D.C., November 1969—July 1971.
11. **Thompson, W. E.,** Field tests of slow release herbicides, rep. no. 15, *Proc. Int. Controlled Release Pesticide Symp.,* Cardarelli, N. F., Ed., University of Akron, Ohio, 1974.
12. **Quinn, S. A., Cardarelli, N. F., and Gangstad, E. O.,** Aquatic herbicide chronicity, *Aquat. Plant Manag. J.,* in press, 1978.
13. **Cardarelli, N. F.,** Aquatic Herbicides Chronicity Study, Final Rep., DACW 73-73-C-0042, Office of the Chief Engineer, Department of the Army, Washington, D. C., November 1, 1972 to October 31, 1973; February 15, 1974.
14. **Janes, G. A., Walker, K. E., and Cardarelli, N. F.,** Controlled Release Herbicides Utilizing Several Elastomeric Carriers, Am. Chem. Soc. Symp., San Francisco, August 29 to September 3, 1976, to be published.
15. **Cardarelli, N. F.,** *Controlled Release Molluscicides,* University of Akron, Akron, Ohio, 1977.
16. **Weast, R. C., Ed.,** *Handbook of Chemistry and Physics,* 57th ed., CRC Press, Cleveland, Ohio, 1977.
17. **Sawyer, A. K., Ed.,** *Organotin Compounds,* Vol. 3, Marcel Dekker, New York, 1972.
18. **Barrer, R. M.,** Diffusion and permeation in heterogeneous media, in *Diffusion in Polymers,* Crank, J. and Park, G. S., Eds., Academic Press, New York, 1968, 165.
19. **Crank, J.,** *The Mathematics of Diffusion,* Oxford University Press, London, 1956.
20. **Roseman, T. J.,** Boundary diffusion layer considerations in the evaluation of controlled release delivery systems, *Proc. Int. Controlled Release Pesticide Symp.,* Gould, R., Ed., University of Oregon, Corvallis, 1977, 403.

21. **Kanakkanatt, S. V.**, The effect of matrix properties on the diffusion coefficients of liquids through rubber, *Macromol. Reprint*, 1, 614, 1971.
22. **Kanakkanatt, S. V.**, Influence of formulation properties on the diffusion-dissolution mechanism, *Proc. Int. Controlled Release Pesticide Symp.*, Cardarelli, N. F., Ed., University of Akron, Ohio, 1976, 8.42.
23. **Cardarelli, N. F. and Kanakkanatt, S. V.**, Matrix factors affecting the controlled release of pesticides from elastomers, in *Controlled Release Pesticides*, Scher, H. B., Ed., American Chemical Society, Washington, D. C., 1977, 60.
24. **Harris, F. W.**, Theoretical aspects of controlled release, *Proc. Int. Controlled Release Pesticide Symposium*, Cardarelli, N. F., Ed., University of Akron, Ohio, 1974, 8.1.
25. **Desai, S. J., Simonelli, A. P., and Higuchi, W. I.**, Investigation of factors influencing release of solid drug dispersed in inert matrices, *J. Pharm. Sci.*, 54, 1459, 1965.
26. **Harris, F. W., Post, L. K. and Field, W. A.**, Investigation of Factors Influencing Release of Herbicides from Polymer Matrices, Q. Rep. 2, October 1972 to December 1972, DACW 73-72-C-0064, Office of the Chief Engineer, Department of the Army, Waashington, D.C., December 1972.
27. **Crank, J. and Park, G. S.**, *Diffusion in Polymers*, Academic Press, New York, 1968.
28. **Cardarelli, N. F.**, Slow release molluscicides and related materials, in *Molluscicides in Schistosomiasis Control*, Cheng, T. C., Ed., Academic Press, New York, 1974, 177.
29. **Bishop, J. H. and Silva, S. R.**, Antifouling paint film structure with particular reference to cross sections, in *ACS Meet. Div. Org. Coat. Plast. Chem.*, 30(1), 364, 1970.
30. **Driscoll, C.**, X-ray fluorescence spectrometry in antifouling coating research, *ACS Meet. Div. Org. Coat. Plast. Chem.*, 29(2), 1969.
31. **Janes, G. A., Bille, S. M., and Cardarelli, N. F.**, Development and Evaluation of Controlled Release Herbicides, Annu. Rep. DACW 39-76-C-0029, Office of the Chief Engineer, Washington, D.C., October 1975 to January 1977.
32. **Bille, S. M. and Cardarelli, N. F.**, Development and Evaluation of Controlled Release Organoleads, Final Rep. Proj. LC-249, International Lead Zinc Research Organization, New York, May 15, 1976.
33. **Walker, K. E. and Cardarelli, N. F.**, Development of Slow Release Copper Sulfate as a Molluscicide, Annu. Rep., Proj. No. 203, International Copper Research Association, New York, July 1, 1974.
34. **Kanakkanatt, S. V.**, Schistosomiasis Control by Slow Release Molluscicides, Rep. to the World Health Organization, Product Development Laboratory, University of Akron, Ohio, September 15, 1972.
35. **Bille, S. M., Mansdorf, S. Z., and Cardarelli, N. F.**, Development of Slow Release Herbicide Materials for Controlling Aquatic P Plants, Annu. Rep. 1, DACW 73-70-0030, AD 879356L, Office of the Chief Engineer, Department of the Army, Washington, D. C., November 1969 to September 1970.
36. **Cardarelli, N. F.**, Method for Dispersing Toxicants to Kill Disease-Spreading Water-Spawned Larva, Trematodes, Mollusks and Similar Organisms, and Products Used in Such Methods, U.S. Patent 3,417,181, 1968.
37. **Cardarelli, N. F.**, Biocidal Elastometric Compositions and Method for Dispersing Biocides therewith, U.S. Patent 3,851,053, 1974.
38. **Carr, D. S.**, Organolead antifouling paints, rep. no. 20, *Proc. Int. Controlled Release Pesticide Symp.*, Cardarelli, N. F., Ed., University of Akron, Ohio, 1974.
39. **Carr, D. S. and Kronstein, M.**, Mechanism of Antifouling Paints, *Proc. Int. Controlled Release Pesticide Symp.*, Harris, F. W., Ed., Wright State University, Dayton, Ohio, 1975, 276.
40. **Kronstein, M.**, Organolead antifouling paint studies, using atomic absorption analysis, *Proc. Int. Controlled Release Pesticide Symp.*, Cardarelli, N. F., Ed., University of Akron, Ohio, 1976, 2.4.
41. Carr, D. S. and **Kronstein, M.**, Improved reactivity of organolead antifouling paints (Abstr.), *Proc. Int. Controlled Release Pesticide Symp.*, Gould, R., Ed., Oregon State University, Corvallis, 1977, 187.
42. **Phillip, A. T.**, Modern trends in marine antifouling paint research, *Proc. Org. Coat.*, 2, 159, 1973/74.
43. **Nishimura, K., Yasunaga, J., and Kaneta, S.**, Nicotinamide as an Antifouling Agent, *JAXXAD 77*, 07053, 3, 1972.
44. **Cardarelli, N. F.**, Controlled release pesticides, in *Pesticide Analysis and Synthesis*, Part 2, Joint National Research Centre-American Chemical Society Workshop, National Research Centre, Cairo, Egypt, 1977, 53.
45. **Cardarelli, N. F.**, Controlled Release Organotin Larvicides, *Mosq. News.*, 38(3), 328, 1978.
46. **Deschiens, R., Brottes, H., and Mvogo, L.**, Applacation sur le terrain, au Cameroun, dans la prophylaxie des bilharziasis de l'oxide de tributyl-etain, *Bull. Soc. Pathol. Exot.*, 59, 968, 1966.
47. **Berrios-Duran, L. A., Ritchie, L. S., and Wessel, H. B.**, Field screening tests on molluscicides against *Biomphalaria glabrata* in flowing water, *Bull. WHO*, 39, 316, 1968.

48. **Cardarelli, N. F.**, Method for Dispersing Toxicants to Kill Disease Spreading Water-spawned Larva, Trematodes, Molluscs, and Similar Organisms; and the Products used in Such Methods, U.S. Patent 3,417,181, 1968.
49. **Cardarelli, N. F. and Walker, K. E.**, Chronic Effects of Ultralow Toxicant Concentrations, Final Rep., 1-R22-AI11861-01A1, National Institutes of Health, Bethesda, Md., 1976.
50. **Fenwick, A.**, Some Observations on Molluscicides in Connexion with the Schistosomiasis Pilot Control and Training Project at Misungwi (Tanzania), World Health Organization Rep. Ser. AFR/BIL-HARZ/14, February 3, 1969, Addendum 1, May 21, 1969.
51. **Upatham, E. S.**, Preliminary results on slow release TBTO pellets and (MT-1E) against St. Lucian *Biomphalaria glabrata* in a marsh and a ravine, *Proc. Int. Controlled Release Pesticide Symp.*, Cardarelli, N. F., Ed., University of Akron, Ohio, 1976, 5.32.
52. **Cardarelli, N. F.**, Laboratory and Field Evaluation of Controlled Release Molluscicides and Schistolarvicides, Prog. Rep. 8, 276-0091, Edna McConnell Clark Foundation, New York, September 1977.
53. **Mansouri, A.**, Brief summary of a study of MT-1E slow release molluscicide, *Proc. Int. Controlled Release Pesticide Symp.*, Cardarelli, N. F., Ed., University of of Akron, Ohio, 1976, 5.52.
54. **Santos, A.**, Field Trials with CBL-9B and MT-1E in Leyte, Research Note, Schistosomiasis Control and Research Service, Department of Health, Manila, Philippines, 1976.
55. **Cardarelli, N. F.**, Controlled release molluscicide activity through an intervening soil barrier, *Controlled Release Molluscicide Newsl. (Univ. Akron)*, 1, 1, 1975.
56. **da Souza, C. P. and Paulini, E.**, Laboratory and Field Evaluations of Some Biocidal Rubber Formulations, World Health Organization Rep. Ser., PD/MOL/69.9, 1969.
57. **Castleton, C. F.**, Brazilian Field Trials of MT-1E, Prog. Rep. 2, Centro de Pesquisas de Produtos Naturais, Rio de Janeiro, Brazil, 1973.
58. **Castleton, C. F.**, Brazilian Field Trials of MT-1E, Final Rep., Centro de Pesquisas de Produtos Naturais, Rio de Janeiro, Brazil, 1973.
59. **Castleton, C. F.**, Brazilian Field Trials of BioMet SRM, rep. presented at the 1st Int. Controlled Release Pesticide Symp., University of Akron, Ohio, September 16—18, 1974.
60. **Ross, A.**, Analysis of BioMet-SRM Sample Pellets from Brazil, M&T Chemicals Incorporated, Rahway, N. J.,; personal correspondence to Cardarelli, N. F., Creative Biology Laboratory., Barberton, Ohio, May 8, 1974.
61. **Gilbert, B., et al.**, Slow release molluscicides in schistosomiasis control, *Proc. Int. Controlled Release Pesticide Symp.*, Cardarelli, N. F., Ed., University of Akron, Ohio, 1976, 5.9.
62. **Shiff, C. J. and Evans, A. C.**, The role of slow release molluscicides in snail control, *Cent. Afr. J. Med.*, Suppl. 23(11), Nov. 1977.
63. **Ritchie, L. S., Lopez, V. A., and Cora, J. M.**, Prolonged applications of an organotin against *Biomphalaria glabrata* and *Schistosoma mansoni*, in *Molluscicides in Schistosomiasis Control*, Cheng, T. C., Ed., Academic Press, New York, 1974.
64. **Deschiens, R. and Floch, H.**, Comparative biological effects of six chemical molluscicides used for bilharziasis prophylaxis, *Bull. Soc. Pathol. Exot.*, 61, 640, 1968.
65. **Shiff, C. J.**, Focal control of schistosome bearing snails, in *Molluscicides in Schistosomiasis Control*, Cheng, T. C., Ed., Academic Press, N.Y., 1974, 241.
66. **Shiff, C. J. and Yiannakis, C.**, Controlled Release Molluscicide Studies, Rep., Blair Research Lab., Salisbury, Rhodesia, 1975.
67. **Cardarelli, N. F., Gingo, P. J., and Walker, W. E.**, Laboratory and Field Evaluations of Controlled Release Molluscicides and Schistosolarvicides, Annu. Rep., 276-0091, Edna McConnell Clark Foundation, New York, July 1, 1977.
68. **Chu, K. Y.**, Effects of environmental factors on the molluscicidal activities of slow-release hexabutyldistannoxane and copper sulfate, *Bull. WHO*, 54, 417, 1976.
69. **Shiff, C. J.**, Field tests of controlled release molluscicides in Rhodesia, *Proc. Controlled Release Pesticide Symp.*, Harris, F. W., Ed., Wright State University, Dayton, Ohio, 1975, 177.
70. **Quick, T.**, BioMet SRM dosage as a function of surface area, *Proc. Controlled Release Pesticide Symp.*, University of Akron, Ohio, 1976, 5.56.
71. **Evans, W.**, BioMet SRM contact studies, *Proc. Int. Controlled Release Pesticide Symp.*, Cardarelli, N. F., Ed., University of Akron, Ohio, 1976, 5.48.
72. **Trachman, H. L., Tyberg, A. J., and Branigan, P. D.**, Atomic absorption spectrometric determination of sub-part-per-million quantities of tin in extracts and biological materials with a graphite furnace, *Anal. Chem.*, 49, 1090, 1977.
73. **Upatham, E. S., Engelhart, J. E., and Seeyave, J.**, Effects of organotin compounds on St. Lucian *Biomphalaria glabrata* and bananas, Report 24, *Proc. Int. Controlled Release Pesticide Symp.*, University of Akron, Ohio, 1974.

74. **Cardarelli, N. F.**, Laboratory and Field Evaluations of Controlled Release Molluscicides and Schistolarvicides, Final Rep., 276-0091, Edna McConnell Clark Foundation, New York, March 1978.
75. **Sheldon, A. W.**, The effects of antifouling coatings on man and his environment, *Am. Chem. Soc. Ser.*, 34(1), 600, 1974.
76. **Monaghan, C. P., et al.**, An Evaluation of Leaching Mechanisms for Organotin Containing Antifouling Coatings, report presented to *Controlled Release Pesticide Symp.*, Oregon State University, Corvallis, Aug. 22—24, 1977.
77. **Anon.**, Evaluation of Slow Release Molluscicides, Final Rep., Creative Biology Laboratory, Barberton, Ohio, World Health Organization, Geneva, Switzerland, July 1976.
78. **Bille, S. and Cardarelli, N. F.**, The development and laboratory evaluation of controlled release organolead molluscicides, Report 5.1, *Proc. Int. Controlled Release Pesticides Symp.*, University of Akron, Ohio, 1976.
79. **Cardarelli, N. F.**, Slow Release Molluscicides: Preliminary Information, Creative Biology Laboratory Misc. Rep. 72-2, Creative Biology Laboratory, Barberton, Ohio, 1972.
80. **Anon.**, Development and Testing of Molluscicidal and Cercariacidal Formulations, 1971 Annu. Rep., Creative Biology Laboratory, Barberton, Ohio, January 26, 1972.
81. **Cardarelli, N. F. and Walker, K. E.**, Slow Release Copper Toxicant Compositions, U.S. Patent 4,012,221, 1977.
82. **Cardarelli, N. F.**, Slow release molluscicides and related materials, in *Molluscicides in Schistosomiasis Control*, Cheng, T. C., Ed., Academic Press, New York, 1974, 177.
83. **Cardarelli, N. F.**, Microenvironmental evaluation of slow release molluscicides, Report 29, *Proc. Int. Conrolled Release Pesticide Symp.*, Cardarelli, N. F., Ed., University of Akron, Ohio, 1974.
84. **Kemper, R. L. and Albright, F. R.**, Chemistry of Copper in Water, Q. Rep. No. 1, INCRA Proj. No. 177-D, April 1, 1975 to June 30, 1975, Lancaster Laboratories Incorporated, Lancaster, Pa., 1975.
85. **Shiff, C. J., Yiannakis, C., and Evans, R. C.**, Further trials with TBTO and other slow release molluscicides in Rhodesia, *Proc. Int. Controlled Release Pesticide Symp.*, Harris, F. W., Ed., Wright State University, Dayton, Ohio, 1975, 177.
86. **Hess, E. H. and Albright, F. R.**, Chemistry of Copper in Water, Annu. Rep., INCRA Proj. 177C, Lancaster Laboratories, Incorporated, Lancaster, Pa., 1975.
87. **Arfaa, F.**, Field trial with slow release molluscicide, *Proc. Int. Controlled Release Pesticide Symp.*, Cardarelli, N. F., Ed., University of Akron, Ohio, 1976, 5.55.
88. **Appleton, C. C. and Stiles, G.**, A preliminary field trial with the slow release molluscicide "INCRACIDE E-51", *Proc. Int. Controlled Release Pesticide Symp.*, Cardarelli, N. F., Ed., University of Akron, Ohio, 1976, 5.42.
89. **Paulini, E.**, Algumas consideracoes sobre o modo de acao dos molluscocidas, *Rev. Bras. Malariol. Doencas Trop.*, 8, 545, 1956.
90. **Paulini, E.**, Copper Molluscicides: Research and Goals, in *Molluscicides in Schistosomiasis Control*, Cheng, T. C., Ed., Academic Press, New York, 1974, 155.
91. **Ritchie, L. S., Lopez, V. A., and Cora, J. M.**, Prolonged application of an organotin against *Biomphalaria glabrata* and *Schistosoma mansoni*, in *Molluscicides in Schistosomiasis Control*, Cheng, T. C., Ed., Academic Press, New York, 1974, 77.
92. **Cheng, T. C. and Sullivan, J. T.**, Biological Studies of Copper Containing Molluscicides, Annu. Rep., INCRA Proj. No. 193, Institute of Pathobiology, Lehigh University, Bethlehem, Pa., 1972.
93. **Walker, K. E. and Cardarelli, N. F.**, Development of Slow Release Copper Sulfate as a Molluscicide, Annu. Rep., INCRA Proj. 203, International Copper Research Association, New York, 1973.
94. **Bille, S. M., Walker, K. E., and Cardarelli, N. F.**, A bait method of destroying *Oncomelania* and *Biomphalaria glabrata*, in Snail Hosts of Schistosomiasis and Related Studies, Annu. Rep. 2, U.S. Army Medical Research Development Command DADA 17-69-C-9116, University of Akron, Ohio, July 1 to June 30, 1971.
95. **Walker, K. E., Quinn, S. A., and Bille, S. M.**, Development of a Toxic Bait for Destruction of *Biomphalaria glabrata* and Other Snail Vectors of Schistosomiasis, Q. Rep. 5, National Institutes of Health NIH-70-2273, University of Akron, Ohio, 1971.
96. **Bille, S. M., Walker, K. E., and Cardarelli, N. F.**, Field Tests of Molluscicidal Baits, Final Rep., U.S. Army Medical Research Development Command, DADA 17-72-G-9351, Creative Biology Laboratory, Barberton, Ohio, October 1 to December 19, 1971.
97. **Walker, K. E.**, Development and field evaluation of frescon baits, *Proc. Int. Controlled Release Pesticide Symp.*, Cardarelli, N. F., Ed., University of Akron, Ohio, 1974.
98. **Cardarelli, N. F.**, Laboratory and field Evaluation of Controlled Release Molluscicides and Schistolarvicides, Progress Report VI, Edna McConnell Clark Foundation, 276-0091, July 1977.
99. **Cardarelli, N. F.**, Laboratory and Field Evaluation of Controlled Release Molluscicides and Schistolarvicides, Prog. Rep. 7, 276-0091, Edna McConnell Clark Foundation, August, 1977.

100. **Cardarelli, N. F.**, Laboratory and Field Evaluation of Controlled Release Molluscicides and Schistolarvicides, Progress Report 10, 276-0091, Edna McConnell Clark Foundation, November 1977.
101. **Cardarelli, N. F.**, Laboratory and Field Evaluation of Controlled Release Molluscicides and Schistolarvicides, Progress Report 11, 276-0091, Edna McConnell Clark Foundation, October 1977.
102. **Thompson, W. E.**, Field tests of slow release herbicides, rep. no. 15, *Proc. Int. Controlled Release Pesticide Symp.*, Cardarelli, N. F., Ed., University of Akron, Ohio, 1974.
103. **Furmidge, C. G. L., Hill, A. C., and Osgerby, J. M.**, Physicochemical aspects of the availability of pesticides in the soil. I. Leaching of pesticides from granular formulations, *J. Sci. Food Agric.*, 17, 518, 1966.
104. **Scott, H. D. and Phillips, R. E.**, Diffusion of selected herbicides in soil, *Soil Sci. Soc. Am. Proc.*, 36, 714, 1972.
105. **Danielson, L. L. and Campbell, T. A.**, Evaluation of latex based herbicide formulation, Report 41, *Proc. Int. Controlled Release Pesticide Symp.*, Cardarelli, N. F., Ed., University of Akron, Ohio, 1974, A-1.
106. **Dinman, B. D.**, "Non-concept" of "no-threshold": chemicals in the environment, *Science*, 175, 495, 1972.
107. **Kraybill, H. F.**, Pesticide toxicity and potential for cancer: a proper prospective, *Pest Control*, 43(12), 1975.
108. **Cardarelli, N. F.**, Hypothesis concerning chronic intoxication, *Proc. Int. Controlled Release Pesticide Symp.*, Cardarelli, N. F., Ed., Wright State University, Dayton, Ohio, 1975, 349.

Chapter 4

MEMBRANE SYSTEMS

W. P. O'Neill

TABLE OF CONTENTS

I. INTRODUCTION

A. Scope

This chapter is concerned with the development of controlled release systems that function by the movement of active molecules from a reservoir to a sink through a polymeric film membrane that offers a predetermined resistance to said movement. The driving force of such systems is the concentration gradient of active molecules between reservoir and sink. The mechanism of movement is diffusion, and the medium may be either the polymer itself or a fluid entrapped in a polymeric network. The resistance provided by the membrane is a function of film thickness and is characteristic of both the film and the migrating species in a given environment. A reliable supply of agent must be available from the reservoir, which may be solid or liquid. The membranes themselves represent a diverse continuum of synthetic compositions that includes homogeneous materials and heterogeneous structures in which one substance fills void spaces in a network of another.

Over the years, polymeric films have been developed for an enormous variety of applications. As protective packaging and coatings, films impede the permeation of some species, e.g., H_2O, and permit the passage of others, e.g., O_2. On the other hand, specific permeation films have been used in reverse osmosis, dialysis, hyperfiltration, and ion exchange to separate mixtures of various species. Only recently, the unique permeation characteristics of polymeric membranes have begun to be applied for controlled delivery of active chemical or therapeutic agents.

A generalized delivery system contains the active agent in sufficient quantity to provide treatment at a predetermined, precisely controlled rate for an extended time period. For some purposes, constancy in concentration may be the goal, while some may require predictable temporal variations.

The factors governing the behavior of membrane systems used for barrier, permselective, or controlled release applications are similar in that the physicochemical composition and structure of the polymeric membrane system govern the solution and transport properties. Extensive investigations in the science and technology of membrane transport are serving as both a theoretical and practical basis for the utilization of membrane systems, especially in the area of controlled delivery. The principles, which shall be interpreted here primarily in the context of the development of drug-delivery systems, are equally applicable for other systems and other purposes, as the succeeding chapter on pesticide and consumer products applications will illustrate. Examples of other delivery technologies such as microencapsulation, which are discussed in detail in Volume II, will be mentioned briefly when it is clear that these make use of the same rate-moderating mechanisms as are involved in macroscopic membrane systems.

B. Pharmaceutical Applications

By far, the principal applications evaluated and/or commercialized with respect to controlled delivery from membrane systems, have been directed toward problems in drug administration. The sort of benefits attainable through this new class of dosage forms, called therapeutic systems, include (1) localization of drug action to tissues requiring treatment, with maximized therapeutic response and minimized exposure of nontarget tissues (2) maintenance of optimal concentrations of drug without undesirable surges or periods of drug insufficiency, (3) avoidance of drug loading into tissue sinks from which elimination is slow and difficult, (4) continuity of therapy for prescribed periods with less drug and decreased risk, (5) ability to utilize drugs with short biological half-lives more safely, (6) greater assurance of compliance of the patient with the prescribed therapeutic regimen, and (7) greater convenience, comfort, and peace of mind for the patient.[1]

In controlled drug-delivery systems, the rate-controlling membrane must usually serve the purpose of interfacing with the biological site, often in intimate contact with delicate tissue. Thus, a high degree of emphasis has been placed upon the biocompatibility of candidate membrane materials, as well as their permeability characteristics.

Therapeutic systems are a new class of pharmaceutical products specified in terms of rate and duration of drug delivery. They provide preprogrammed, unattended delivery of drug at a rate, and for a time period, established to meet a specific therapeutic need. Each therapeutic system consists of a drug component, a delivery module, a platform, and a therapeutic program.

The drug component is selected on the basis of its established therapeutic effectiveness, appropriate pharmacokinetic behavior, and pattern of duration at the active site, if known.

The drug delivery module is subdivided into four elements:

1. The *drug reservoir* is a single- or multicomponent element which stores the drug in a stable form in the required amount for the execution of a prescribed drug program.
2. The *rate controller* establishes and maintains the prescribed rate of drug administration through the operational life of the system.
3. The *energy source* effects the transfer of the drug molecules from the drug reservoir to their selected point of entry in the body.
4. The *delivery portal* provides egress for the drug from the delivery module so that it can reach the target tissue.

The platform, which must be compatible with local tissues, houses the drug and delivery module and couples the therapeutic system to the selected body site. The body site dictates the size, shape, and mechanical properties necessary for patient comfort. The platform may either be fixed to the body site or mobile within a defined area.

The therapeutic program, which is executed by the drug-delivery module, accomplishes the basic purpose of the therapeutic system: to present the drug in the most beneficial and reliable manner.

II. THEORY

A. Principles of Permeation

1. Diffusion in General

a. Fick's Law Applied to Membranes

If a fluid mixture is brought in contact with one surface of a membrane under conditions where the other surface of the membrane is in contact with another fluid mixture, and the chemical potentials of the components common to both fluids differ across the barrier, there will generally be a transfer of components through the membrane in the direction of declining chemical potential. If a homogeneous film of uniform thickness separates two fluid phases of differing composition, and the boundary conditions on both sides of the membrane are maintained constant, a steady-state flux of one or more components through the membrane will ultimately be established as given by Fick's first law of diffusion:

$$J_i = -D_i \frac{d\,C_i}{dx} \tag{1}$$

where J_i is the mass flux, D_i is the local diffusivity, C_i is the local concentration of component i, and x is the distance through the membrane. The units of J_i are mass/area × time.

When the permeating component is in equilibrium with the respective surface layers of the membrane, the local concentration in the membrane surface can be related to the concentration in the solution contacting the membrane:

$$C_i = k\,C \tag{2}$$

where k is defined as the partition coefficient.

Assuming the diffusivity to be constant, under steady-state conditions, Equation 1 can be integrated to give:

$$J_i = D \frac{\Delta C_i}{\ell} \tag{3}$$

where ℓ is the membrane thickness, and D is the diffusion coefficient.

Since the local concentration within the membrane is seldom known, Equation 3 is usually written:

$$J = \frac{D\,k\,\Delta C}{\ell} \tag{4}$$

Where

$$\Delta C = C^{*} - C^{\ell}$$

the differences in concentration across the membrane. (Figure 1)

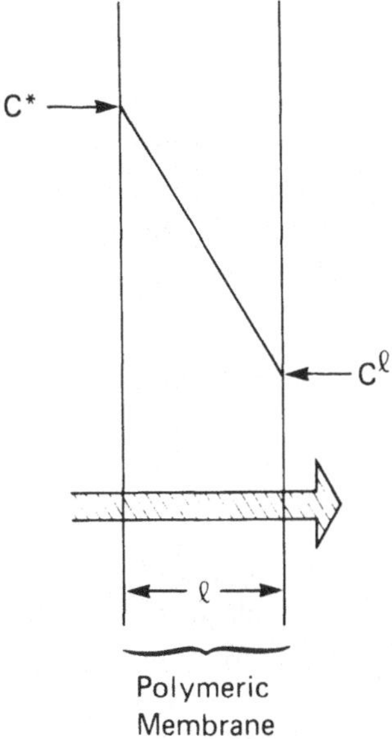

FIGURE 1. Concentration profile across rate-controlling membrane. (Illustration provided by ALZA Corporation, Palo Alto, California.)

The flux of the permeant is more properly recognized to be proportional to the gradient in thermodynamic activity rather than in concentration. The distinction between concentration and activity becomes significant when different solvents are compared. If the solubilities differ, a given solute will permeate a given membrane at different rates from solutions at the same concentration in different solvents. Regardless of absolute concentration, all saturated solutions produce the same flux.

The implications of Fick's Law for many delivery systems, including the membrane situations of interest here, have been the subject of detailed treatments by Baker and Lonsdale,[2] Flynn et al.,[3,4] and Rogers[5] — who elaborated the principles articulated by Barrer, Crank, and Park in their texts on diffusion.[6-8] The general topic of polymeric membrane permeation has been surveyed by Michaels[9] and Hopfenberg.[10]

b. Measurement of Diffusion Coefficients

Techniques for the measurement of diffusion coefficients in polymers have been well discussed by Crank and Park.[7] The commonly used methods for the determination of diffusivity include the measurement of permeation rates and sorption/desorption kinetics. For steady-state permeation occurring through a membrane of thickness, ℓ, the diffusivity can be deduced from a single observation of the flux using Equation 4, provided the solubility in the membrane is known. In the case of transient permeation measurements, the diffusivity can be deduced from the time lag.

If one plots the amount of permeant diffused through a membrane of thickness ℓ versus the time elapsed, a typical curve is observed in which the instantaneous rate of appearance of the permeant increases gradually until the steady-state rate is established, i.e., the plot assumes a straight-line form. The intercept of the projection of that line back to zero is defined as the time lag, t_1, which corresponds to the time needed for the first diffusing molecules to be transported across the thickness of the membrane. The diffusion coefficient, D, is simply the solution of Equation 5:

$$D = \frac{\ell^2}{6\, t_{lag}} \tag{5}$$

Alternatively, one can follow the desorption of permeant from a saturated membrane. By plotting the amount desorbed at time t, M_t, as a fraction of the total desorbed, M_∞, versus the square root of t/ℓ^2, the diffusion coefficient can be obtained, since as long as M_t/M_∞ is less than 0.6 Equation 6 will fit the data.

$$\frac{M_t}{M_\infty} = 4\left(\frac{Dt}{\pi \ell^2}\right)^{1/2} \tag{6}$$

2. *Diffusion Through Dense Polymers*

a. *Influences Upon Diffusivity*

The medium for diffusion in a homogeneous polymeric membrane is its amorphous phase. The diffusion of a solute molecule within an amorphous polymer matrix is an activated process involving the cooperative movements of the penetrant and of the polymer-chain segments that surround it. In effect, thermal fluctuations of chain segments must allow sufficient local separation of adjacent chains to permit the passage of a penetrant molecule. It is by this stepwise process that diffusion occurs. The frequency with which a diffusion step will occur depends upon (1) the size and shape of the diffusing molecule, (2) the tightness of packing and force of attraction between adjacent polymer chains, and (3) the stiffness of the polymer chains. Mathematical expressions have been derived relating the diffusivity and activation energy for diffusion of simple molecules within polymer matrices to the molecular dimensions of the penetrant and the configuration of the polymer chains.[9] Diffusivity relates to the free-volume of the amorphous phase, which for a given polymer decreases with increasing molecular weight, but is best appreciated in terms of the glass-transition temperature.

Most polymers undergo discontinuous changes in several second-order thermodynamic properties at their glass-transition temperature, T_g. Such polymers in membrane form display corresponding changes in properties that govern transport, e.g., heats of solution and diffusion activation energy, at or near T_g.[9]

Experiments, such as those of Michaels et al.[9] with polyethylene terephthalate, above and below T_g support the suggestion that polymers in the glassy state contain frozen-in submicroscopic voids that entrap penetrant molecules and contribute little to the diffusive process. Apparently, below the T_g there are regions of virtually immobilized chains through which activated diffusion is impossible. In general, the farther below its glass transition a polymer is employed as a diffusion barrier, the less its free-volume, and the less permeable its amorphous phase. Since T_g increases with polymer molecular weight, diffusion coefficients are inversely related to the degree of polymerization. Plasticizers in effect lower the T_g, increase free-volume, and promote polymer-segment mobility. Thus, diffusivity generally increases with plasticizer incorporation.[5]

Diffusivity, along with other properties, can be modified by copolymerization. Random insertion of comonomer units in a polymer chain generally decreases T_g and increases the amorphous fraction. Grafts, depending on their nature and length, may increase or decrease permeability.[5]

It has long been recognized that cross-linking reduces chain mobility and diffusivity and, thus, can be used to modify the membrane characteristics of nonthermoplastic polymers. Generally, fillers simply decrease the volume fraction of membrane available for diffusion, but well-dispersed reinforcing fillers may decrease diffusivity further by reason of their cross-linking activity.[3]

By definition, thermoplastics do not contain permanent cross-links. However, there are two important sources of chain constraint in such polymers, crystallinity and strain-induced orientation. Polymeric crystallites are regions of high molecular order and packing. Penetrant molecules are generally insoluble in such crystallites, and hence, their diffusion is confined to the remaining amorphous regions. The crystallites reduce the available volume of the polymer for penetrant solution, and they constrain diffusion to take place through irregular, tortuous pathways between crystallites. Furthermore, within the amorphous phase, the diffusing molecules, if they are large enough, encounter intercrystalline passages which are too narrow to accommodate them. Thus, the fraction of the amorphous polymer phase accessible for diffusion of a given penetrant decreases as the molecular size of the penetrant increases. The literature on perm-separations provides much evidence that such semicrystalline materials can be discriminating molecular sieves.[9]

Michaels[9] and others[7] have demonstrated that the thermal history of a semicrystalline polymer membrane markedly affects permeability. Linear polyethylene slowly cooled from the melt has much lower gas permeability than the same polymer quenched from the melt and subsequently annealed at a high temperature, even though both membranes have the same level of crystallinity. Presumably, this is a consequence of the presence of well-formed, thin crystallites in the former polymer and of thick, defective lamellae in the latter.

The nature of the permeant, however, can apparently reverse the consequences. Klein and Briscoe[11,12] have shown recently that for relatively long permeant species, e.g., a series of paraffin esters, diffusion is faster in the slowly cooled matrix than in a rapidly quenched polymer. This trend, contrary to that observed for simple gases, is the case for long penetrants even though the degree of crystallinity of the quenched polymer is somewhat less than that of the slowly cooled material.

The effects of mechanically induced orientation of crystalline polymers upon their transport characteristics are also dramatic. If a polymer film is uniaxially stretched well below its melting point, the drawn film, despite negligible change in the amount of crystallinity consequent to the deformation, displays permeability characteristics orders of magnitude lower than its undrawn counterpart. On cold drawing, the crystalline structure of the polymer is fragmented, and the amorphous polymer chains are pulled into parallel alignment. Apparently, this fragmentation and alignment greatly increases the number of intercrystalline tie chains, substantially decreasing the mobility of individual chain segments.[10]

When the penetrant solvates the polymer, the effects of microstructure are greatly exaggerated. While the crystalline phase is not swollen by the penetrant, the amorphous phase may swell within the limits imposed by the elastic deformation of the chain segments that bridge between crystallites. Low-crystallinity polymer may swell much more than a sample with slightly greater crystallinity, and as a result, the less crystalline material may be many times more permeable to solvating species than the difference in their amorphous contents alone would suggest. For example, low-crystallinity polyethylene (50% amorphous phase) is 10 times more permeable to liquid hydrocarbons than high-crystallinity polyethylene (30% amorphous phase). Annealing facilitates even greater swelling and, thus, can further enhance permeability to solvating species.[9]

Several points are illustrated by Figure 2 which presents a comparision of the variation with solute molecular weight of the diffusivity in polymers and water.[2] The diffusants plotted range from gases to complex organic species. Diffusion in polymers is more sensitive to molecular weight of the diffusing species than is diffusion in liquids such as water. Furthermore, the difference between the diffusivity in liquids and polymers increases with increasing size of the penetrant molecule. This can be understood as a consequence of the fact that diffusion of a molecule in a polymer requires the

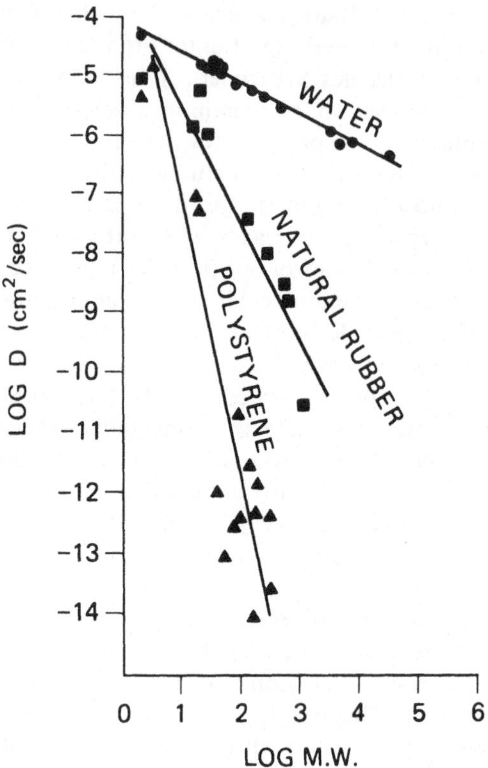

FIGURE 2. Variation of diffusion coefficients with molecular weight for solutes diffusing in water, natural rubber, and polystyrene. (Reprinted with permission of the University of Akron from *Controlled Release Pesticide Symposium,* September 1974, 11.26, and with permission of Plenum Publishing Corporation from Baker, R. W. and Lonsdale, H. K., in *Controlled Release of Biologically Active Agents,* Tanquary, C. A. and Lacey, R. E., Eds., Plenum Press, New York, 1974, 21.)

cooperative movement of several polymer backbone segments for each molecular jump. The larger the molecule, the larger the movement required. The stiffer that polymer backbone, the greater the energy required for said movement, and the lower will be the diffusion coefficient. The steepness of the slopes in Figure 2 reflects these requirements.

Natural rubber (polyisoprene) and polystyrene are representative of the two major classes of amorphous materials, rubbers and glasses. Diffusivities of solutes in most of the polymers of interest lie between the value in rubbers (relatively permeable polymers, used above their glass-transition temperatures) and the value in polystyrene (a relatively stiff and impermeable glass since it is well below T_g under normal conditions).

b. Influences Upon Partition

Permeation is a consequence not only of diffusivity, but also of the distribution of the solute between the polymer and the media surrounding it. The partition coefficient (k) is the ratio of the solubility of the permeant in the membrane to its solubility in a given liquid solvent.

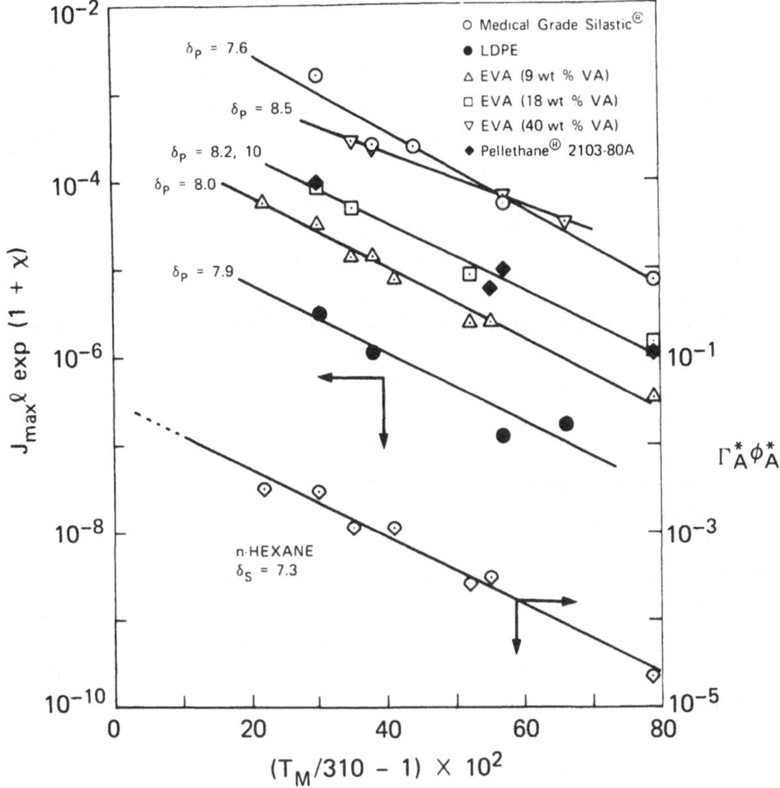

FIGURE 3. Correlation of permeabilities of steroids in various polymers with steroid melting temperature. (Reprinted with permission of AIChE from Michaels, A. S., Wong, P. S. L., Prather, R. and Gale, R. M., *AIChE J.*, 21, 1073, 1975.)

When a liquid containing a solute is brought into contact with a polymer, the extent to which the polymer will sorb the solute will be governed primarily by the similarity of chemical constitution of the solute and polymer. A convenient measure of compatibility of a given solute in a polymer is the difference in the solubility parameters of the solute and polymer. Techniques for this determination have been extensively described in the literature, and the data for many substances have been consolidated in handbooks.[13]

Recently, Michaels and Wong[14] have applied Hildebrand's theory of the solubility of microsolutes in ordinary solvents, and Flory-Huggins theory, to the solubility of steroids in polymers. They have derived a predictive correlation between polymer permeability and steroid crystalline melting temperature. The other correlating parameters are the entropy of fusion of the steroid and the solubility parameters of steroid and polymer (Figure 3). If the permeability and diffusivity of one steroid in a given polymer are experimentally measured, and the melting temperature of that steroid is known, then with this correlation, the permeability of any steroid in that polymer can be predicted from knowledge only of its melting temperature. Furthermore, if the diffusivity of a steroid molecule in a given polymer can be estimated by interpolation or extrapolation of a diffusivity vs. molecular-size correlation for that polymer, it may even be possible to predict the permeability of any steroid in that polymer in the complete absence of experimental data relating to steroid permeation in the material.

c. Implications for Release Rates

Baker and Lonsdale[2] have pointed out that the relatively low diffusivity and solubility of large molecules in polymers constrain the release rates practically attainable with delivery systems that employ dense polymer membranes. Even silicone rubber, which ranks highly with respect to both diffusion and partition coefficients for many organic molecules of interest, rarely permits a maximum daily flux greater than 2.5 mg/cm^2 for a membrane 0.1mm in thickness. For less permeable polymers, release rates two orders of magnitude lower are typical. Hence, it is no surprise to find that the applications have generally involved potent physiological compounds and the more powerful synthetic drugs.

Permeation through porous media occurs by diffusion within the fluid entrapped in the void space of the polymer network, and as the line for diffusion in water plotted in Figure 2 suggests, the diffusivity in porous membranes is several orders of magnitude greater than that for dense polymer membranes.

3. Diffusion Through Entrapped Fluids

a. Hydrophilic Membranes

It was noted above that plasticizers promote solute diffusion by increasing chain mobility. In effect, plasticizer molecules increase the relative volume of the amorphous phase through which the permeant species can migrate. Hence, combination of a polymer with an appropriate plasticizer provides a significant degree of freedom with respect to membrane design. Of course, the plasticizer concentration must remain fixed to get constant permeability.

A special case is the swelling interaction between water and certain polymers composed of hydrophilic monomer units. These materials, called hydrogels, imbibe relatively huge proportions of water (0.1 to 0.9 volume fraction), swelling to an equilibrium level that is related to the osmotic properties of the surrounding fluid. This structured water, as it is termed,[15] really constitutes a novel continuous phase with permeability quite distinct from the polymer itself. Some water molecules are bound to the polymer network through hydrogen bonding and relatively immobilized, but the bulk of the water of hydration is free to move. Most applications of interest involve interfacing a controlled delivery system with an aqueous recipient environment. The boundary between the aqueous phase of the membrane and its environment must be rather vague, with rapid exchange of water. Water-soluble substances, too, should freely diffuse from such a membrane into its environs. Diffusion coefficients through such membranes have been observed to span up to five orders of magnitude for a simple polar species such as sodium chloride, increasing with water content.[16] Analysis of hydraulic permeability, assuming Poiseulle flow through tortuous cylindrical capillaries, yields estimated submicroscopic pore sizes between 10 and 100 Å. The theory has evolved in terms of "fluctuating pores" filled with water. Diffusivity of small molecules thus approaches that in water and can be simulated by diffusion in aqueous polymer solutions at the extreme. As in the case of dense polymer membranes, permeation is a function of the free volume, which in high swollen hydrogels is equivalent to the volume fraction of water. Yasuda, Lamaze, and Peterlin[17] showed that for highly swollen membranes, the following relationship holds:

$$\ln \frac{D}{D_o} = \frac{-\beta X (1 - \alpha)}{1 + \alpha X} \tag{7}$$

where D/D_o is the ratio of the diffusion coefficient in the hydrogel to that in pure water, X is a measure of hydration, α is the ratio of the free volume in a unit of polymer to the free volume in a unit of pure water, and β is a constant that relates the volume of the permeant species to the free volume in a unit volume of water.

It has turned out that substantial discrepancies arise, however, when the volume fraction of polymer and/or degree of cross-linking increase in a series of hydrogels. There may well be a smooth transition from transport by solution in the polymer to diffusion through micropores, but there are large gaps between theory and practice which are matters of controversy that cannot be treated here. Some are touched upon in Section III, and recent publications from the University of Utah provide a current perspective and review.[18-20]

At the extreme of this family of hydrophilic membranes, exist structures with microscopic pores, such as the cellulosic dialysis membranes. Permeation through the aqueous phase, pervading both the environment and the pores, is primarily governed by the molecular volume of the solute.[21-22] For comparisons of porous and nonporous hydrophilic membranes see References 23 and 24.

b. Hydrophobic Microporous Membranes

Hydrophilic microporous membranes, especially cellulosics, have been in use for a long time. Microporous structures have now been introduced in several hydrophobic polymers and can be employed to immobilize a liquid phase different from the, usually aqueous, environment. Microporous membranes have well-defined pores connecting the two surfaces of the membrane. The dimensions of these micropores can vary from hundredths of microns to several microns. Small-molecule permeation through these membranes occurs principally by diffusion, the liquid phase filling the tortuous microporous network (Figure 4). To express the transport flux through the membrane, Equation 4 is modified to read:

$$J = E\,D\,\frac{k\,\Delta C}{\tau\,\ell} \tag{8}$$

where E is the porosity and τ is the tortuosity of the microporous membrane. This method of immobilizing lipophilic liquids may enable the delivery of substances of low water solubility at rates comparable to those achievable with water-soluble species and hydrogels or dialysis membranes. Previous publications treating transport in porous media, in addition to those cited in the preceding section, include References 25 to 27.

B. Design Considerations

1. Membrane Parameters

Synthetic polymeric films are attractive candidates for controlling release by virtue (1) of the dissolution and diffusion of small molecules in the dense polymer or immobilized liquid phase, and (2) the tractability and reproducibility of these materials. In the design and development of controlled-delivery systems, it is desireable to have a thorough understanding of the physicochemical and molecular kinetic factors influencing transport of agents through candidate flux-moderating materials.

The total rate of solute transport through a membrane composed of a specific polymeric material is controlled by adjusting the area and thickness of the membrane and/or the activity or chemical potential gradient across the membrane. Since integrity of the membrane is a *sine qua non,* the mechanical properties of the rate-controlling element create specific limitations to which the overall design must respond. The effects of geometry on release kinetics in membrane-moderated systems have been analyzed. Therefore, theoretical patterns can be predicted for assemblies such as spheres, cylinders, and laminated, sandwich-type systems.[2]

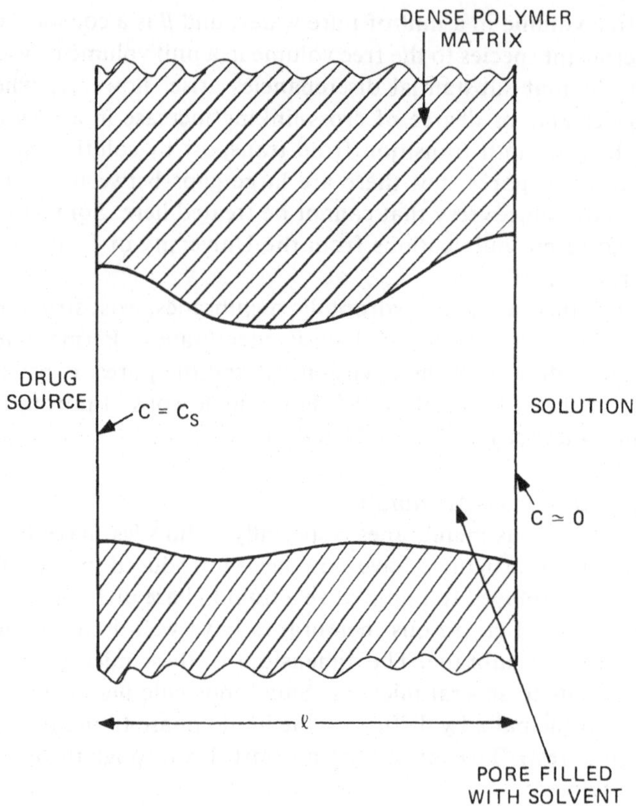

FIGURE 4. Schematic representation of transport through microporous membranes. (Reprinted with permission of Plenum Publishing Corporation from Chandrasekaran, S. K., and Shaw, J., in *Contemporary Topics in Polymer Science*, Vol. 2, Pearce, E. M. and Schaefgen, J. R., Eds., Plenum Press, New York, 1977, 298.)

Stability and susceptibility to modification by adsorption or imbibition of constituents of the environment are major concerns in selection among alternative materials for a given application. It is essential that membranes can be created reproducibly and will retain the desired characteristics throughout the period of storage and use.

2. Reservoir Variables

In most instances, since infinite-sink conditions of negligible diffusant activity are maintained downstream, all other things being equal, the delivery rate is determined by the thermodynamic activity of the diffusant maintained upstream of the membrane. Under these conditions, Equation 4 can be modified to read:

$$J = D\,k\,\frac{C^*}{\ell} \tag{9}$$

where C^* is the concentration or activity maintained upstream of the membrane. If this activity is maintained constant, the delivery rate from the system will also remain constant.

Change in volume of the reservoir as a function of time can vary the area of contact between the reservoir medium and the membrane. This, of course, will modify the overall delivery rate of the device. This is especially a concern when a solid agent has been simply packed into a tubular membrane. After a portion of the agent has been delivered, the remaining solid often settles, reducing the area of membrane in contact with solid agent and decreasing the overall delivery rate. Employment of a liquid medium in the reservoir makes possible continuous contact between the supply of agent and the entire membrane. This overcomes the depletion problem. Although this problem is recognized, the use of packed solid persists in many experiments. Many patents have been issued dealing with technology for preparing stable and reliable reservoirs,[28-30] and they relate primarily to the maintainance of unit activity on the entire upstream side of the membrane.

The quantity, $J\ell$, is referred to as the normalized flux. When a saturated solution is maintained upstream of the membrane and an infinite sink lies downstream, the maximum flux J_{max} is obtained, and the product $J_{max}\ell$ is termed transference, $\overline{T}$. Transference completely describes the permeation process, and drug solubilities in the upstream solvent are not required to compute fluxes through the membrane. Transference is a reproducible intensive property of the solute/membrane system independent of the solvent.[31]

The pattern of release of a membrane-controlled system could be made to vary in a predictable manner by employing two or more cosolvents in the reservoir.[32] If one of the solvents diffuses from the system and leaves behind a solution with higher thermodynamic activity than that initially present, the rate will predictably increase. The opposite trend of activity and rate could also be programmed.

3. Issues Related to Recipient Environment

It should be understood that the availability of active agent to modify a physiological state is almost always influenced by the permeability, thickness, etc. of barriers not part of the delivery system, but of the recipient organism. That aspect of membrane control will not be treated here as it is a broad biological topic common to all sorts of delivery systems, regardless of their own release and rate-control mechanisms. A good introduction to this physiologically and anatomically specific problem, written from the standpoint of vaginal absorption of steroids, is provided by Flynn, Higuchi et al.[33]

Because of relatively slow transport of agent away from the membrane surface, agent concentration can reach appreciable levels at the surface and in the fluid immediately surrounding the device. Should the concentration in the boundary layer exceed the solubility of the agent, delivery will actually cease. This has frequently caused poor agreement between in vitro and in vivo results, especially with drugs of low water solubility in devices that release at high rates. A release rate in vivo that is virtually independent of membrane thickness, but is proportional to membrane surface area, indicates that such a rate-limiting boundary phenomenon is in operation. Baker and Lonsdale[2] analyzed boundary layer effects on transmembrane diffusion rates and have shown that rates are dependent on the quotient J_{max}/C_s; where J_{max} is the maximum flux under infinite sink conditions, and C_s is the solubility in the recipient compartment.

Membrane-controlled delivery is quite temperature sensitive, so that the temperature of the operating environment is an important element of design. Fortunately for drug delivery purposes, warm-blooded organisms maintain constant temperatures within relatively narrow limits. Rogers[5] reports that, over a reasonable temperature range, the temperature dependence of permeability can be represented by an Arrhenius-type equation:

$$P = P_o e^{-E_p/RT} \tag{10}$$

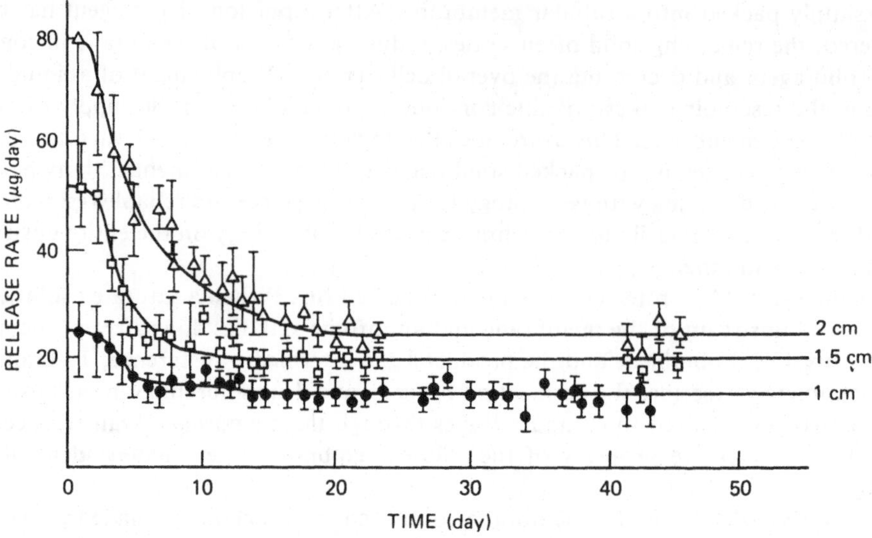

FIGURE 5. Release in vitro of megesterol acetate from silicone-rubber cylinders of the lengths indicated, showing burst effect and steady state. (Reprinted with permission of the editor and author from Kincl, F. A. and Rudel, H. W., *Acta Endocrinol. (Copenhagen) Suppl.*, 151, 14, 1971.)

where E_p is the activation energy for permeation, equal to the sum of the apparent activation energy for diffusion and the heat of solution.

4. Temporal Idiosyncrasies

During early delivery-time periods, systems will exhibit release rates higher or lower than the steady-state value, depending upon the history of the system. When systems are used immediately after fabrication, they will require a certain time period to establish the steady-state concentration gradient within the membrane. This time period is termed "time lag". On the other hand, when systems are stored before use, solute will saturate the membrane, and the delivery rate will initially be higher than the steady-state value. This phenomenon is termed the "burst effect". The magnitude of these two effects is dependent upon the solute diffusivity in the membrane, the membrane thickness and the relative temperatures of storage and use.

Mathematical equations have been derived by Crank[7,8] and Baker et al.[2] to quantify this phenomenon at constant temperatures. In the case of the burst effect, we have

$$\frac{J}{J_{max}} = 1 + 2\,e^{-D\pi^2 t/\ell^2} \qquad (11)$$

where J_{max} is the steady state flux, and J is the flux during the burst period. Figure 5 depicts a typical burst phenomenon.

For time lag, the equation describing the phenomenon is

$$\frac{J}{J_{max}} = 1 - 2\,e^{-D\pi^2 t/\ell^2} \qquad (12)$$

5. Manufacturing Perspectives

All of the prior concerns come together in considering the possibility of producing membrane-controlled systems for sale and use. The specifics have been alluded to, and for any given product are the subject of much proprietary art. In general, the concerns are the following:

1. Purity and stability of raw materials
2. Forming and assembly processes
3. In-process monitoring
4. Packaging and provision for placement
5. Sterilization (where necessary)
6. Stability

This topic has been covered at some length, especially from a quality-assurance point of view.[34]

III. APPLICATIONS

A. Diffusion Through Dense Polymers

1. Silicone-Rubber Membranes

a. Background

The era of controlled release commenced with a serendipitous collaboration between two physicians working at the U.S. Naval Medical Research Center in Bethesda.[35] Judah Folkman was studying the effect of thyroid hormone on heart block induced in animals. Donald Long was photographically investigating the turbulence induced by artificial heart valves made from silicone rubber. Folkman needed a noninflammatory vehicle for prolonged release of thyroid hormone, and Long observed that certain dyes permeated silicone rubber, and others did not. It was found that powders of oil-soluble dyes placed in silicone-rubber tubing diffused through the walls and released for months. A series of experiments followed in which drugs were tested in commercially available polydimethylsiloxane (PDMS) tubing. The results were disclosed in a series of publications, beginning in 1964, which described the release of anesthetics, thyroid hormone, and certain cardiovascular agents in animals.[36-40] Generally, they found that drugs of low molecular weight (<1000 daltons) and lipophilic character diffused through silicone rubber, and that high molecular weight and polar species did not. This work, on which a patent was granted in 1966,[41] had a catalytic effect in many areas of pharmacology. The early work was reviewed in 1968.[42]

Stimulated by Folkman et al., and encouraged by the emerging evidence of the biocompatibility of PDMS, others soon conducted sustained-release experiments of therapeutic, prophylactic, and physiologic compounds with PDMS in animals. Among the latter, the steroid hormones quickly became the major focus of development, because of their enormous potency and the desirability of sustained, chronic effects that might be produced by "artificial glands" secreting μg/day of the physiologic messengers, or their analogues.

In 1966, Dzuik and Cook reported the release in vitro of several steroids through the walls of PDMS tubing for extended periods of time, and they demonstrated the possibility of prolonged hormonal influence in vivo by modifying the estrus cycle in sheep with implanted subcutaneous PDMS packets containing melangestrol acetate.[43] About the same time, Martinez-Manitou et al. showed that minute quantities of orally administered progestins could control human conception.[44] As a result, Segal and Croxatto put forth the concept of combining these effects to create long-term contraceptive systems.[45]

Kincl and his co-workers at the Population Council in New York compared PDMS implants with injections in animals. They confirmed the amplification and prolongation of steroid action obtainable by this route.[46-48] With the support of the Population Council and the Ford Foundation, workers in Chile and Brazil commenced clinical studies with subcutaneous implants.[49-52] Simultaneously, others began to investigate devices for placement in the vagina,[53,54] the uterine cervix,[55] or the uterine cavity.[56] Membrane systems for the first and last of these applications will be the subject of further discussion below (Section III.A.1.C).

Polydimethylsiloxane was the first synthetic polymer studied as a rate-controlling membrane, and its adoption was predicated primarily upon the biocompatibility that was being demonstrated in the process of developing surgical prostheses and devices. Fortuitously, silicone rubbers are highly permeable to a variety of drug substances. Although supplanted for many commercially oriented developments by other agents that offer more favorable processing characteristics, mechanical properties, or cost, silicone rubber remains the substance upon which a greater variety of work on controlled release applications has been conducted and published than for any other polymer.

For several years, Dow Corning Corporation was the only source of silicone polymers. As a result, much of the work reported in the literature refers to the tradename Silastic® silicone rubber and to specific medical-grade cured tubing, sheet, or compounded fluid suitable for curing (cross-linking or vulcanization). Typically, the fluids consist of hydroxy-terminated dimethylsiloxane polymer and a cross-linking agent that can be cured by addition of a catalyst. Cured membranes frequently contain 20 to 30% by weight of an inorganic filler. The filler is usually a reinforcing silica in those formulations that are cured with peroxide or trimethylacetoxysilane and water vapor. However, nonreinforcing diatomaceous earth is the customary filler in the fluids that are cured at room temperature in the presence of stannous octanoate.

The thermoset characteristic of the vulcanized system and the stability of the siloxane backbone make these devices easy to sterilize and resistant to solvents and to thermal or mechanical stress. Silicone rubber is readily tractable on a laboratory scale, but is generally less attractive for mass production than thermoplastics, which more conveniently can be cast into thin films, extruded, injection molded, and joined rapidly into multicomponent assemblies. Low molecular-weight siloxane fluids analogous to PDMS have often been employed as suspension media for the reservoir component of systems controlled by silicone rubber or other membranes.

Background information on silicone rubber for biomedical purposes has been surveyed by Braley.[57] Fabrication of silicon rubber drug delivery systems has recently been reviewed, in a comprehensive chapter on implants, by Chien.[214]

b. Membrane Characteristics

The transport properties and characteristics of PDMS membranes have been extensively studied.[58-66] The variable of polymer-backbone modification in the polysiloxane series has been explored as well.[67] In connection with their interest in the contraceptive applications, by 1968 Kincl, Benagiano, and Argee had demonstrated the relatively high permeability to progesterone of PDMS, compared to that of a variety of other polymers, e.g., it is 100 to 1000 times as permeable as nylon and polystyrene, respectively.[58] They also showed (as Table 1 illustrates) the range over which the daily release rates vary for a variety of steroids permeating through a common polydimethylsiloxane (PDMS) membrane, and demonstrated (with the examples in Table 2) the influence of different substituents along the polysiloxane-polymer backbone upon the permeation of specific steroids.

TABLE 1

Diffusion of Steroids Across PDMS Membrane[a]

Steroid	Flux(μg/cm^2 day)
19-Norprogesterone	1350
Progesterone	470
Testosterone	320
Megestrol acetate	235
Norethindrone	75
Estradiol	60
Mestranol	45
Corticosterone	20
Cortisol	6

[a] Membrane thickness, 0.1 mm.

From Kincl, F. A., Benagiano, G., and Angee, I., *Steroids*, 11, 673, 1968. With permission.

TABLE 2

Diffusion of Steroids Across Various Polysiloxane Membranes

	Permeability Constants (μg/cm^2 hr × 10^{-4})		
Membrane	Norprogesterone	Dehydrocortisol	Cortisol
Poly(dimethyl siloxane)	650	0.4	0.2
Poly(methyl phenyl siloxane)	600	1.1	0.6
Poly(trifluoro propyl methyl siloxane)	75	—	0.06

From Friedman, S., Koide, S. S., and Kincl, F. A., *Steroids*, 15, 679, 1970. With permission.

In the late 1960s, many laboratories verified the applicability of Fick's law to PDMS with a variety of compounds. They demonstrated the low diffusivity of ionized species in PDMS. Most of the parameters relevant to steroid transport in PDMS were measured, or the data was gathered together on a comparable basis, by Cowsar and Lacey.[66] The marked effect of steroid structure on diffusivity stimulated the development of theory and methods.Transport rates were observed to vary directly with lipophilicity and with solubility in PDMS. Compounds with few or no hydroxyl groups diffuse at much higher rates than polar species.

Roseman and Flynn[68] pointed out that the reinforcing silica filler (typically 20 to 30% by weight) commonly incorporated in PDMS may increase tortuosity and decrease the volume fraction of the polymer matrix, slowing transport. Adsorption of drug onto filler particles may contribute to overstatement of the partition coefficient and delay establishment of steady-state flux, with consequent underestimation of the diffusion coefficient.[65,68-70] However, the permeability (the product of distribution coefficient and diffusivity) is little affected by the presence of filler.[63,65,68-70] Taubert and Schaumann were unsuccessful in their attempts to achieve rate modification by addition of nonreinforcing fillers such as kaolin.[70]

The diffusion coefficients for many steroids of interest in PDMS are in the range of 10^{-6} cm^2/sec. This is orders of magnitude greater than those of much smaller molecules in thermoplastics, such as polyethylene, which of course, relates to the easy formation of passages for the diffusing species resulting from the high segmental chain mobility of the rubber.[3,69] Chien[71] noted that these progestins have diffusion coefficients approximately equal to the self-diffusion coefficient of a dimethylsilicone oligomer, containing 10 silicon atoms, suggesting that an opening in the polymer sufficiently large to permit a steroid molecule to jump from one hole to another would require the motion of a chain segment of the order of 10 monomer units in length. To some extent, this high level of permeability proved disadvantageous, in that the membrane often turned out not to be the constituent of the entire system which, in reality, determined the rate of release.

In many of the early experiments, the limited solubility of steroids in the recipient sink gave rise to less than theoretical rates because of boundary-layer phenomena,[72] and to in vivo release rates that exceeded in vitro rates.[70] For example, in preparation for one of the first clinical trials, Lifchez and Scommegna[73] observed that doubling the wall thickness of cylindrical silicone-rubber delivery devices only decreased by 10 to 30% the release rate in a subcutaneous rat-implant model. The release rates in vivo turned out to be about three times the rates predicted from the in vitro release into water. However, when plasma, which contains solubilizing proteins, was used as the recipient fluid in vitro, the in vitro rates did correspond to the release rate measured in vivo.[59]

Just the opposite was encountered with respect to vaginal devices, where rates diminished by a factor of two were observed in vivo, and a boundary layer with a thickness of about 500 μm was hypothesized to rationalize the in vivo results.[69,74] In 1974, Chien and Lambert published a method to overcome these discrepancies by enhancing the solubility in the in vitro sink several hundredfold through the addition of a water-miscible cosolvent, such as polyethylene glycol, and assuring adequate stirring.[71,75] Caution with respect to the relevance of in vitro test methods for each biologic situation is still very much in order. Chien has reviewed this topic recently.[214]

c. Major Applications

All of the principal developments relying upon silicone-rubber membranes have involved steroids, primarily for contraception. Subcutaneous implantation of thin cylinders of silicone rubber filled with synthetic progestational steroids, especially megestrol acetate (6-methyl-17α-acetoxypregna-4,6-diene-3,20-dione) in women represented the first membrane-delivery effort to reach clinical trial. Two reviews by Kincl provide a thorough account of the events prior to 1970[76] and a briefer view of subsequent events.[77] Segal,[78] too, has reviewed the early work on implants for contraception, which he is credited as having originated.[45] Continuing reports appear in the journal, *Contraception.*

Progesterone itself proved ill-suited for subcutaneous delivery because it sometimes provoked the formation of abcesses.[79] Furthermore, relatively large doses of progesterone would have been required for systemic treatment, since it is very rapidly metabolized. Figure 5 illustrated how readily the release rate of megestrol acetate from PDMS tubing could be varied, simply by selecting different lengths of tubing.The first couple of years of clinical studies with PDMS capsules containing megestrol acetate implanted under the skin in the forearm showed that human fertility control could be achieved at delivery rates roughly in the 100 μg/day range.[49-52,80] Subsequent studies indicated some problems: (1) in the pattern of delivery (with the rate, as measured by ^{14}C-megestrol acetate excreted, diminishing to 50% of initial and in vitro rate by the

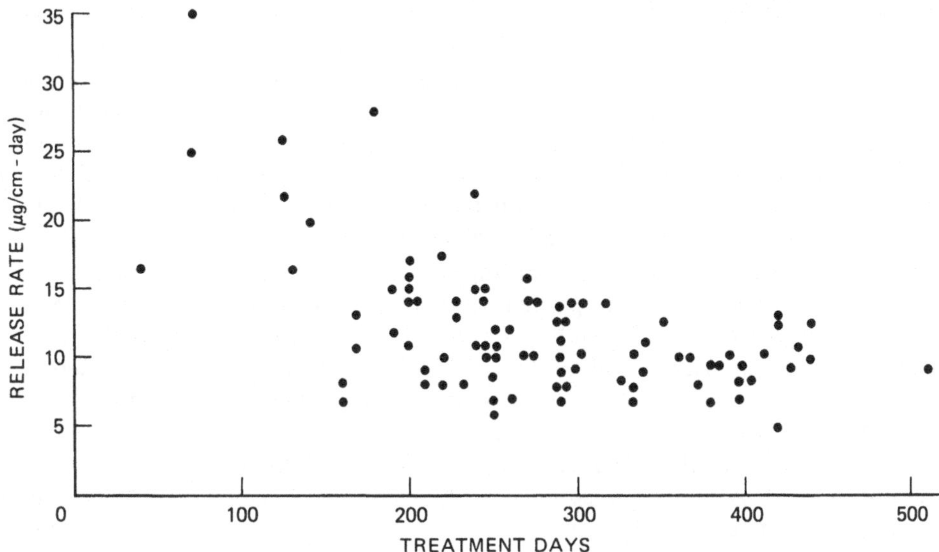

FIGURE 6. Average release rates in vivo of norethindrone from implanted silicone-rubber membrane systems, based on residual drug after clinical trial. (Reprinted with permission of Finnish Medical Society Duodecim from Weise, J., Marker, I. I., Holma, P., Vartiainen, E., Osler, M., Pyorala, T., Johansson, E., and Laukkainen, T., *Ann. Clin. Res.*, 8, 95, 1976.)

third month[81]), and (2) the response (with irregular and prolonged menstrual bleeding occurring frequently[82,83]). Infertility was generally achieved without suppressing ovulation.

Typical of recent clinical trials, one study[84] described implantation of six capsules (3.5 mm long and 2.4 mm OD) in each subject's left forearm, with an 11-gauge trocar needle and a local anesthetic. Each capsule contained 35 mg of megestrol acetate and released approximately 50 μg/day. The plasma levels of megestrol acetate varied between subjects and from day to day for a given subject, but stabilized after about 3 months at about 0.6 ng/mℓ plasma. At this rate of administration, which prior studies suggest may be three times that required to prevent conception, the normal ovarian hormone patterns were disturbed, and ovulation was inhibited. Absorption from the implanted PDMS capsules in vivo has been reported by a number of observers to vary significantly. The in vivo measurements reported in such studies employ radioimmunoassay, which makes possible safer and more convenient determination of the functionality of delivery systems in vivo than was feasible, even with radioactively labeled drug, during the earlier days of controlled-release development.[215]

PDMS capsules containing norgestrienone,[85] norethindrone,[86] progestin R2323,[87] and *d*-norgestrel,*[88] have been implanted subdermally in women. The Population Council in New York is preparing and supplying capsules of these drugs.[215] The last, if successful, could provide 5-year contraceptive activity. Figure 6 depicts average release-rate data for a large trial with norethindrone PDMS capsules in gluteal (buttock) implant sites. This degree of data scatter is quite normal and, apparently, does not impair efficacy.[89] Capsules of other progestational steroids, such as delmadinone acetate, have been prepared, tested, and in some cases, used for estrus control in animals.[90]

* Levonorgestrel is the name recommended by WHO for the optically active enantiomorph of *d,ℓ*-norgestrel. It was previously designated as d-norgestrel.[216]

The second application for silicone-rubber membrane-controlled steroidal contraception involved intrauterine delivery. The endometrial tissue that lines the uterus is the natural target for progesterone, and it undergoes substantial modification under the influence of progesterone in the course of the menstrual cycle that makes the modified endometrium resist implantation and maturation of a fertilized ovum. Controlled delivery offered the opportunity to achieve the desired endometrial effect with minute doses of hormone that would not inhibit the higher centers in the brain or produce other side effects common to orally administered steroidal contraceptives.

Scommegna et al.[56] modified an intrauterine device, known as the Tatum T, by attaching a length of PDMS tubing filled with milled, crystalline progesterone (35 mg), closed and attached with silicone adhesive. The delivery capsules were 30 mm in length, 3.2 mm in diameter, with a wall thickness of 0.8 mm. Such devices were inserted in volunteer women for periods of up to 200 days (norm = 6 months). Under the influence of this local hormonal delivery system, the women continued to ovulate and menstruate normally, since the systemic hormone levels that influence these functions remain undisturbed. A mean release rate of 130 μg/day was calculated on the basis of the amount of progesterone remaining in devices after withdrawal from the subjects 161 to 200 days postinsertion. Inhibition of fertility and the parameters of success in contraception are difficult to characterize, and we will not get into them here. Suffice it to say, no pregnancies were observed with an active progesterone-delivering PDMS device in the uterine cavity through 1662 woman-months of that initial test. Further development of T-shaped intrauterine delivery systems for progesterone, culminating in the commercialization of the Progestasert® system by ALZA Corporation, involved ethylene vinylacetate-polymer membranes and will be discussed in section A.2.c. of this chapter. Animal studies have shown that other progestational steroids, such as *d*-norgestrel[91] also exert a local effect when delivered within the uterus.

Intrauterine PDMS devices containing crystalline *d*-norgestrel were prepared by el-Mahgoub and tested in vitro and in the clinic.[92,93] They were horseshoe shapes made from 2.4 mm OD tubing. Release rates of 5 and 13 μg/day were measured in the initial experiments. Then a study was conducted on 100 women with T-shaped devices containing 5 mg *d*-norgestrel supposed to release at about 5 μg/day. No pregnancies had occurred within 728 woman-months at the time of Mahgoub's report late in 1974. Similar favorable results were reported in 1977 by workers in Finland, albeit with a smaller clinical group.[94]

Other shapes have been examined. For example, spheres were evaluated for intrauterine delivery of the progestins, progesterone, and chlormadinone acetate.[95] Hollow silicone-rubber spheres (0.6 cm diameter) were filled with the micronized steroids in a silicone-rubber binder. Varying the wall thickness (core diameter) produced systems with the range of in vitro release rates and characteristics shown in Figure 7. There appears to have been a slight, unaccounted for, increase in release rate with time.

Vaginal contraceptive devices shaped like the outer ring of a conventional contraceptive barrier diaphragm (a toroid), and referred to as vaginal rings, have been much investigated, especially by workers associated with the Upjohn Company[96] and the World Health Organization Population Control Program.[217] However, most of this work has involved steroids dispersed or impregnated within silicone rubber, rather than membrane-controlled systems.

The administration of the male hormone, testosterone, to rats was investigated early in the development of silicone-rubber delivery technology,[97] but several years elapsed before it was evaluated in the clinic.[98] While work was underway elsewhere oriented toward therapeutic or contraceptive administration of testosterone to humans, PDMS-membrane-controlled delivery of testosterone was recognized as having potential value

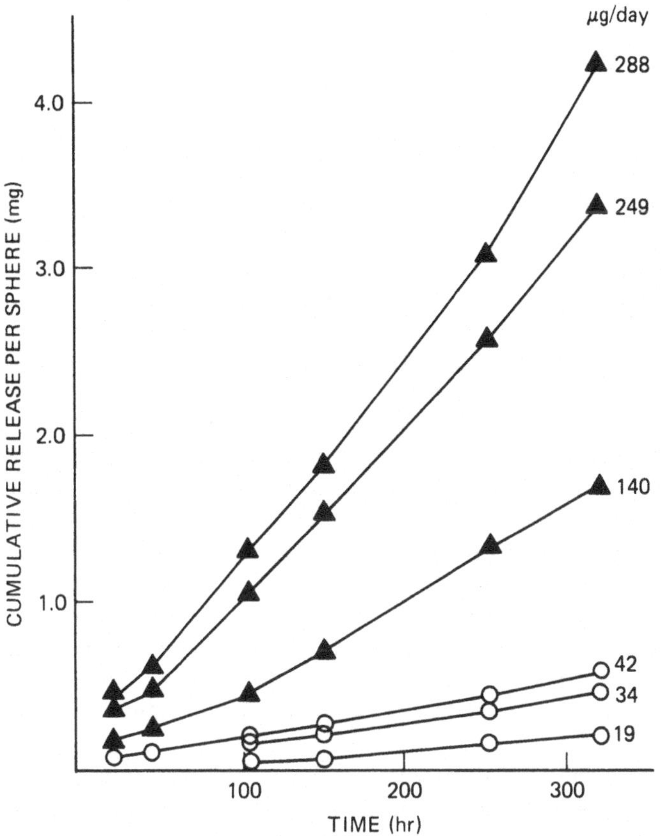

FIGURE 7. Release in vitro of chlormandinone acetate (O) and progesterone (▲) from spherical silicone-rubber membrane systems. (Reprinted with permission of the author from Vickery, B. H., Erickson, G. I., Bennett, J. P., Mueller, N. S., and Haleblian, J. K., *Biol. Reprod.*, 3, 160, 1970.)

to inhibit estrus, heat, in the canine bitch. Simmons and Hamner[99] used tubing with a wall thickness of 0.65 mm to release testosterone, or androstenedione, subcutaneously. A dose of 170 μg/kg/day or more of testosterone was observed to inhibit estrus throughout studies lasting up to 28 months. Reversibility was demonstrated by whelping healthy pups after withdrawal of the implants. Androstenedione was ineffective at the delivery rates tested. The steroids were simply packed into the tubing, and released in vivo at rates (measured by difference) between 28% and 185% of the release rate anticipated on the basis of literature data.[58] Masculinization in anatomy, but not behavior, was reported. The development of similar products for inhibition of estrus in pet animals appears to be approaching commercialization.

The utility of androgen delivery systems for contraception in humans is doubtful for reasons given in the next section, but therapeutic applications have reached the clinical stage. Frick et al.,[98] prepared PDMS capsules delivering testosterone at 55 μ g/day for subcutaneous implantation (in the chest) to treat castrated men and victims of hypogonadism. Dosage was individually determined, with three to five cylindrical capsules (2 cm by 2.4 mm diameter) generally proving adequate to restore libido, potency, beard growth, etc. The same workers report using similar capsules to release ethinyl estradiol at 45 μ g/day for treating patients with cancer of the prostate. Altogether, more than 200 patients have been treated in these studies without complications.

d. Exploratory Studies

In efforts directed toward the development of a contraceptive system for men, Ewing et al. utilized PDMS capsules releasing testosterone at various rates to demonstrate with male rabbits that sperm production can be upset for months at a time without permanent loss of fertility. The rate of administration required to achieve the desired effect in rabbits,[100,101] along with preliminary results of rhesus-monkey studies,[102] suggest testosterone alone will not be sufficient for contraception in man.

An interesting aspect of work by Shippy and his colleagues[103] on testosterone delivery, was their admixture of drug to the silicone rubber before forming the membrane in order to overcome a peak in release rate, which otherwise occurred a few days after tests were initiated (Figure 8). This phenomenon, which they attributed to water uptake, has since been explained[2,104] indirectly in terms of storage/test conditions. Namely, testing (or using) at a different temperature than that at which the system equilibrated during storage can cause a transient peak.

Knobil et al.[105] employed PDMS systems to deliver estradiol and progesterone in studies of menstrual physiology. They determined the delivery rate corresponding to threshold levels triggering events that, in effect, reset the biologic clock for this cyclic mechanism. Delivery of estrogens, especially estradiol, at therapeutic rates was investigated early in the exploration of PDMS steroid systems,[106] but has never been commercially exploited. Other studies of sex hormone implants have been reported recently.[218]

The high permeability of PDMS to steroids is inappropriate for some applications, and yet its biocompatibility record has made PDMS the first choice for the biological-interface role. One way to deal with this limitation has been the preparation of laminated structures. To control the release of estradiol from subcutaneous implants in rats, Bloch et al.[107] applied PDMS elastomer to a polyethylene film. Since the steroid diffuses through PDMS 1000 times as fast as it does through the polyethylene, only the latter substantially influences the rate of release (if the thicknesses of the layers are comparable).

Analogously, Colter et al.[108] reported preparation of laminates by polymerization with high-energy plasma of monomers, such as ethylene and tetrafluoroethylene, into films (400 to 2000Å in thickness) on the surface of PDMS membranes. They also created a densely cross-linked skin within the PDMS itself with plasma. The technique is referred to as CASING (crosslinking by activated species in inert gas). Curiously, they report that for a given film thickness, plasma-polymerized tetrafluoroethylene is more permeable to progesterone than the film obtained from ethylene. Up to a 40-fold reduction in progesterone flux was attained with plasma-polymerized ethylene compared to untreated silicone rubber.

Table 3 provides a summary of the nonsteroid agents that have been delivered, or studied for possible delivery, by means of a silicone-rubber membrane-controlled system. None of these applications appears to have resulted in a controlled release product. Release through PDMS from complexes and micelles is reported for a number of combinations of drug and adjuvant.[221,222]

2. Ethylene-Vinylacetate Copolymer Membrane Systems

a. Background

Consideration of the processing and performance characteristics of candidate materials appropriate for delivery of drugs directly to target organs in the body led to rec-

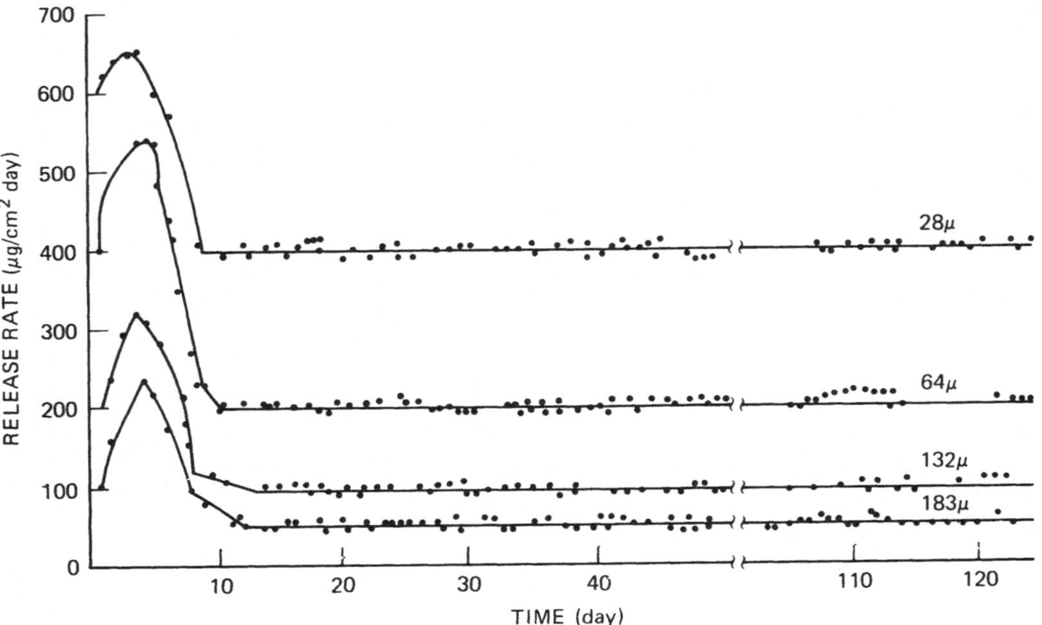

FIGURE 8. Release in vitro of testosterone through cylindrical silicone-rubber membranes of the thickness (in μg) indicated. (Reprinted with permission of Plenum Publishing Corporation from Baker, R. W. and Lonsdale, M. K., in *Controlled Release of Biologically Active Agents*, Tanquary, C. A., and Lacey, R. E., Eds., Plenum Press, New York, 1974, 40, and of John Wiley & Sons Inc. from Shippy, R. L., Hwang, S., and Bunge, R. G., *J. Biomed. Mater. Res.*, 7, 102, 1973.)

TABLE 3

Other Agents Studied with Silicone-Rubber Membranes

Type of agent	Workers	Year	Lead Ref.
Anesthetics			
Ether, nitrous oxide	Folkman et al.	1966—68	38—40
Halothane and cyclopropane	Folkman et al.	1966—68	42
Benzocaine	Most; Nakano et al.	1972, 1977	64, 109
Tranquilizers			
Chlorpromazine	Nakano et al.	1971	110
Butamben	Nakano et al.	1976—78	109, 111
Chlordiazepoxide	Lovering et al.	1974—76	112
Diazepam	Lovering et al.	1974—76	112
Sedatives			
Phenobarbital, amobarbital	Garrett, Lovering	1968, 1973	62, 113
Antiparkinsonism			
L-Dopa	Siegel and Atkinson	1971	114
Anti-secretory			
Atropine	Bass et al.	1965	115
Stimulants			
Histamine	Bass et al.	1965	115
Caffeine	Nakano et al.	1970—76	111
Nicotine	Gaginella et al.	1974	116
Isoproterenol	Folkman et al.	1964—68	37, 42
Cardioregulatory			
Digitoxin	Folkman and Long	1964	37
Triiodothyronine	Folkman and Long	1964	37
Antiinflammatory			
Eleven agents	Jones and Datko	1977	220
Phenylbutazone	Lovering and Black	1974	117
Prostaglandins	Nuwayser, Spilman,	1974, 1977	118, 119
	Roseman	1979	219
Antineoplastics			
BCNU (nitrosourea)	Schmidt et al.	1972	120
5-Fluorouracil and derivatives	Nakano et al.	1976	121
Antimicrobials			
Chloramphenicol, nitrofurantoin	Mangelson and Cockett	1967	122
Oxytetracycline, and penicillin G	Wepsic	1971	123
Antiparasitics			
Malaria and schistosomiasis	Powers	1965	124
Microfilaria	Collins	1974	125
Lice	Clifford et al.	1967	126

ognition by ALZA Corporation of the ethylene vinylacetate (EVA) copolymers as a family of materials with a number of promising advantages.[127] For certain applications, these thermoplastic materials were found to be far superior to silicone rubber in terms of both membrane function and formability.

Because of the high permeability of silicone rubber relative to the rate at which drugs with low water-solubility, such as progesterone, are cleared from the surface of a drug-delivery device in a body site, such as the uterus, it was desirable to select less permeable membrane materials in order to retain control of release within the delivery system. In 1974, Zaffaroni[128] disclosed the invention of combining a highly permeable reservoir matrix, containing excess solid agent, with a membrane 10 times less permeable than the matrix, and cited as an example one copolymer of ethylene and vinylacetate (16% vinylacetate) that is an order of magnitude less permeable to progesterone than polydimethylsiloxane.

EVA was also attractive because it lends itself to high-speed thermoplastic processes such as extrusion, film casting, and injection molding. The capability of forming reliable fused welds can be of great value in component assembly. Certain members of the polymer family are flexible, even elastic, and require no plasticizers to achieve these properties, or for processing.

On account of the latter characteristic, EVA copolymers were candidates for the replacement of plasticized polyvinylchloride in blood containers and tubing. The prior toxicological record was favorable, and upon it ALZA built an extensive data base with respect to the compatibility of specific EVA copolymers within certain biologic environments, including several species, and in contact with uterine or ocular tissues.

Utilization of copolymer materials allows the variation of properties, individually or collectively, by small changes in monomer ratios. The introduction of the comonomer into the basic polyethylene structure reduces the crystallinity of the polymer, which has multiple consequences. At extremely low vinylacetate levels, EVA approaches low density polyethylene in characteristics, but at higher levels, it more closely resembles plasticized polyvinylchloride. As the vinylacetate concentration is increased, stiffness, tensile strength, and softening point decrease, while toughness, permeability, and flexibility increase. Its pseudoelastic characteristics are presumably the result of crystalline polyethylene regions acting as cross-links in a matrix in which the glass transition occurs below room temperature. The average molecular weight, reflected in bulk and solution viscosities, is variable and independent of the comonomer ratio. Judicious selection of vinylacetate content and melt-flow index provides membrane materials with a broad spectrum of processing and performance properties. A thorough review of the structure/property relationships of general interest within the EVA series has been presented by Salyer.[129]

A wide range of EVA materials (with vinylacetate content from 4 to 60%) is now available from various sources, although none of these are supplied as medical-grade polymers. The EVA polymers are relatively clean, chemically stable, and can be sterilized with either ethylene oxide or radiation.

b. Membrane Characteristics

Ethylene vinylacetate copolymers vary in permeability as a function of comonomer ratio. Michaels et al.[14] and Baker et al.[104] reported normalized fluxes for various steroids in specific samples of EVA with 9, 18, and 40% vinylacetate content. The former source is represented here by Figure 3, and the data from the latter appears in Table 4. Permeability to progesterone has been reported to double when the vinylacetate content of EVA membranes is increased from 9 to 16%.[28,128]

In general, the solubility parameter increases as the proportion of vinylacetate mon-

TABLE 4

Permeability to Estriol of Polyurethanes and Reference Polymers[104]

Polymer	Supplier Tradename	Normalized flux (μg-mm/cm²-day)
Ether-based Polyurethanes	Goodrich Estane 5714®	3.3—3.6
	Goodrich Tuftane TF-410®	5.5
	Upjohn Pellethane 80A®	0.3—2.4
	Upjohn Pellethane 55D®	1.4
	Cyanamid Cyanaprene 9341®	1.3
	Uniroyal Roylar A-895®	6.3
	Uniroyal Roylar 84N®	1.9
	Hooker Rucothane P-602-3®	2.9
	Quinn PE-49®	16
	Quinn PE-55®	5.4—5.8
	Quinn PE-18®	3.7
	Quinn PE-19®	7.7
Ester-based Polyurethane	Goodrich Tuftane 80D®	<0.1
	Goodrich Estane 5702®	2.7
	Goodrich Estane 5710®	0.3
	Uniroyal Roylar E9®	0.4
Silicone rubber	Dow-Corning (3 grades)	<0.1
Ethylene vinyl-acetate		
9% VA	USI UE 638®	<0.1
18% VA	USI UE 630®	~0.1
40% VA	DuPont Elvax 40®	0.8

omer rises, e.g., from 8.0 to 8.5 within the range from 9 to 40% vinylacetate content, while crystallinity diminishes from 47% to zero. (The crystallinity of low-density polyethylenes is typically in the 50 to 60% range.[14]) As a result, both the diffusion coefficient and partition coefficient change markedly with monomer ratio.

c. Major Applications

The efforts of ALZA Corporation leading up to commercial introduction of an ethylene vinylacetate-membrane-controlled intrauterine progesterone contraceptive system, with the tradename Progestasert®, took place parallel to the evaluation of the concept involving PDMS-membrane systems described earlier (Section A.1.c). The ALZA work was distinctive with respect to the emphasis on achieving precisely controlled, predictable, and uniform delivery rates verifiable by in vitro testing,[130] as well as the employment of EVA and the scale of the effort. Before the product was approved for sale (March 1976), over 45,000 women-months of clinical testing had been logged.

ALZA clinical studies showed that delivery of 65μg/day of progesterone from the stem of a T-shaped unit was sufficient to produce conception rates comparable to those ordinarily obtained with oral contraceptives.[131] A system was designed which would maintain that rate for 400 days and could be inserted conveniently by a physician annually.

The total annual dosage of 24 mg of progesterone is less than the amount secreted on a single day in the normal course of the menstrual cycle by the mature corpus luteum of the ovary. The only detectable physiologic action of the system involves the lining of the uterus (the endometrium), which shows evidence of suppression similar to that seen in pregnancy and in users of oral contraceptives. Women wearing the Progestasert® intrauterine progesterone system ovulate and menstruate regularly.

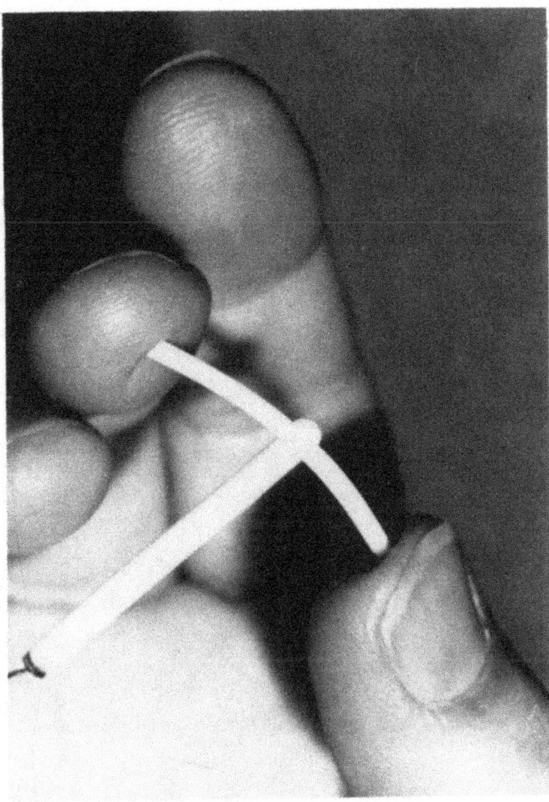

FIGURE 9. The Progestasert® intrauterine progesterone contraceptive system. (Illustration provided by ALZA Corporation, Palo Alto, California.)

Thus, the endometrium, with which the Progestasert® lies in contact, is sloughed off and renewed normally each month.

Unlike the intrauterine devices that preceded it, the Progestasert® system does not depend upon physical configuration, size, stiffness, or presence of a foreign element (such as copper) to prevent conception. Thus, the platform selected has a T-shape. As an inert plastic device, this was recognized to be a relatively ineffective, but well-tolerated and well-retained, intrauterine device. In addition to the delivery module and platform (Figure 9), which were engineered carefully to be compatible with the uterine environment, a novel device was developed to facilitate insertion through the uterine cervix and ascertainment of proper placement (Figure 10).

The stem of the Progestasert-T® is an extruded ethylene vinylacetate tube 3.5 cm in length with a wall thickness of approximately 250 μm.[127] The extrusion of such a membrane required the development of sophisticated process controls.[34] The stem tube contains 35 mg of micronized progesterone suspended in silicone oil. The delivery profile of the Progestasert® is depicted in Figure 11.

Another membrane-controlled therapeutic system employing an EVA copolymer was developed and commercialized by ALZA Corporation during the period that the Progestasert® system was under development. This development arose through the inventions of Ness[132] and many at ALZA. In October 1974, two forms of Ocusert® pilocarpine ocular therapeutic systems were approved and commercially introduced

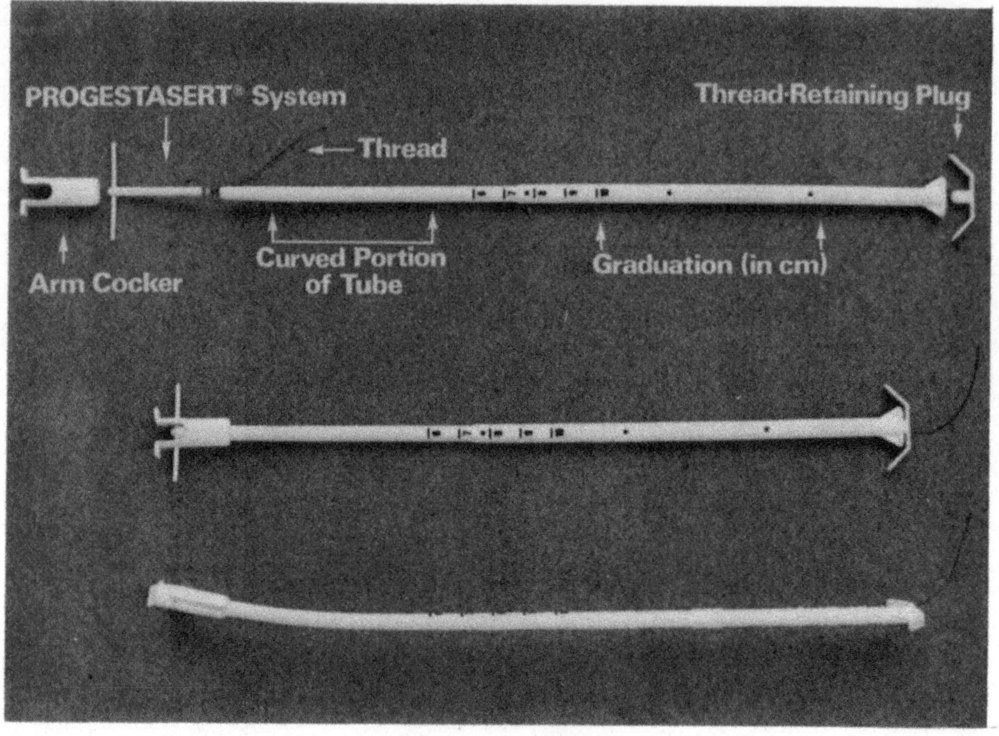

FIGURE 10. Inserter device for Progestasert® system. (Reprinted with permission of the editor from Pharriss, B. B., *J. Reprod. Med.*, 17, 93, 1976.)

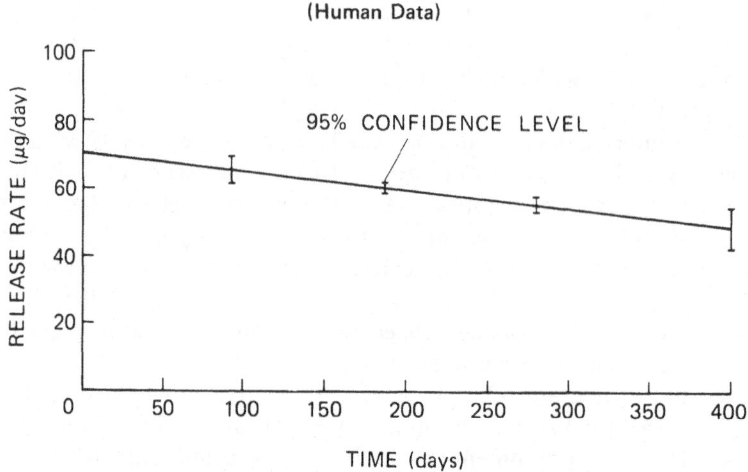

FIGURE 11. Release kinetics of Progestasert® system. (Reprinted with permission of Academic Press from Yates, F. E., Benson, H., Buckles, R., Urquhart, J., and Zaffaroni, A., in *Advances in Biomedical Engineering,* Vol. 5, Brown, J. H. and Dickson, J. F., III, Eds., Academic Press, New York, 1975, 19. Pharriss, B. B., *J. Reprod. Med.*, 17, 94, 1976.)

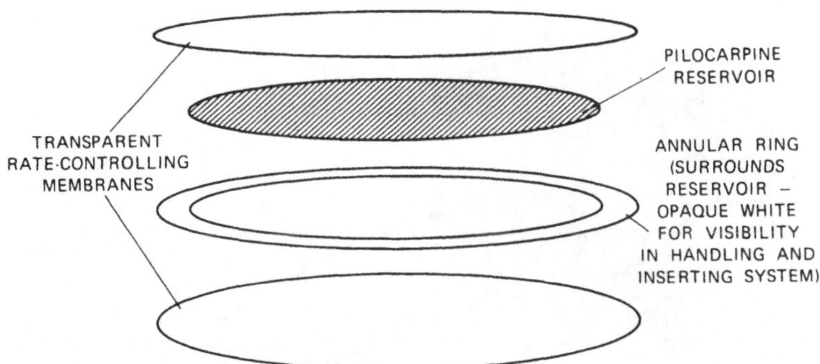

FIGURE 12. Schematic diagram of Ocusert® pilocarpine ocular therapeutic system. (Reprinted with permission of Academic Press from Yates, F. E., Benson, H., Buckles, R., Urquhart, J., and Zaffaroni, A., in *Advances in Biomedical Engineering*, Vol. 5, Brown, J. H. and Dickson, J. F., III, Eds., Academic Press, New York, 1975, 15.)

the treatment of elevated intraocular pressure associated with chronic, open-angle glaucoma.

Chronic, open-angle glaucoma customarily requires lifetime treatment with a miotic agent, like pilocarpine, for control of intraocular pressure, and conventionally, this involves instillation of eye drops at least four times a day. For suitable patients, Ocusert® systems permit once-per-week treatment. Like the Progestasert® system, Ocusert® systems are target-organ specific, but they are designed to deliver at much faster rates and over shorter periods of time. Ocusert® systems are thin, flexible, lamillar ellipses (Figure 12) that deliver pilocarpine to the tear fluid continuously for a week after placement behind the upper or lower eyelid. Figure 13 shows an Ocusert® system in place, floating on the tear fluid, and revealed by drawing down the eyelid.

Ocusert® systems with pilocarpine release rates of 20 μg/hr and 40 μg/hr were developed by ALZA to permit titration of a patient's intraocular pressure, just as a wide range of eyedrop concentrations are utilized to allow for the condition and response of individual patients.[133] Each system consists of a pilocarpine-reservoir core laminated between two EVA copolymer membranes. The reservoir is surrounded by an annular ring, also EVA, which seals the edges so that release occurs primarily through the membranes. Titanium dioxide pigment is incorporated into the annular ring to make it white, thus rendering the otherwise clear system easier to locate when it must be manipulated. The completely assembled and packaged systems are sterilized by irradiation. Ocusert® systems measure 6 mm by 13 mm axially and 0.5 mm in thickness. It was not necessary to add preservatives, although they are required in pilocarpine eyedrop solutions.

The drug reservoir is a film of pilocarpine in a natural polymer (alginic acid, commonly employed in pharmaceutical formulations) that serves as a carrier medium and permits the reservoir component to be handled as a film during manufacture. In fact, all components are prepared as films and assembled by lamination.[34] In proportion to the release rates of the Pilo-20 and Pilo-40 systems, the reservoirs contain 5 mg or 11 mg of pilocarpine as the free base. Actually, only 3.4 or 6.7 mg are delivered during a week of use in the eye, significantly less than the 28 mg of drug that a patient would instill in each eye in the course of a week with 2% pilocarpine solution eyedrops.

Delivery in vitro and in vivo by the Pilo-20 system are compared in Figure 14. After an initial surge, caused by the relatively high concentration built up in the membrane

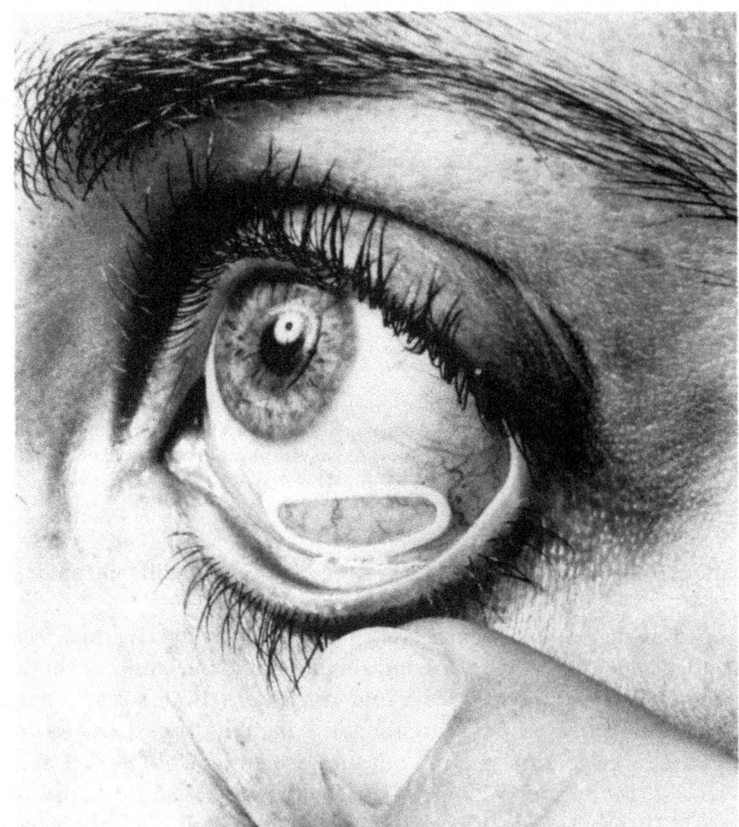

FIGURE 13. Ocusert® system in place for use, eyelid retracted to show system. (Reprinted with permission of Academic Press from Yates, F. E., Benson, H., Buckles, R., Urquhart, J., and Zaffaroni, A., in *Advances in Biomedical Engineering*, Vol. 5., Brown, J. H. and Dickson J. F., III, Eds., Academic Press, New York, 1975, 15.)

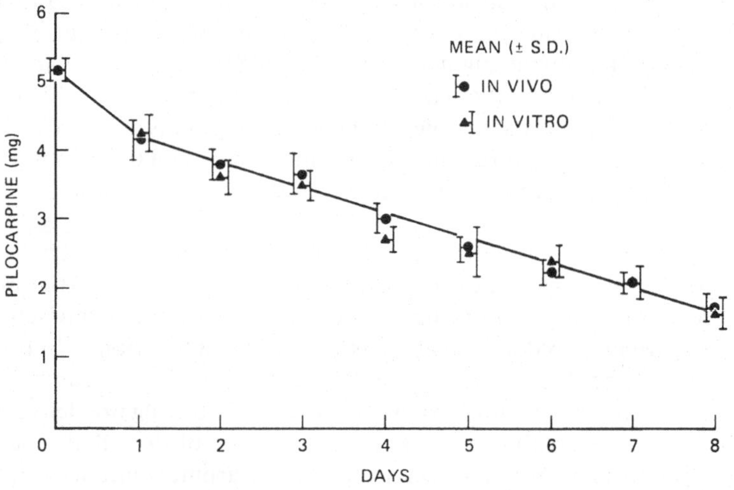

FIGURE 14. Comparison of in vitro and in vivo release for the Ocusert® pilocarpine system, residual drug after indicated periods. (Illustration provided by ALZA Corporation, Palo Alto, California.)

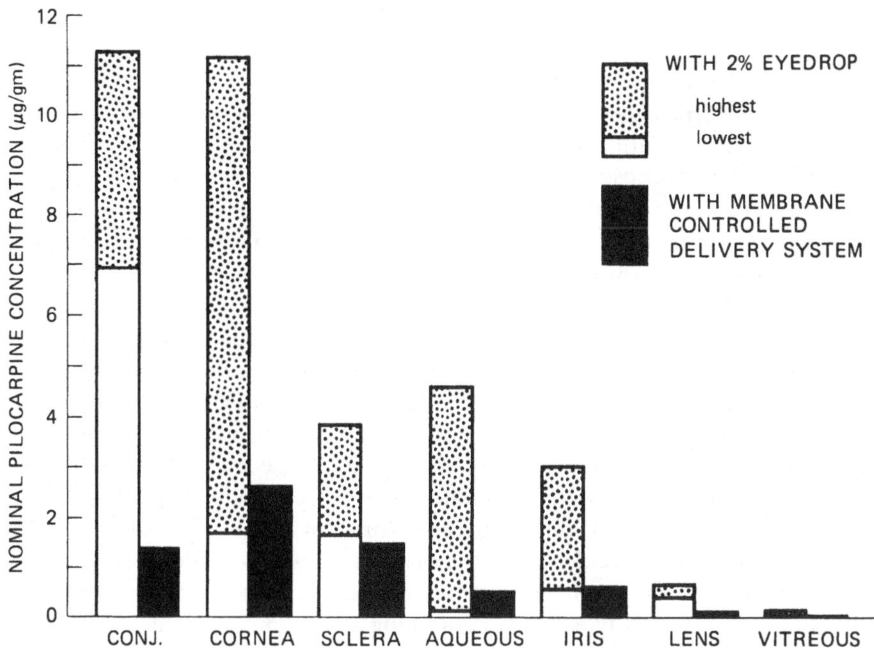

FIGURE 15. Levels of ^{14}C in rabbit ocular tissues after administration of ^{14}C-pilocarpine via eyedrop and membrane-controlled delivery system. (Reprinted with the permission of The Ophthalmic Publishing Co. from Sendlebeck, L., Moore, D., and Urquhart, J., *Am. J. Opthalmol.*, 80, 274, 1975.)

during storage, the release rate remains nearly constant over a seven day period (17.9 ± 2.9 μg/hr in vivo or 18.3 ± 1.3 μg/hr in vitro).[134]

The rate-controlling membranes of both Ocusert® systems are each approximately 100 μm thick, and during manufacture the thickness must be controlled within ± 10%. A plasticizer, di(2-ethylhexyl)phthalate, is incorporated into the EVA copolymer to obtain the higher release rate of the Pilo-40 system.[135]

In recent years, a number of publications have described the use of soft contact lenses or similar matrices presoaked in dilute pilocarpine solutions to prolong therapeutic action. However, such approaches produce release rates that decline rapidly, e.g., 90% of the drug is released within ½ hr.[2]

One benefit of the Ocusert® system is the fact that administration of drug continues even through the hours of sleep, when eyedrop medication is known to lose effect and ocular pressure rises.[136] It is somewhat surprising to find that continuous delivery of drug actually results in lower drug concentration in tissues such as the lens and cornea, which are believed to be unrelated to pilocarpine's hypotensive effect, than occurs when the drug is given by eyedrop (Figure 15).[137] Troublesome side effects, such as myopia and miosis, and discomfort are reduced, presumably because the Ocusert® systems expose the eye to less drug and have less impact on the pH of the tear fluid than pilocarpine eyedrops.[138,139]

d. Exploratory Studies

A modest number of publications have appeared disclosing exploratory investigations in which EVA membranes have been evaluated as to their control of release rates. Two examples relate to the interest within the past few years in delivery systems suitable for releasing prophylactic agents in the oral cavity to prevent tooth decay. The third example suggests the potential of extending EVA-membrane-controlled release into the realm of macromolecular agents, which could have significant fundamental and practical implications.

Workers at the University of Kentucky[140] have reported the evaluation of EVA membranes to release the antibacterial substance hexylresorcinol at rates on the order of 50 μg/cm²-hr. The purpose of these experiments was to determine the effects of continuous presentation of hexylresorcinol on *Streptococcus mutans*, a bacterium involved in the formation of dental caries. The initial results suggested that inhibition, if not elimination, of the bacterium might be feasible by this means.

As part of a feasibility and dose-range-finding exercise, workers at Polysciences, Inc.[141] reported the evaluation of a number of polymers for the purpose of releasing sodium fluoride at rates of 1 mg/day or less from devices that could be semipermanently attached to the teeth. The principal focus of this work came to rest on polymethacrylate (methyl and *n*-butyl) membranes around varous hydrophilic polymer reservoirs containing NaF or Na_2PO_3F. However, with respect to EVA, they report that when hydrophilic matrices were dip-coated with an ethylene vinylacetate copolymer (20 μm thickness), the rate of release became constant in the range of 0.5 mg/cm²-day for more than 30 days. Incidentally, they report relatively rapid release (~0.15 mg/cm²-hr) from a NaF/EVA matrix, and we note recent issuance of a patent on an athletic mouthguard, formed from EVA containing a fluoride salt, which provides topical fluoride dental prophylaxis along with the more conventional function of such mouthguards.[142]

Folkman and Langer[35,143,144] have reported observations that appear to involve the slow release through EVA membranes of larger molecules than one would expect to permeate a dense polymer. They find that proteins are released at near-constant rates from EVA copolymer matrices which are formed by dispersion of the proteins in a polymer solution followed by solvent evaporation, if those matrices are coated with a thin layer of the same polymer. Macromolecules with molecular weights between 10^3 and 10^7 daltons were included in the experiments, and a relatively polar ethylene vinylacetate copolymer (40% vinyl acetate) was employed. Release rates for enzymes such as lysozyme (mw 14,400), alkaline phosphatase (mw 88,000), and catalase (mw 250,000) surprisingly approach zero-order kinetics over periods of 20 to 100 days, as shown in Figure 16. Generally, more than 50% of the protein is released rapidly from such systems, as the figure shows, but the residual material is released at a rate equivalent in biochemical activity to about a μg/day over 100 days from a 1 mm³ pellet. The burst phase appears to reflect molecular weight, but beyond the first 20 days, the near-constant release rates seem to be independent of the molecular weight of the agent. The technique was also found to apply to insulin, and the nonprotein biopolymers heparin and DNA, with each being released at about 1 μg/day for up to 60 days. The in vivo functionality was demonstrated through the use of such EVA "sandwiches" (as the inventors refer to the matrices coated with EVA) containing a proteinaceous substance extracted from cancer cells called tumor angiogenesis factor (TAF), and which induces the formation of blood vessels.

The mechanism involved in the release of large molecules from such EVA systems is not clear, but cannot depend upon simple diffusion since these macromolecules do not diffuse through films of pure polymer under customary test conditions. Nevertheless, a useful mechanism for prolonged release of extremely low levels of pharmacologically potent entities from a tiny noninflammatory source has been identified. It should prove useful for animal experiments, and perhaps, for therapeutic or prophylactic applications.

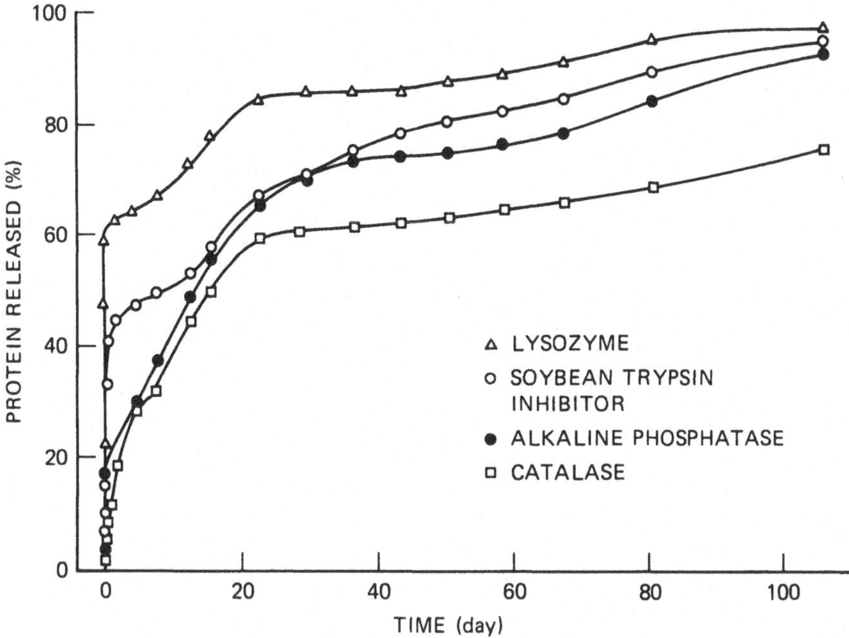

FIGURE 16. Release in vitro of four proteins from ethylene vinylacetate copolymer reservoirs coated with ethylene vinylacetate copolymer membranes. (Reprinted with permission of Macmillan Journals Ltd. from Langer, R. and Folkman, J., *Nature (London)*, 263, 799, 1976.)

3. Polyurethane Membranes

a. Background

Polyurethanes have been the object of a great deal of attention for biomedical applications for a number of years, but until very recently, little had been published on their utility as rate-controlling membranes.

The family of polyurethanes encompases a diverse variety of compositions and physical characteristics. Both cross-linked and soluble forms are available from many manufacturers, and various isocyanate (hard segment) and polyether or polyester (soft segment) combinations are offered. The ester-based polyurethanes slowly hydrolize under physiologic conditions and are, thus, less attractive for biologic applications.[145] The polyether compounds have been used in various artificial-organ developments, and there is an excellent data base, both physical and biologic, upon which to build.[146]

b. Membrane Characteristics

Lyman[24] did a good deal of work with copolyether-urethane membranes directed toward creating artificial-kidney elements that would separate molecules on the basis of chemical structure as well as physical size, the determinant in conventional dialysis.

Michaels et al.[14] reported data on a copolyether-urethane (Pellethane® 2103-80A) in the work correlating steroid properties with polymer characteristics, indicating its permeability to be intermediate between silicone rubber and polyethylene. Zentner et al.[19] report permeation data for progesterone through two polyurethanes that have received much attention in artificial-organ developments.

Recently, a wide variety of members of this class were surveyed by Baker et al. of Bend Research as membranes to control the delivery of estriol in the uterus.[104] A portion of the data is reproduced in Table 4. With respect to estriol, a relatively polar steroid, the polyurethanes are more permeable than either silicone rubber or ethylene vinylacetate, and an order-of-magnitude range of permeabilities was observed among the ether-based polyurethanes tested.

*c. **Major Applications***

For their purpose, the Bend Research workers concluded that one copolyether-urethane, Estane® 5714, offered the best combination of properties and favorable preliminary evidence with respect to toxicology for testing in rabbit uteri. Unlike some of the other polyurethanes, notably the Pellethane® 80A, the membrane characteristics of the selected polymer appeared to be independent of the technique of membrane preparation. Cylindrical systems releasing at rates from 1.25 to 12.5 μg/day were prepared from the chosen polymer. One cm lengths of tubing with specified wall thickness, created by dip-coating steel rods, were filled with crystalline estriol and sealed.

The influences of reservoir-solid depletion and of the limiting constraints with respect to membrane thickness are considered in a manner of general interest.[104] A useful graphical method was presented for taking into account design constraints such as:

1. Wall thickness—rigidity
2. Wall thickness—fragility and precision
3. Outside diameter—site limit
4. Internal diameter—capacity
5. Reservoir depletion—linear dependence of release rate

*d. **Exploratory Studies***

Workers at the University of Utah[19], as part of a multipolymer screening series, have recently examined two copolyether-urethanes and a blend of copolyether-urethane with polydimethylsiloxane for progesterone release-rate-controlling membranes. It is interesting to note that these workers also report that the prior treatment of Pellethane® membranes had a much more marked influence on the permeability of those membranes than prior treatment did for the other materials tested. They report evidence of irreproducibility between films, and inhomogeneities within a given film, that swamp out other experimental variables, such as the effect of environmental lipids on the Pellethane® membranes. For the other copolyether-urethane tested, Ethicon's Biomer®, they found that the permeability to progesterone decreased for the first week of exposure to blood plasma, and increased thereafter. The effect was attributed to lipid uptake by the hydrophobic regions of the copolymer, followed by hydrolytic breakdown. After the initial dip, the permeability actually rose higher than that prior to exposure to plasma. At least over the short (1 week) term, the blend of copolyether-urethane and PDMS was found to provide a more stable membrane than either of the copolyether-urethanes.

4. Other Dense Polymers

Comparatively little work has been reported with other hydrophobic polymers as release-rate-moderating membranes. Gonzales et al.[147] verified the conformance to Fick's Law of polyethylene with a variety of organic compounds in aqueous solution. Kalkwarf et al.[148] identified a grade of polyethylene sufficiently permeable to progesterone to be considered for contraceptive purposes, and a polyethylene vaginal ring, containing medroxyprogesterone dispersed in a silicone-rubber matrix, has been described.[28]

A great deal of attention has been given to the barrier properties of hydrophobic polymers such as polyethylene with respect to small molecules, e.g., O_2, H_2O,[10] but little has been reported with regard to controlled release because of the low permeability to most organic species of interest. Michaels et al. present data on the permeation of several steroids through a low-density polyethylene with 53% crystallinity.[14] It is three orders of magnitude less permeable than silicone rubber, and we have noted earlier (in the section on silicone rubbers) that this characteristic has been exploited by Bloch,[107] in the form of laminates, to deliver estradiol very slowly. Polymerization with high-energy plasma, discussed in the silicone rubber and hydrogel segments of this chapter, provides some examples of exploratory investigations of very thin (0.2 to 0.5 μm) films of polyolefins as rate-controlling membranes.[108,149]

Polyvinylchloride polymers, especially plasticized, flexible formulations, are widely used in biomedical applications, including matrices from which active agent can be leached,[150] but they have yet to find use as a rate-controlling membrane.

Polyamides have been tested for many controlled release membrane applications (e.g., References 32 and 151—153), but the only practical application seems to be the microcapsular insecticide marketed by Pennwalt,[154] and related polyamide microencapsulates. Polyester membrane materials that should disintegrate after performing their membrane function in vivo are being developed.[223]

Thin (15 to 150 μm) polymethyl methacrylate and poly-*n*- butyl methacrylate membranes, applied by dip-coating, turned out to be the most promising release-rate moderators identified by Halpern et al.[141] in their exploratory work on fluoride delivery. A 10-μm-thick cellulose-acetate coating was also reported in that work, but the data are scant.

Acylated and alkylated cellulosics provide a sort of bridge into the hydrophilic-polymer realm, since the affinity for water is inversely related to the degree of hydrophobic acylation or alkylation. Cellulose acetates have been exhaustively investigated as membranes with ultrathin, dense-barrier segments.[155] Furthermore, various cellulosic derivatives are employed for dialysis, including hemodialysis in artificial kidneys.

The relatively high water permeability and solute reflectivity of cellulosic esters of lower aliphatic acids has made these polymers unattractive for release via diffusion, but their utility for release via osmotic pumping is being exploited by ALZA Corporation.[156]

Bearing on the subject of the transport of organic species through relatively hydrophobic cellulosic membranes is the work of Short and Rhodes.[157-159] They investigated the effect of nonionic surfactants on the diffusion of steroids, such as testosterone and hydrocortisone, across cellulose-acetate membranes. Withington and Collett[160] describe investigations of the effect of surfactants on the transport of salicylic acid across cellulose-acetate membranes in studies modeling gastrointestinal absorption. Nuwayser and Williams[118] reported on release of prostaglandin from microcapsules of five cellulose derivatives.

Without getting into the microencapsulation aspect, that is the subject of a chapter in Volume II, we feel obliged to call attention to the fact that cellulose-acetate-butyrate spherical membranes have shown promise for controlled release of steroids, such as progesterone and estrone, from microreservoirs.[161] Gardner et al. of Battelle have reported examples of near-zero-order release, as shown in Figure 17. The objective of their work is the preparation, for contraceptive and medicinal purposes, of steroid delivery systems that will migrate, in response to the observed natural flux of microscopic particles, from the vagina into the uterus and fallopian tubes. If successful, this will be an exquisite, target-seeking, membrane-controlled drug-delivery system!

By and large, the cellulosics have not proved useful for controlled release applications, but much basic data relevant to potential applications is available, some of which is discussed later in this chapter (Section B.1.d).

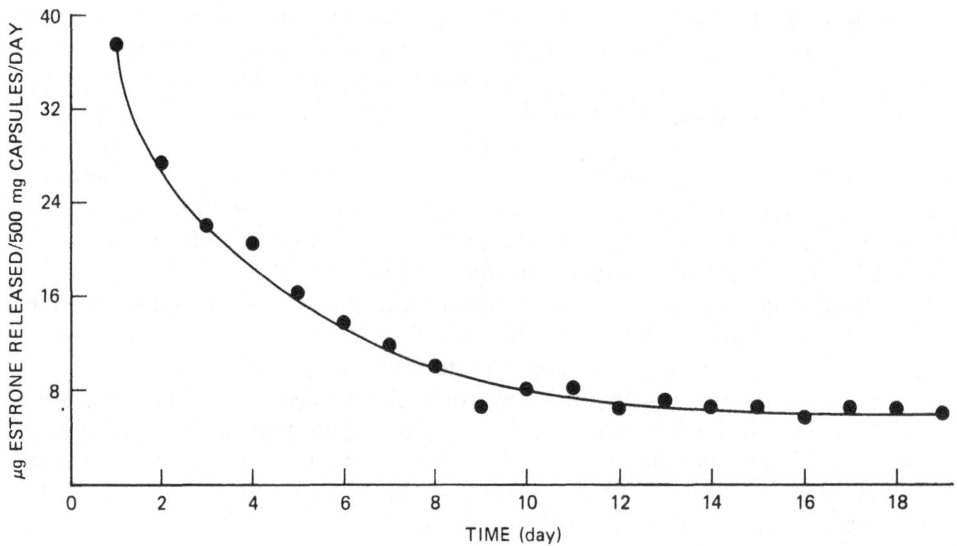

FIGURE 17. Release in vitro of estrone microencapsulaed with cellulose acetate butyrate. (Reproduced with permission of American Chemical Society from Gardner, D. L., Fink, D. J., Patanus, A. J., Baytos, W. C., and Hassler, C. R., in *Controlled Release Polymeric Formulations,* Paul, D. R. and Harris, F. W., Eds., American Chemical Society, Washington, D. C., 1976, 178.

B. Diffusion Through Entrapped Fluids

1. Hydrogel Membranes

a. Background

Certain hydrophilic monomers form polymeric materials that imbibe substantial proportions of water, generating stable three-dimensional networks with an equilibrium water content. Such materials are referred to as hydrogels. Natural polymers in the category have received little attention as rate-controlling membranes for delivery systems. Kiraly and Nose[162] provide a recent review of the natural hydrogels. Some synthetic polymer series, especially the hydroxyalkyl acrylates, offer great versatility, since the degree of hydration and other properties influencing permeability can be varied by altering comonomer ratios, cross-linking agent concentration, and conditions of polymerization. As rate-controlling membranes, these materials can be considered intermediate between dense and microporous structures. Transport across a hydrogel membrane depends largely upon the water solubility of the agent. It involves primarily, the entrapped, continuous, aqueous phase rather than dissolution of the agent in the polymer itself. Synthetic hydrogels were reviewed recently by Ratner and Hoffman[163] within a symposium on biomedical applications of hydrogels in general.[164]

The first preparation and characterization of hydrogels from 2-hydroxyethyl methacrylate (HEMA) was reported by Wichterle and Lim of the Czech Academy in 1960.[165] Workers at National Patent Development Corporation, the U.S. licensee of the Czech inventions, and many others have elucidated the relationships between the physical properties of the gels, their structure (including water content and cross-links), and conditions of polymerization.[166-169] Much effort to establish the biocompatibility of these materials has related to their utility in soft contact lenses, as catheter coatings, and for burn dressings.

Homopolymers and random copolymers can be prepared readily, in water or mixed aqueous solvents, with redox or azo catalysts. Since it is very difficult to prepare HEMA — monomer free from ethylene glycol dimethacrylate, which gives rise to cross-linking, assiduous purification is necessary to obtain the noncross-linked homopolymer (PHEMA). PHEMA itself is not water soluble, but the homopolymers of most of the other monomers are, so that difunctional acrylates, such as ethylene glycol dimethacrylate, are commonly added to obtain cross-linked gels.

The homopolymers of other hydroxyalkyl methacrylates have not received nearly as much attention as PHEMA, but the trends in properties are generally in accord with chemical intuition. For example, poly(hydroxypropyl methacrylate) (HPMA) hydrogels contain less water than PHEMA and have better mechanical properties,[169] whereas poly(glyceryl methacrylate) (GMA) hydrogels have much higher equilibrium water contents and are weaker than PHEMA.[170]

b. Membrane Characteristics

The ease of synthesis of copolymers from hydrophilic acrylic monomers simplifies the preparation of membranes with a wide range of properties potentially suitable to control the release of water soluble species. Yasuda et al.[16] determined the diffusional parameters for sodium chloride in a series of hydrated acrylic copolymer gels. Methyl methacrylate (MMA), glycidyl methacrylate (GdMA) and the three previously mentioned monomers (HEMA, HPMA, and GMA) were employed to produce a series of hydrated membrane materials having equilibrium water contents ranging from 10 to 70%. The diffusion coefficients for NaCl were observed to vary over five orders of magnitude with changes in the equilibrium water content. However, the diffusivity was independent of the specific chemical structure of the copolymers and primarily a function of membrane water content. Water soluble species as diverse as salts, sugars, urea, and fluorescein diffuse freely.

Yasuda, Lamaze, and Peterlin[17] developed theoretical relationships interpreting the permeability of hydrogels to water and solutes in terms of the free volume (water content by volume) in the membrane. Straight-line relationships were obtained when a semilog plot was constructed of the diffusion coefficient in the membrane (divided by the diffusion coefficient in pure water) vs. the volume fraction of water in the various membranes. However, there is a break in slope between 30 and 40% hydration, denoting a transition in the factors influencing transport (Figure 18).

Chen[171] observed that the concentration of cross-linking agent has an effect upon water uptake, and Wisniewski et al.[18] have added support to Chen's hypothesis of transition from porous to dissolution transport. Work continues at the University of Utah[19-20] on phenomena related to the nature and concentration of cross-linking agents with respect to transport of solutes, including hydrophobic species such as progesterone, in PHEMA and related copolymers. Most of the work involves gels with water content in the anomalous 30 to 40% region, and the general implications are not clear.

The behavior of water and solutes in uncross-linked PHEMA and other hydroxyalkyl methacrylate polymers with low cross-link density can be explained in terms of a very loose network that is porous on a submicroscopic level. Pore radii ranging from 3 to 50 Å have been calculated.[172] Wisniewski et al.[18] found that the diffusion coefficient of water approached a limiting value as the concentration of cross-linking agent exceeded approximately 5 mol%, which happens to occur in the region in which equilibrium water content is 35 to 40%. Transition to a partition-type mechanism, even for water, was proposed to explain the observation. Subsequently, the work of Zentner et al. with progesterone has been interpreted in the same terms.[19,20] That is, the permeation of progesterone is primarily through "pores", except at high concentration of one (the shorter) cross-linking agent. Otherwise, the diffusion coefficient shows a straight-line inverse correlation with cross-link concentration and is proportional to water content in accord with the porous-transport mechanism of Yasuda et al.[17]

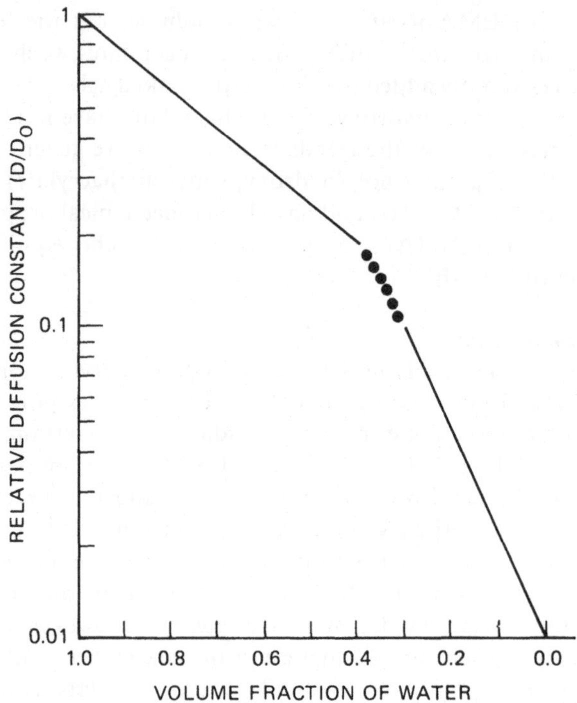

FIGURE 18. Relationship between relative diffusivity to water and the specific water content by volume of hydrogels. (Reproduced with permission of John Wiley & Sons, Inc. from Refoio, M., in *Encycl. Polym. Sci. Tech.*, Suppl. 1, 1976, 214.)

c. Major Applications

The most extensive work reported to date in which hydrogels are employed as rate-moderating membranes has been in connection with the development of systems for the controlled release of fluoride salts in the mouth over extended periods of time to retard the formation of dental caries. The most successful approach has been to fabricate solid reservoirs in which the fluoride compound is dispersed in one hydrogel, and to surround the reservoirs, by coating, with another hydrogel through which the fluoride permeates more slowly. The purpose of the initial work was to develop technology that would permit delivery at rates from 0.02 mg/day to 1 mg/day over intervals of at least 6 months. This would facilitate determination of the optimal delivery rate required to achieve the desired prophylactic effect.

This is part of a major effort, sponsored by the National Institute of Dental Health, that is still in the stage of animal studies. Evaluation of continuous delivery of fluorides in the oral cavity is, of course, an outgrowth of the observed benefits achieved through topical application of fluorides to the teeth and by the maintenance of a fluoride level of approximately 1 ppm in the drinking water of many communities. The present study should certainly be regarded as a dose-response study, rather than an attempt to anticipate the form which a practical intraoral fluoride-delivery system might take. We have made reference earlier to the exploratory work of Halpern et al. in this field.[141] Though they did much of their work with hydrophilic matrices, most of the membranes were of a hydrophobic nature.

Cowsar, and his colleagues at Southern Research Institute,[173] did an exemplary piece of work to select membrane and reservoir components appropriate for desired delivery

TABLE 5

Membrane Parameters of HEMA/MMA Copolymers and Sodium Fluoride

Monomer ratio	Equilibrium water content (%)	$D(\times 10^{-8})$ (cm^2sec^{-1})	$C_s(\times 10^{-4})$ ($g/cm^3\ F^-$)	K	$P = DK(\times 10^{-10})$ ($cm^2\ sec^{-1}$)
30/70	11.1	8	2.5	0.012	9.6
40/60	13.8	26	3.1	0.015	39
50/50	20.3	63	3.3	0.017	107

rates and system dimensions. Building upon the findings of Yasuda,[16] with respect to the diffusion of NaCl in various methacrylate copolymer hydrogels, they prepared HEMA/MMA copolymers in ratios ranging from 80:20 to 40:60 as candidates for fluoride delivery-system components. The equilibrium water content varied in direct proportion to the HEMA/MMA molar monomer ratio up to 50% HEMA. On the basis of the diffusion coefficients (D), partition coefficient (K), and saturation solubilities (C_s), the 50:50 copolymer was selected for the core and the 30:70 copolymer was chosen for coating. Table 5 presents a small portion of the data considered. Comparable data was obtained against water, synthetic saliva, and human saliva.

The limiting factor with respect to device dimensions was the necessity to contain sufficient NaF for 6 months activity (over a 50-fold range), plus a 20% excess to prevent rate loss as the reservoir becomes depleted. Cores, consisting of 62 or 80% by weight powdered NaF dispersed in a HEMA/MMA copolymer, were made by placing a mixture of copolymer solution and NaF in a mold and allowing the solvent (acetone-dioxane) to evaporate. The mold dimensions were derived by considering the minimum core volume sufficient for a given NaF concentration and the area of a rhombus corresponding to that volume. The systems were all ribbon shaped, approximately 0.5 cm wide and 0.4 to 4 cm in length. The membranes used were thin enough (0.1 to 0.3 mm) that the core surface area approximated the membrane area after the latter had been coated on the core. Application of Fick's Law to the estimated area and measured permeability permitted calculation of the membrane thickness necessary to obtain the target rate. If too thin a membrane would be necessary at the minimum core volume, then the area and volume could be increased until a practical membrane thickness would suffice. Since the membrane is generated by coating the core, the geometry and precision of the cores for such systems are critical, Cowsar et al. describe in some detail (1) their use of stainless-steel replicas and silicone-rubber molds for this purpose, and (2) the dip-coating process. Each dip in the 12% polymer solution increased coating thickness about 10 μm.

Excellent conformance of in vitro release to that anticipated has been reported for at least two months, and the release rates were shown to be independent of the composition and pH of the receiving fluid ($<\pm 0.03$ mg/day variation for a device with a mean daily release rate of 0.47 mg). The delivery of fluoride into the saliva is being evaluated in the mouths of beagle dogs, where the systems are attached via stainless-steel wires to a special orthodontic appliance that retains the systems beneath the upper lip and above the teeth. Early results indicated that a system designed to deliver 0.2 mg/day produced a level of about 0.24 ppm, and a 0.5 mg/day system produced 0.75 ppm fluoride in the dog's saliva. Increased salivation, induced with the drug pilocarpine, decreased the fluoride level in the saliva as one would expect for a system in which the release rate is membrane controlled. Preclinical toxicological testing of such systems is underway.

Abrahams and Ronel[174] of Hydromed (a subsidiary of National Patent Development) have described the preparation of three types of hydrogel-membrane-controlled systems for releasing the narcotic antagonist cyclazocine, which is a synthetic analog of morphine. Narcotic antagonists prevent the expression of the opiate effects of illicit drugs and are being studied, in connection with a number of delivery systems, as one component in the rehabilitation of narcotic addicts. Capsular devices filled with powdered cyclazocine (free base) were prepared from spun-cast tubes with a wall thickness of 1 mm composed of HEMA copolymers containing low levels of methacrylic acid (MAA). Minor changes in the ionogenic comonomer (MAA) concentration have a dramatic effect on delivery. The release rates through copolymers with 0.3 and 1.0 mole % MAA content were constant over at least 150 days, the rate increasing from about 0.3 mg/day to 0.7 mg/day with the increased MAA level. At 2.4 mol% MAA level, the rate rose to 2 mg/day, but fell steadily, presumably due to depletion of the reservoir. The results are reportedly quite reproducible. A similar effect, although not as marked, was noted when the cross-link density was varied through the range of 0.02 to 3.0% ethylene glycol dimethacrylate.

Tablets of the free-base form of cyclazocine, weighing about 300 mg, were dip coated with pure PHEMA or with HEMA copolymers containing methacrylic acid. With the former (membrane thickness unspecified), the release rate was constant at 0.7 mg/day for at least six months — during which time 40% of the contents were delivered. With the copolymer containing 2.4 mol% of the ionogenic MAA, the rate rose to 5 mg/day and then decreased as depletion occurred.

Lastly rods formed from a solution of cyclazocine hydrochloride in HEMA that was polymerized *in situ*, were dip coated with a copolymer of 80% ethoxyethyl methacrylate and 20% hydroxyethyl methacrylate. These latter systems provided release at about 1 mg/day in excess of one month, while the uncoated cores released 50% of their drug content in 2 days and were essentially empty in 6 days.

d. Exploratory Studies

Wichterle and his colleagues[175] report that they have explored the diffusion of several anticancer drugs through HEMA/butyl methacrylate (BMA) copolymers over the range from zero to 20% BMA. Permeation data for four drugs (5-FU, Cytembena, Butocin, and Ftorafur), as well as sodium chloride, are reported. The objective was said to be pouch-shaped devices for placement directly on tumors.

Some studies involving diffusion of drugs through hydrogels, although not involving them as rate-limiting membranes per se, provide useful data for consideration of that approach. Anderson, et al.[176], in devising a method to administer hydrocortisone sodium succinate to chick embryos in order to assess an effect of steroids on embryonic differentiation, determined diffusion coefficients for this drug in polyHEMA (Hydron-type N) cross-linked to various degrees. Good[177] recently reported on the simultaneous hydration of and drug release from PHEMA for oral dosage of tripelennamine and other drugs.

Several groups have incorporated antibiotics (cephalosporins, penicillins, tetracyclines, aminoglycosides) into HEMA and ethylene glycol methacrylate polymer gels for sustained release.[178-181] Hydrogel contact lenses have been saturated with drug solutions (pilocarpine, epinephrine, polymixin antibiotic, idoxuridine) for prolonged release to the eye.[182-183] However, the principle of interposing a barrier membrane has not been investigated with hydrogels for such applications.

An investigation which is directed toward combination of membrane-moderation with the capacities of hydrogel technology is underway at the University of California.[149] It involves CASING (cross-linking by active species of inert gases), by means of which the surface of a hydrogel is modified to create a barrier membrane *in situ*. The initial attempts to cross-link the surface of PHEMA itself did not produce the anticipated retardation of pilocarpine-hydrochloride-release rate, but the related technique of generating a polyolefin film *in situ* did yield favorable results. Films formed by plasma polymerization of ethylene, tetrafluoroethylene, or ethane on the surface of water-swollen hydrogels of HEMA homopolymer and of HEMA/MA copolymer did modify the release kinetics from first order to zero order. The rates appear to be inversely proportional to the film thicknesses, which are fractions of a μm. The problems one can anticipate for such laminates, with respect to adhesion and fracture of the relatively stiff surface film, as the substrate swells and shrinks, makes the practicality of this modification of hydrogels doubtful.

Another example of high-energy physics applied in this field involves the surface polymerization of hydroxyalkyl acrylates to create hydrogel coatings on other polymeric substrates. This was developed at the Franklin Institute. In 1973, Scott et al.[184] described its use to coat polyethylene rods wrapped with copper wires. A 0.5 mm thick, continuous, water-swellable film of PHEMA was applied. Earlier workers had demonstrated the contraceptive effectiveness of intrauterine devices incorporating copper wire, which release copper ions slowly as the constituents of the uterine milieux corrode the wire. The purpose of the hydrogel grafting work was primarily to enhance the biocompatability of the copper-bearing device. Because the hydrogel-forming monomer does not graft to metals, the polyethylene backbone of the device was grooved to allow the hydrogel graft polymer to bridge the gaps and cover the entire device. The copper beneath the hydrogel coat was observed to degrade in rabbit uteri just as that on uncoated devices, so far as studies with scanning electron microscopy, X-ray, and X-ray diffraction could reveal. The copper-release rate (33 $\mu g/day/cm^2$) of the hydrogel-coated devices in the rabbit uterus was considered comparatively low by the workers, but was too close to the rates of release reported[185-186] for bare copper-bearing IUD's (45 to 60 $\mu g/day/cm^2$) to support the inference that the coating exerts a rate-controlling effect in the sense we have discussed for other applications. However, the coating maintained its integrity well in the uterine environment and could be expected to prevent the release of gross amounts of copper in the form of particles, which have been observed to flake off as the embrittled metal corrodes. Scott et al. concluded that these coatings could thus prolong the activity of copper-bearing IUD's as well as contribute to *in utero* tolerance.

Although other synthetic hydrogel polymers, such as polyacrylamide and polyvinylpyrrolidone, have been investigated for prolonged-release applications,[187-189] there have been no reports on the sort of membrane control that is the subject of this chapter, presumably because the high water content of such gels makes them extremely permeable and mechanically weak. Poly(*N*-vinyl-2-pyrrolidone) is a water-soluble substance frequently employed in pharmaceutical and veterinary formulations, but the highly cross-linked gel-forming version has received relatively little attention. Indicative of its membrane properties is the work in which polyvinylpyrrolidone gels have been evaluated as hemodialysis membranes.[190] Grafts of polyacrylamide and polyvinylpyrrolidone, as well as other hydrophilic monomers, can be prepared by several routes, and may prove useful in the creation of membrane-controlled systems by conferring biocompatability on a material otherwise more suitable from the standpoint of permeability and mechanical properties. Other hydrophilic polymers, such as polyvinylalcohol, ionic polymers, and their complexes, have been examined as hemodialysis membranes and for other biomedical applications, but not yet in controlled release membranes. Ratner and Hoffman's review[163] provides lead references on all of these polymers.

Natural hydrogels include polysaccharides, such as cross-linked dextrans (Sephadex®) and cellulose (Cuprophan®), and polypeptides, such as collagen. The regener-

TABLE 6

Some Commercial Microporous Films

Base polymer	Range of pore size (Å)	Tradename	Source
Regenerated cellulose	>1000	Millipore®	Millipore
Cellulose nitrate/acetate	>1000	Diapor®	Amicon
Cellulose triacetate	15—250	Poroplastic®	Moleculon
Polypropylene	200 × 2000	Celgard®	Celanese
Polytetrafluoroethylene	>1000	Gore-tex®	W. L. Gore Associates
Polycarbonate	>300	Nucleopore®	General Electric

ated cellulose film called Cuprophan® has become the most commonly employed material for hemodialysis membranes, and thus, a voluminous literature is available with respect to its permeability to molecules of biomedical interest. It is interesting to note that in addition to its use to counteract renal insufficiency, the so-called artificial kidney is utilized to detoxify victims of all sorts of poisoning. We are not aware of the inverse application in which such membranes are employed to release drugs or other agents. However, the annual review of dialysis and hemoperfusion of poisons and drugs that appears in the *Transactions of the American Society for Artificial Internal Organs* provides access to data on the interaction with membranes of molecules, especially drugs, that might be considered for delivery. Of course, most of the membranes employed for blood purification are, as the word "dialysis" implies, porous to some extent. The pores are intermediate in size between those of hydrogels like PHEMA and the microporous structures we will next discuss. Below a critical molecular size, corresponding to the mean pore size, the rate of permeation of solutes through a membrane like Cuprophan® is inversely proportional to the molecular weight of the hydrated diffusant. For Cuprophan®, rejection on the basis of molecular size commences with the hydrated species of molecules of only a few hundred daltons.[21-25] For microporous structures, rejection only occurs for molecules above 10,000 daltons.

Collagen, the material that forms connective tissue and is the most common protein in the body, is receiving increasing attention as a film-forming polymer. It has been used for applications as diverse as surgical sutures and sausage casings. A recent publication by Bradley and Wilkes[191] compares the properties of reconstituted collagen prepared by nine different routes. Collagen suspensions were prepared by enzymatic degradation of beef tendons, and films were cast from the suspension. The films were cross-linked or tanned by various means. The Princeton workers determined the rates of release (into ethanol) of the contraceptive steroid medroxyprogesterone after that drug had been dispersed in the collagen prior to cross-linking. A wide range of release patterns was observed as conditions of preparation were varied. A collagen membrane per se was not reported.

2. Microporous Membranes

a. Background

In recent years, a number of films have become available that provide discrete, continuous, very fine pores. Sheet materials a few mils in thickness and with pore sizes ranging from several microns to a few angstroms are considered microporous membranes. Some examples of the base polymers and commercial sources in the U.S. are shown in Table 6. All are highly permeable to gases and water vapor, although the hydrophobic films effectively impede liquid water at atmospheric pressure because of high surface tension.

b. Membrane Characteristics

Each film type is unique, with the intrinsic characteristics of the pores being very much dependent upon method of preparation and subsequent history. Furthermore, the barrier properties of these films depend largely upon the medium with which the pores are filled. The film itself provides the network in which the transport medium is entrapped. The stability and reproducibility of that network differs among the candidate materials, with crystallinity and resistance to attack, by either the enclosed medium or constituents of the environment in which it is to be used, ranking high among the criteria for acceptability. Not only effective pore diameter, but percent porosity and the side-to-side average path length or tortuosity are critical characteristics influencing transmission rate of a given molecular species.

Thus far, microporous films have found greatest utility as filters, as supports for ultrafiltration or reverse-osmosis membranes, and as separators for battery compartments. They are being investigated for a number of biomedical barrier applications, from wound dressings to blood oxygenators. Some have been evaluated as matrices that can be impregnated with an agent for slow release by diffusion and evaporation or leaching, as described in a later chapter. Relatively little has been published on microporous membranes as rate-controlling barriers between a reservoir and a sink. This, coupled with the idiosyncrasies peculiar to each family of materials, makes generalization almost impossible. There are only a single major application and a few exploratory studies in the field of microporous-membrane systems, as we have defined them. With respect to general references, the literature on membrane separation[192] and the promotional publications of the membrane suppliers are the best sources.

c. Major Applications

ALZA Corporation has employed a microporous rate-controlling membrane in the first of a product line called transdermal therapeutic systems (TTS), which are designed to deliver drug through the skin at a controlled rate.[193] Understanding of the barrier properties of human skin itself made it possible to consider the controlled administration of drugs through the unbroken surface of the skin into the systemic circulation for action at a remote site. The pattern of drug so obtained in the body corresponds to that achieved by infusion. Controlled, unattended administration of a drug to the skin surface could, in theory, also be beneficial for local treatment of certain topical skin lesions as well, but interest at this time focuses primarily upon the continuous delivery of systemic medication.

The initial system, Transiderm® Scop-210, has been developed to prevent and treat nausea, a condition for which oral medication is often unsatisfactory. The permeability of human skin to scopolamine is sufficiently high that systems of a practical size can deliver adequate drug for therapy while maintaining drug levels low enough to minimize side effects (such as dryness of the mouth, drowsiness, and impairment of eye accommodation) that normally accompany oral or intramuscular administration of the drug. Clinical research has been reported which was conducted to determine if the transdermal-delivery system for scopolamine is efficacious in treating and preventing motion-induced nausea.[194,195]

The Transdermal Therapeutic System-scopolamine, shown schematically in Figure 19, is a multilayer laminate. It consists of a drug reservoir, containing scopolamine dispersed as a separate phase within a highly drug-permeable matrix, laminated between the rate-controlling membrane and an external metallic foil which is impermea-

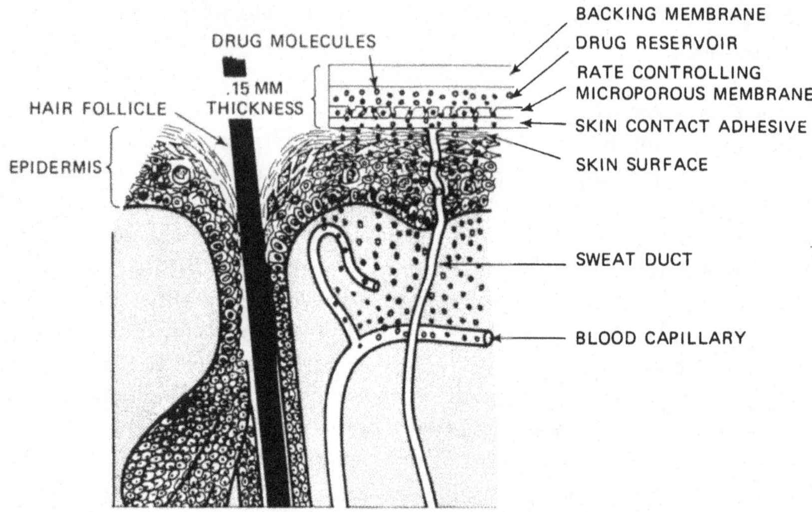

FIGURE 19. Schematic diagram of Transdermal Therapeutic System in place on the surface of human skin. (Reprinted with the permission of Plenum Publishing Corporation from Chandrasekaran, S. K. and Shaw, J., in *Contemporary Topics in Polymer Science*, Vol. 2, Pearce, E. M. and Schaefgen, J. R., Eds., Plenum Press, New York, 1977, 296.)

ble to drug and moisture. The specific nature of the microporous membrane is not disclosed. Its pores are filled with fluid highly permeable to scopolamine.

On the side of the microporous membrane, opposite the reservoir is a gel which serves as both a pressure-sensitive adhesive to secure the system on the skin surface and as an auxiliary reservoir that provides an initial priming dose of drug upon attachment to the skin. A protective strippable film covers the adhesive layer until just before the system is to be applied to the skin. The amount of scopolamine incorporated in the skin-contact layer was determined in light of the degree of drug immobilization in the skin and the pharmacokinetic characteristics of scopolamine.[196] That is, the delivery system must provide the appropriate net input to the body to achieve the necessary drug level for effect, given the processes of sorption, distribution, metabolism, and elimination that the body imposes.

The delivery rate of scopolamine from the system is governed by diffusion of drug molecules through the various elements of the multilayer laminate and the skin. At steady state, the rate limiting component is the microporous membrane. Actual and theoretical profiles of scopolamine release in vitro from a transdermal therapeutic system into an infinite sink under isotonic and isothermal conditions are shown in Figure 20. The Transiderm® — Scop 210, with a surface area of 2.5 cm^2, is designed to release 200 μg as a priming dose and 10$\mu g/hr$ for 72 hr at steady-state.

A great deal of effort has gone into the comprehension of the kinetics and variability of drug permeation through skin, so that indeed it can be assured that control will be vested in the delivery system. The principal resistance to penetration of drugs and other small molecules through intact human skin resides within the stratum corneum, which is comprised of dead, keratinized, partially desiccated, epidermal cells. The stratum corneum is a heterogeneous structure containing about 40% protein, about 40% water, and 20% lipid in a layered, closely-packed array of flattened, interlocking cells.[197] Skin appendages (glands and hair follicles) could contribute to skin permeation, but in fact, have little importance, at least in the cases presently of interest. The lower layers of the epidermis and the dermis offer little resistance to penetration of drugs, but appar-

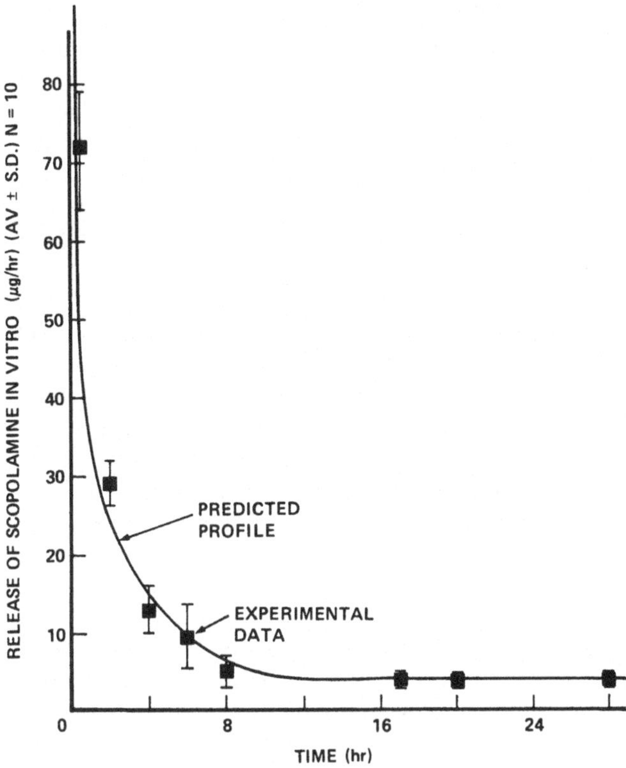

FIGURE 20. Release of scopolamine in vitro from Transdermal Therapeutic System — scopolamine. (Reprinted with permission of Plenum Publishing Corporation from Chandrasekaran, S. K. and Shaw, J., in *Contemporary Topics in Polymer Science,* Vol. 2, Pearce, E. M. and Schaefgen, J. R., Eds., Plenum Press, New York, 1977, 302.)

ently bind certain substances in a manner analogous to protein binding.[198,199] The permeation of drug through this barrier depends upon the physico-chemical properties of the drug, such as its molecular size, oil/water partition coefficient, and polarity. The vehicle in which the drug is dissolved may modify its partition into the stratum corneum and permeation through the skin. Occlusion of the skin entraps moisture and raises the water content of the stratum corneum, which facilitates absorption and permeation of water-soluble agents. A useful model of skin permeation has been developed by the workers at ALZA and described from several perspectives.[200]

For scopolamine, skin permeability is strongly pH dependent in that the nonionic (more lipophilic) form of the drug is decidedly more permeable than the ionic form. The ALZA model for skin permeation, which invokes the coexistence of dissolved and mobile sorbed molecules in equilibrium with site bound and immobile molecules within the tissue, quite accurately correlates experimental sorption data and transient transport measurements.[196,197]

The evidence for the correspondence between delivery profile and levels achieved and maintained in vivo is too complex to present here because of the extensive metabolism that scopolamine undergoes. However, it has been clearly demonstrated[197] that about 4 hr following application of the system, the excretion of the drug by the kidneys approaches the level corresponding to the input rate determined in vitro.

d. Exploratory Studies

ALZA is investigating other applications of the principles of transdermal drug delivery and microporous-membrane-controlled release, but nothing further has been published with respect to the microporous membrane aspect, outside of the patent literature.[201-203] A few other examples of release moderated by porous materials are worthy of citation.

Certain microencapsulation processes give rise to microporous structures capable of releasing by the sort of process we have described.[204,205] The EURAND Corporation in Milan manufactures a series of microencapsulated products, called Diffucap® dosage forms, that are reported to release by a dialysis mechanism. These are encapsulated with a Diffulac® membrane of undisclosed composition that is said to become microporous in water and permits dissolution of the contents at the ambient pH. The dissolved agent then diffuses through the pores over an 8 to 12 hr period.

Sciarra et al.[206,207] investigating the possibilities of incorporating various drugs in ethyl cellulose and polyamide films for use as wound dressings, and Donbrow et al.,[208-211] who focused on ethyl-cellulose films, have demonstrated the possibility of spontaneously generating porous films for the purpose of controlled drug delivery. Employment of water-leachable plasticizers, such as tributyl citrate in the former case or polyethylene glycol in the latter case, gave rise to formation of pores in the relatively hydrophobic films when they were placed in an aqueous environment. In the first case cited, the delivery system was of the monolithic variety, albeit in film form, as have been some of Donbrow's systems,[210] but the significance of the work of Donbrow's group in the present context relates especially to the use of this approach to create *in situ* a microporous controlling membrane. Laminates were described in which drug (salicylic acid, caffeine, or barbital) was dispersed in a reservoir matrix of hydroxypropyl cellulose and enclosed within membranes composed of ethyl cellulose and polyethylene glycol (PEG 4000). The PEG was shown to be displaced by water in a matter of minutes. The concentration of PEG 4000 was varied from 10 to 60% by weight in the ethyl cellulose. The drug release rate to water increased linearly with PEG content, over about a fivefold range for caffeine, and less for the other drugs. However, the membranes are impermeable to inorganic species, such as sodium hydroxide, and the permeation of organic species correlates with their solubility in ethyl cellulose.[211] Thus, it appears that, at least when the PEG-plasticized ethyl cellulose is created as described, the porosity is of a closed-cell variety.[209] Nevertheless, the kinetics are zero order as one expects for a membrane-controlled system.

IV. CONCLUSION

The principles of membrane-controlled release have been elucidated and applied with a wide variety of membranes, differing in composition and morphology, to deliver a number of agents. Although only pharmaceutical applications have been discussed, the succeeding chapter will present examples in quite another field.

ACKNOWLEDGMENT

The author wishes to acknowledge the fact that this review was made possible through the privilege of his having served in various management posts over five of the formative years of ALZA Corporation and through the cooperation of his colleagues there. In particular, the contribution of S. Kumar Chandrasekaran and the guidance and assistance of Sharon Hamrick, Susan Laird, and Jim Yuen deserve grateful recognition.

REFERENCES

1. **O'Neill, W. P. and Wolman, A. J.**, Markets for therapeutic systems, *Drug Cosmet. Ind.*, 120, 28, 1977.
2. **Baker, R. W. and Lonsdale, H. K.**, Controlled release: mechanisms and rates, in *Controlled Release of Biologically Active Agents*, Tanquary, C. A. and Lacey, R. E., Eds., Plenum Press, New York, 1974, chap. 2.
3. **Flynn, G. L.**, Influence of physico-chemical properties of drug and system on release of drugs from inert matrices, in *Controlled Release of Biologically Active Agents*, Tanquary, C. A. and Lacey, R. E., Eds., Plenum Press, New York, 1974, chap. 3.
4. **Flynn, G. L., Yalkowsky, S. H., and Roseman, T. J.**, Mass transport phenomena and models: theoretical concepts, *J. Pharm. Sci.*, 63, 479, 1974.
5. **Rogers, C. E.**, Structural factors governing controlled release, in *Controlled Release Polymeric Formulations*, Paul, D. R. and Harris, F. W., Eds., American Chemical Society, Washington, D.C., 1976, chap. 2.
6. **Barrer, R. M.**, *Diffusion In and Through Solids*, Cambridge University Press, Cambridge, 1955.
7. **Crank, J. and Park, G. S., Eds.**, *Diffusion in Polymers*, Academic Press, London, 1968.
8. **Crank, J.**, *The Mathematics of Diffusion*, 2nd ed., Clarendon Press, Oxford, 1975.
9. **Michaels, A. S. and Bixler, H. J.**, Membrane permeation: theory and practice, in *Progress in Separation and Purification*, Interscience, New York, 1968, 143, and references therein.
10. **Hopfenberg, H. B., Ed.**, *Polymer Science and Technology*, Vol. 6, Plenum Press, New York, 1974.
11. **Klein, J.**, Diffusion of long molecules through solid polyethylene. I. Topological constraints, *J. Polym. Sci.*, 15, 2057, 1977.
12. **Klein, J. and Briscoe, B. J.**, Diffusion of long molecules through solid polyethylene. II. Measurements and results, *J. Polym. Sci.*, 15, 2065, 1977.
13. **Burrell, H. and Immergut, B.**, Solubility parameters, in *Polymer Handbook*, Brandrup, J. and Immergut, E. H., Eds., Interscience, New York, 1967.
14. **Michaels, A. S., Wong, P. S. L., Prather, R., and Gale, R. M.**, A thermodynamic method of predicting the transport of steroids in polymer matrices, *AIChE J.*, 21, 1073, 1975.
15. **Jhon, M. S., Ma, S. M., Hattori, S., Gregonis, D. E., and Andrade, J. D.**, The role of water in the osmotic and viscoelastic behavior of gel networks, in *Hydrogels for Medical and Related Applications*, Andrade, J. D., Ed., American Chemical Society, Washington, D.C., 1976, chap. 4.
16. **Yasuda, H., Lamaze, C. E., and Ikenberry, L. D.**, Permeability of solutes through hydrated polymer membranes. I. Diffusion of sodium chloride, *Macromol. Chem.*, 118, 1935, 1968.
17. **Yasuda, H., Lamaze, C. E., and Peterlin, A.**, Diffusive and hydraulic permeabilities of water in water-swollen polymer membranes, *J. Polym. Sci. Part A*, 9, 1117, 1971.
18. **Wisniewski, S. J., Gregonis, D. E., Kim, S. W., and Andrade, J. D.**, Diffusion through hydrogel membranes. I. Permeation of water through poly (2-hydroxy ethyl methacrylate) and related polymers, in *Hydrogels for Medical and Related Applications*, Andrade, J. D., Ed., American Chemical Society, Washington, D.C., 1976, chap. 6.
19. **Zentner, G. M., Cardinal, J. R., and Kim, S. W.**, Progestin permeation through polymer membranes. I. Diffusion studies on plasma soaked membranes, *J. Pharm. Sci.*, 67, 1347, 1978.
20. **Zentner, G. M., Cardinal, J. R., and Kim, S. W.**, Progestin permeation through polymer membranes. II. Diffusion studies on hydrogel membranes, *J. Pharm. Sci.*, 67, 1352, 1978.
21. **Craig, I. C. and Chen, H.**, On a theory for the passive transport of solute through semipermeable membranes, *Proc. Natl. Acad. Sci. U.S.A.*, 67, 702, 1972.
22. **Farrell, P. C. and Babb, A. L.**, Estimation of the permeability of cellulosic membranes from solute dimensions and diffusivities, *J. Biomed. Mater. Res.*, 7, 275, 1973.
23. **Muir, W. M., Gray, R. A., Courtney, J. M., and Ritchie, P. D.**, Perm-selective dialysis membranes. II. Films based on acrylic acid-*n*-butyl methacrylate copolymers — a critical comparison with cellulosic films, *J. Biomed. Mater. Res.*, 7, 3, 1973.
24. **Lyman, D. J. and Kim, S. W.**, Aqueous diffusion through partition membranes, *J. Polym. Sci. Polym. Symp.*, 41, 139, 1973.
25. **Craig, L. C. and Konigsberg, W.**, Dialysis Studies. III. Modification of pore size and shape in cellophane membranes, *J. Phys. Chem.*, 65, 166, 1961.
26. **Lyman, D. J. and Kim, S. W.**, Membranes in artificial kidney devices: past, present and future, *Biomater. Med. Devices Artif. Organs*, 1, 431, 1973.
27. **Yasuda, H. and Tsai, J. T.**, Pore size of microporous polymer membranes, *J. Appl. Polym. Sci.*, 18, 805, 1974.
28. **Zaffaroni, A.**, IUD having a Replenishing Drug Reservoir, U.S. Patent 3,896,819, 1975.
29. **Zaffaroni, A.**, IUD having a Replenishing Drug Reservoir, U.S. Patent 3,845,761, 1975.

30. **Zaffaroni, A.**, Intrauterine Contraceptive Device Containing Pharmaceutically Acceptable Steroids, U.S. Patent 3,895,103, 1975.
31. **Theeuwes, F., Gale, R. M., and Baker, R. W.**, Transference: a comprehensive parameter governing permeation of solutes through membranes, *J. Membr. Sci.*, 1, 3, 1976.
32. **Theeuwes, F., Ashida, K., and Higuchi, T.**, Programmed diffusional release from encapsulated cosolvent system, *J. Pharm. Sci.*, 65,648, 1976.
33. **Flynn, G. L., Ho, N. F. H., Hwang, S., Owada, E., Molokhia, A., Bahl, C. R., Higuchi, W. I., Yotsuyanagi, T., Shah, Y., and Park, J.**, Interfacing matrix release and membrane absorption — analysis of steroid absorption from a vaginal device in a rabbit doe, in *Controlled Release Polymeric Formulations*, Paul, D. R. and Harris, F. W., Eds., American Chemical Society, Washington, D.C., 1976, chap. 7.
34. **Michaels, A. S., Mader, W. J., and Manning, C. R.**, New concepts and standards of quality control as applied to controlled drug delivery systems, in *The Quality Control of Medicines*, Deasy, P. B. and Timoney, R. F., Eds., Elsevier, Amsterdam, 1976, 45.
35. **Langer, R. and Folkman, J. M.**, Sustained release of macromolecules in *Polymeric Delivery Systems, Midland Macromolecular Monograph*, Vol. 5, Gordon & Breach, New York, 1978.
36. **Folkman, J. M. and Long, D. M., Jr.**, The use of silicone rubber as a carrier for prolonged drug therapy, *J. Surg. Res.*, 4, 139, 1964.
37. **Folkman, J. and Long, D. M.**, Jr., Drug pacemakers in the treatment of heart block, *Ann. N.Y. Acad. Sci.*, 111, 857, 1964.
38. **Folkman, J. M., Long, D. M., Jr., and Rosenbaum, R.**, Silicone rubber: a new diffusion property useful for general anesthesia, *Science*, 154, 148, 1966.
39. **Folkman, J., Winsey, S., and Moghul, T.**, Anesthesia by diffusion through silicone rubber, *Anesthesiology*, 29, 410, 1968.
40. **Folkman, J., Mark, V. H., Ervin, F., Suematsu, K., and Hagiwara, R.**, Intravenous gas anesthesia by diffusion through silicone rubber, *Anesthesiology*, 29, 419, 1968.
41. **Long, D. M., Jr. and Folkman, J.**, Polysiloxane carrier for controlled release of drugs and other agents, U.S. Patent 3,279,996, 1966.
42. **Folkman, J. and Mark, V. H.**, Diffusion of anesthetics and other drugs through silicone rubber: therapeutic implications, *Trans. N.Y. Acad. Sci.*, 30, 1187, 1968.
43. **Dzuik, J. and Cook, B.**, Passage of steroids through silicone rubber, *Endocrinology*, 78, 208, 1966.
44. **Martinez-Manatou, J., Giner, A., Cortez, V., Azner, R., Casasola, J., and Rudel, H. W.**, Low doses of progestogen as an approach to fertility, *Fertil. Steril.*, 17, 49, 1966.
45. **Segal, S. J. and Croxatto, H. B.**, Single administration of hormones for long-term control of reproductive function, presented at the 23rd Meet. Am. Fertility Society, Washington, D.C., April 14—16, 1967.
46. **Chang, C. C. and Kincl, F. A.**, Sustained hormonal preparations. III and IV. Biological effectiveness of steroid hormones, *Steroids*, 12, 689, 1968; *Fertil. Steril.*, 21, 134, 1970.
47. **Zbuzkova, V. and Kincl, F. A.**, Sustained release hormonal preparations. XII. Plasma levels of 6-methyl-17α-acetoxy-4,6-pregnadiene-3,20 dione in hamsters, *Steroids*, 16, 447, 1970.
48. **Kincl, F. A., Angee, I., Chang, C. C., and Rudel, H. W.**, Sustained release hormonal preparations. IX. Plasma levels and accumulations in the various tissues of 6-methyl-17α-acetoxy-4,6-pregnadiene-3,20-dione after oral administration or absorption from PDS implants, *Acta Endocrinol. (Copenhagen)*, 64, 508, 1970.
49. **Croxatto, H. B., Diaz, S., Vera, R., Etchart, M., and Atria, P.**, Fertility control in women with a progestagen released in microquantities from subcutaneous capsules, *Am. J. Obstet. Gynecol.*, 105, 1135, 1969.
50. **Tatum, H. J., Coutinho, E. M., Adeodato Filho, J., and Sant' Anna, A. R. S.**, Acceptability of longterm contraceptive steroid administration by subcutaneous Silastic capsules, *Am. J. Obstet. Gynecol.*, 105, 1139, 1969.
51. **Coutinho, E. M., Mattos, C. E. R., Sant' Anna, A. R. S., Adeodato Filho, J., Silva, M. J., and Tatum, H. J.**, Longterm contraception by subcutàneous Silastic capsules containing megestrol acetate, *Contraception*, 2, 313, 1970.
52. **Croxatto, H. B., Diaz, S., Atria, P., Cheviakoff, S., Rosatti, S., and Oddo, H.**, Contraceptive action of megesterol acetate implants in women, *Contraception*, 4, 155, 1971.
53. **Mishell, D. R., Jr. and Lumkin, M. E.**, Contraceptive effect of varying dosages of progestogen in Silastic vaginal rings, *Fertil. Steril.*, 21, 99, 1970.
54. **Mishell, D. R., Jr., Lumkin, M. E., and Stone, S.**, Prolonged inhibition of ovulation with progestogen impregnated intravaginal devices, *Am. J. Obstet. Gynecol.*, 113, 927, 1972.
55. **Cohen, M. R., Pandya, G. N., and Scommegna, A.**, The effect of an intracervical steroid-releasing device on the cervical mucus, *Fertil. Steril.*, 21, 134, 1970.

56. **Scommegna, A., Pandya, G. N., Christ, M., Lee, A. W., and Cohen, M. R.**, Intrauterine administration of progesterone by a slow releasing device, *Fertil. Steril.*, 21, 201, 1970.
57. **Braley, S.**, The chemistry and properties of the medical grade silicones, in *Biomedical Polymers*, Rembaum, A. and Shen, M., Eds., Marcel Dekker, New York, 1971.
58. **Kincl, F. A., Benagiano, G., and Angee, I.**, Sustained release hormonal preparations. I. Diffusion of various steroids through polymer membranes, *Steroids*, 11, 673, 1968.
59. **Sundaram, K. and Kincl, F. A.**, Sustained release hormonal preparations. II. Factors controlling the diffusion of steroids through dimethyl polysiloxane membranes, *Steroids*, 12, 517, 1968.
60. **Kratochvil, P., Benagiano, G., and Kincl, F. A.**, Sustained release hormonal preparations. VI. Permeability constants of various steroids, *Steroids*, 15, 505, 1970.
61. **Garrett, E. R. and Chemburkar, P. B.**, Evaluation, control and prediction of drug diffusion through polymeric membranes. I. Methods and reproducibility of steady state diffusion studies. II. Diffusion of aminophenones through Silastic membranes, *J. Pharm. Sci.*, 57, 944, et seq., 1968.
62. **Garrett, E. R. and Chemburkar, P. B.**, Evaluation, control and prediction of drug diffusion through polymeric membranes. III. Diffusion of barbiturates, phenyl alkyl amines, dextromethorphan, progesterone and other drugs, *J. Pharm. Sci.*, 57, 1401, 1968.
63. **Most, C. F., Jr.**, Some filler effects on diffusion in silicone rubber, *J. Appl. Polym. Sci.*, 14, 1019, 1970.
64. **Most, C. F., Jr.**, Co-permeant enhancement of drug transmission rates through silicone rubber, *J. Biomed. Mater. Res.*, 6, 3, 1972.
65. **Roseman, T. J.**, Release of steroids from a silicone polymer, *J. Pharm. Sci.*, 61, 46, 1972.
66. **Lacey, R. E. and Cowsar, D. R.**, Factors affecting the release of steroids from silicones, in *Controlled Release of Biologically Active Agents*, Tanquary, A. C. and Lacey, R. E., Eds., Plenum Press, New York, 1974, chap. 5.
67. **Friedman, S., Koide, S. S., and Kincl, F. A.**, Sustained release hormonal preparations. VII. Permeability of three types of silicone rubber to steroids, *Steroids*, 15, 679, 1970.
68. **Flynn, G. L. and Roseman, T. J.**, Membrane diffusion. II. Influence of physical adsorption on molecular flux through heterogeneous dimethyl polysiloxane barriers, *J. Pharm. Sci.*, 60, 1788, 1971.
69. **Roseman, T. J.**, Silicone rubber: a delivery system for steroids, in *Controlled Release of Biologically Active Agents*, Tanquary, A. C. and Lacey, R.E., Eds., Plenum Press, New York, 1974, chap. 4.
70. **Schaumann, R. and Taubert, H. D.**, Long-term applications of steroids enclosed in dimethyl polysiloxane (Silastic): *in vitro* and *in vivo* experiments, *Acta Biol. Med. Ger.*, 24, 897, 1970.
71. **Chien, Y. W., Lambert, H. J., and Grant, D. E.**, Controlled drug release from polymeric devices. I. Technique for rapid *in vitro* release studies, *J. Pharm. Sci.*, 63, 365, 1974.
72. **Flynn, G. L., Carpenter, O. S., and Yalkowsky, S. H.**, Total mathematical resolution of diffusion layer control of barrier flux, *J. Pharm. Sci.*, 61, 312, 1972.
73. **Lifchez, A. S. and Scommegna, A.**, Diffusion of progestogens through Silastic rubber implants, *Fertil. Steril.*, 21, 426, 1970.
74. **Roseman, T. J. and Higuchi, W. I.**, Release of medroxyprogesterone acetate from a silicone polymer, *J. Pharm. Sci.*, 59, 353, 1970.
75. **Chien, Y. W. and Lambert, H. J.**, Controlled drug release from polymeric delivery devices. II. Differentiation between partition-controlled and matrix-controlled drug release mechanisms, *J. Pharm. Sci.*, 63, 515, 1974.
76. **Kincl, F. A. and Rudel, H. W.**, Sustained release hormonal preparations, *Acta Endocrinol. (Copenhagen) Suppl.*, 151, 5, 1971.
77. **Kincl, F. A.**, Permeation of drugs through Silastic rubber membranes, *Acta Pharm. Suec.*, 13, 29, 1976.
78. **Segal, S.**, Contraceptive implants in human fertility, in *Human Reproduction*, Hafez, E. S. F. and Evans, T. N., Eds., Harper & Row, New York, 1973.
79. **Kincl, F. A. and Rudel, H. W.**, Sustained release hormonal preparations, a survey of the field, *Excerpta Med. Int. Congr. Ser.*, 273, 977, 1973.
80. **Tejuja, S.**, Use of subcutaneous silastic capsules for longterm steroid contraception, *Am. J. Obstet. Gynecol.*, 107, 954, 1970.
81. **Benagiano, G., Ermini, M., Carenza, I., and Rolfini, G.**, Studies on sustained contraceptive effects with subcutaneous polydimethyl siloxane implants, *Acta Endocrinol.*, 73, 335, et seq., 1973.
82. **Coutinho, E. M., Mattos, C. E. R., Sant' Anna, A. R. S., Adeodato Filho, J., Silva, M. C., and Tatum, H. J.**, Further studies on longterm contraception by subcutaneous Silastic capsules containing megestrol acetate, *Contraception*, 5, 389, 1972.
83. **Croxatto, H. B., Diaz, S., Quinteros, E., Simoneti, L., Kaplan, E., Rencoret, R., Leixelard, P., and Martinez, C.**, Clinical assessment of subdermal implants of megestrol acetate, *d*-norgestrel, and norethindrone as a longterm contraceptive in women, *Contraception*, 12, 615, 1975.

84. **Weiner, E. and Johannson, E. D. B.**, Contraception with megestrol acetate implants. Megestrol acetate levels in plasma and the influence on the ovarian function, *Contraception*, 13, 685, 1976.
85. **Coutinho, E. M. and da Silva, A. R.**, One year contraception with norgestrienone subdermal Silastic implants, *Fertil. Steril.*, 25, 170, 1974.
86. **Tejuja, S., Malhotra, U., and Bhinder, G.**, A preliminary report on the contraceptive use of subdermal implants containing norethindrone, *Contraception*, 10, 361, 1974.
87. **Coutinho, E. M., da Silva, A. R., Carreira, C. M., Chaves, M. C., and Adeodato Filho, J.**, Contraceptive effectiveness of Silastic implants containing the progestin R 2323, *Contraception*, 11, 625, 1975.
88. **Weiner, E. and Johansson, E. D. B.**, Plasma levels of *d*-norgestrel, estradiol and progesterone during treatment with Silastic implants containing *d*-norgestrel, *Contraception*, 14, 81, 1976.
89. **Weise, J., Marker, I. I., Holma, P., Vartiainen, E., Osler, M., Pyorala, T., Johansson, E., and Laukkainen, T.**, Long-term contraception with norethindrone subcutaneous capsules, *Ann. Clin. Res.*, 8, 93, 1976.
90. **Kent, J. S.**, Controlled release of delmadinone acetate from silicone polymer tubing: *in vitro — in vivo* correlations to diffusion model, in *Controlled Release Polymeric Formulations*, Paul, D. R. and Harris, F. W., Eds., American Chemical Society, Washington, D.C., 1976, chap. 11.
91. **Jones, R., Cohen, M., and Bell, J.**, Local progestational effect of norgestrel in an intrauterine Silastic capsule, *Contraception*, 8, 439, 1973.
92. **el-Mahgoub, S.**, *d*-Norgestrel slow-releasing T device as an intrauterine contraceptive, in *Proc. Int. Conf. Intrauterine Contraception, 3rd Cairo, 1974*, Hefnawi, F. and Segal, S., Eds., and Segal, S., *Analysis of Intrauterine Contraception*, Elsevier-North Holland, New York, 1976.
93. **el-Mahgoub, S.**, *d*-Norgestrel slow-releasing T device as an intrauterine contraceptive, *Am. J. Obstet. Gynecol.*, 123, 133, 1975.
94. **Nilsson, C. G. and Laukkainen, T.**, Improvement of a *d*-norgestrel-releasing IUD, *Contraception*, 15, 315, 1977.
95. **Vickery, B. H., Erickson, G. I., Bennett, J. P., Mueller, N. S., and Haleblian, J. K.**, Antifertility effects in the rabbit by continuous low release of progestin from an intrauterine device, *Biol. Reprod.*, 3, 154, 1970.
96. **Pharriss, B. B. and Hendrix, J. W.**, Absorption of steroids through Silastic in animals and humans — results of experiments in animals, *Excerpta Med. Int. Congr. Ser.*, 207, 146, 1969.
97. **Moon, K. H. and Bunge, R. G.**, Silastic testosterone capsules: observations in the castrated male rat, *Invest. Urol.*, 6, 329, 1968.
98. **Frick, J., Marberger, M., and Marberger, H.**, Hormonal therapy with steroid-filled Silastic rubber implants, *Urol. Int.*, 29, 81, 1974.
99. **Simmons, J. G. and Hamner, C. E.**, Inhibition of estrus in the dog with testosterone implants, *Am. J. Vet. Res.*, 34, 1409, 1973..
100. **Ewing, L. L., Desjardins, C., and Stratton, L. G.**, Testosterone polydimethyl siloxane implants and contraception in male rabbits, in *Temporal Aspects of Therapeutics*, Urquhart, J. and Yates, F. E., Eds., Plenum Press, New York, 1973, 165.
101. **Ewing, L. L., Stratton, L. G., and Desjardins, C.**, Effect of testosterone polydimethylsiloxane implants upon sperm production, libido and accessory sex organ function in rabbits, *J. Reprod. Fertil.*, 35, 245, 1973.
102. **Ewing, L. L., Schambacher, B., Desjardins, C., and Chaffee, V.**, The effect of subdermal testosterone-filled polydimethylsiloxane implants on spermatogenesis in rhesus monkeys, *Contraception*, 13, 583, 1976.
103. **Shippy, R. L., Hwang, S., and Bunge, R. G.**, Controlled release of testosterone using silicone rubber, *J. Biomed. Mater. Res.*, 7, 95, 1973.
104. **Baker, R. W., Lonsdale, H. K., Tuttle, M. E., and Ayres, J. W.**, Intrauterine release of estriol for contraception. II. Device fabrication and *in vitro* release rates, in Proc. Drug Delivery Workshop, National Institute of Child Health and Development, U.S. Department of Health, Education, and Welfare, Washington, D.C., 1976, 49.
105. **Karsch, F. J., Weick, R. F., Kotchkiss, J., Dierschke, D. J. , and Knobil, E.**, An analysis of the negative feedback control of gonadotropic secretion utilizing chronic-implantation of ovarian steroids in ovariectomized Rhesus monkeys, *Endocrinology*, 93, 478, 1973, and references therein.
106. **Carnette, J. C. and Duncan, G. W.**, Release, excretion, tissue uptake, and biological effectiveness of estradiol from Silastic devices implanted in rats, *Contraception*, 1, 339, 1970.
107. **Bloch, R., Kraicer, P. F., Binder, H., and Lobel, E.**, Composite membrane estradiol implant, *J. Pharm. Sci.*, 64, 832, 1975.
108. **Colter, K. D., Shen, M., and Bell, A. T.**, Reduction of progesterone release rate through silicone membranes by plasma polymerization, *Biomater. Med. Devices Artif. Organs*, 5, 13, 1977.

109. **Juni, K., Nakano, M., and Arita, T.**, Comparative permeability and stability of butamben and benzocaine, *Chem. Pharm. Bull.*, 25, 1098, 1977.
110. **Nakano, M.**, Effects of interaction of surfactants, adsorbants and other substrates on the permeation of chlorpromazine through dimethyl polysiloxane membrane, *J. Pharm. Sci.*, 60, 571, 1971.
111. **Nakano, M., Juni, K., and Arita, T.**, Controlled drug permeation. I. Controlled release of butamben through silicone membrane by complexation, *J. Pharm. Sci.*, 65, 709, 1976.
112. **Lovering, E. G., Manville, C. A., and Rowe, M. L.**, Drug permeation through membranes. V. Interaction of diazepam with common excipients, *J. Pharm. Sci.*, 207, 1976.
113. **Lovering, E. G. and Black, D. B.**, Drug Permeation through membranes. I. Effect of various substituents on amobarbital permeation through polydimethylsiloxane, *J. Pharm. Sci.*, 62, 602, 1973.
114. **Siegel, P. and Atkinson, J. R.**, *In vivo* chemode diffusion of L-Dopa, *J. Appl. Physiol.*, 30, 900, 1971.
115. **Bass, P., Purdon, R. A., and Wiley, J. N.**, Prolonged administration of atropine or histamine in a silicone rubber implant, *Nature (London)*, 208, 591, 1965.
116. **Gaginella, T. S., Welling, P. G., and Bass, P.**, Nicotine base permeation through silicone elastomers: comparison of dimethylpolysiloxane and trifluoropropylmethylpolysiloxane systems, *J. Pharm. Sci.*, 63, 1849, 1974.
117. **Lovering, E. G. and Black, D. B.**, Drug permeation through membranes. III. Effect of pH and various substances on permeation of phenylbutazone through polydimethyl siloxane, *J. Pharm. Sci.*, 63, 671, 1974.
118. **Nuwayser, E. S. and Williams, D. L.**, Development of a delivery system for prostaglandins, in *Controlled Release of Biologically Active Agents*, Tanquary, A. C. and Lacey, R. E., Eds., Plenum Press, New York, 1974, 145.
119. **Spilman, C. H., Beuving, D. C., Forbes, A. D., Roseman, T. J., and Larion, L. J.**, Evaluation of vaginal delivery systems containing 15[*s*] 15-methyl $PGF_2\,\alpha$ methyl ester, *Prostaglandins, 12, (Suppl.)* 1, 1976.
120. **Schmidt, V., Zapol, W., Prensky, W., Wonders, T., Wodinsky, I., and Kitz, R.**, Continuous cancer chemotherapy: nitrosourea diffusion through implanted silicone rubber capsules, *Trans. Am. Soc. Artif. Organs*, 18, 45, 1972.
121. **Nakano, M., Arakawa, Y., Juni, K., and Arita, T.**, Permeation of 5-fluorouracil and 1-(2-tetrahydrofurfuryl)-5-fluorouracil) through cellophane, collagen, and silicone membranes, *Chem. Pharm. Bull.*, 24, 2716, 1976.
123. **Wepsic, J. G.**, Catheter Having Antibacterial Substance Therein Provided with Means Permitting Slow Release of Said Substance, U.S. Patent 3,598,127, 1971.
124. **Powers, K. G.**, Release of antimalarial agents from silicone rubber capsules, *J. Parasitol.*, 51, 53, 1965.
124. **Powers, K. G.**, Release of antimalarial agents from silicone rubber capsules, *J. Parisitol.*, 51, 53, 1965.
125. **Collins, R. C.**, Implant chemotherapy for experimental filariasis, *Am. J. Trop. Med. Hyg.*, 23, 880, 1974.
126. **Clifford, C. M., Yuker, C. E., and Corwin, M. D.**, Control of the louse *Polyplax serrata* with systemic insecticides administered in Silastic rubber implants, *J. Econ. Entomol.*, 60, 1210, 1967.
127. **Leeper, H. and Benson, H.**, The role of polymers in optimizing therapeutic effectiveness of drugs, *Polym. Eng. Sci.*, 17, 42, 1977.
128. **Zaffaroni, A.**, Drug Delivery System, U.S. Patent 3,854,480, 1974.
129. **Salyer, I. O. and Kenyon, A. S.**, Structure and property relationships in ethylene-vinylacetate copolymers, *J. Polym. Sci. Part A*, 9, 3083, 1971.
130. **Zaffaroni, A.**, Special requirements for hormone releasing intrauterine devices, *Acta Endocrinol., Suppl.*, 75(Suppl. 185), 423, 1974.
131. **Swanson, D. R., Wong, P., and Pharriss, B. B.**, Drug delivery systems in contraception, *Excerpta Med.*, 45, 1975.
132. **Ness, R. A.**, Ocular Insert, U.S. Patent 3,618,604, 1971.
133. **Quigley, H. A., Pollack, I. P., and Harbin, T. S.**, Pilocarpine Ocuserts, long-term clinical trials and selected pharmacodynamics, *Arch. Ophthalmol.*, 93, 771, 1975, and references therein.
134. **Chandrasekaran, S. K., Benson, H., and Urquhart, J.**, Methods for achieving and assessing controlled drug delivery — the biomedical engineering approach, in *Sustained and Controlled Release Drug Delivery Systems*, Robinson, J., Ed., Marcel Dekker, New York, in press.
135. The Ocusert® (pilocarpine) Pilo-20/Pilo-40 Ocular Therapeutic System, a monograph written and published by ALZA Corporation, Palo Alto, California, 1974, 7.
136. **Franfelder, F. T., Shell, J. W., and Herbst, S. F.**, Effect of pilocarpine ocular therapeutic systems on diurnal control of intraocular pressure, *Ann. Ophthalmol.*, 8, 1031, 1976.

137. **Sendlebeck, L., Moore, D., and Urquhart, J.,** Comparative distribution of pilocarpine in ocular tissues of the rabbit, *Am. J. Ophthalmol.*, 80, 274, 1975.
138. **Brown, H. S., Meltzer, G., Merrill, R. C., Fisher, M., Ferre, C., and Place, V. A.,** Visual effects of pilocarpine in glaucoma: a comparative study of administration by eyedrops or by ocular therapeutic systems, *Arch. Ophthalmol.*, 94, 1716, 1976.
139. **Longwell, A., Birss, S., Keller, N., and Moore, D.,** Effect of topically applied pilocarpine on tear film pH, *J. Pharm. Sci.*, 65, 1654, 1976.
140. **Gregorio, S. B., Kraal, J. H., Houssain, A. A., and Akaho, E.,** An experimental continuous release device for hexylresorcinol, *J. Dent. Res.*, 56, B183, 1977.
141. **Halpern, B. D., Solomon, O., Kopec, L., Korostoff, E., and Ackerman, J. L.,** Release of inorganic fluoride ion from rigid polymer matrices, in *Controlled Release Polymeric Formulations*, Paul, D. R. and Harris, F. W., Eds., American Chemical Society, Washington, D.C., 1976.
142. **Jacobs, A. G.,** Athletic Mouthguard, U.S. Patent 4,044,762, 1977.
143. **Langer, R. and Folkman, J.,** Polymers for the sustained release of proteins and other macromolecules, *Nature (London)*, 263, 797, 1976.
144. **Langer, R., Brenn, H., Falterman, K., Klein, M., and Folkman, J.,** Isolation of a cartilage factor that inhibits tumor neovascularization, *Science*, 193, 70, 1976.
145. **Boretos, J. W., Detmer, D. E., and Donachy, J. H.,** Segmented polyurethane: a polyether polymer. II. Two years experience, *J. Biomed. Mater. Res.*, 5, 373, 1971.
146. **Doyle, E. N.,** *The Development and Use of Polyurethane Products*, McGraw-Hill, New York, 1971.
147. **Gonzales, M. A., Nematoblaki, J., Guess, W. L., and Autian, J.,** Diffusion, permeation, and solubility of selected agents in and through polyethylene, *J. Pharm. Sci.*, 56, 1288, 1967.
148. **Kalkwarf, D. R., Sikof, M. R., Smith, L., and Gordon, N.,** Release of progesterone from polyethylene devices *in vitro* and in experimental animals, *Contraception*, 6, 424, 1972.
149. **Colter, K. D., Bell, A. T., and Shen, M.,** Control of the pilocarpine release rate through hydrogels by plasma treatment, *Biomater. Med. Devices Artif. Organs*, 5, 1, 1977.
150. **Rowe, R. C., Elworthy, P. H., and Ganderton, D.,** Effect of sintering on the pore structure and strength of plastic matrix tablets prepared from poly(vinyl choride), *J. Pharm. Pharmacol.*, 25, (Suppl.) 112P, 1973.
151. **de Gennaro, M. D., Thompson, B. B., and Luzzi, L. A.,** Effect of cross-linking agents on the release of sodium pentobarbital from nylon microcapsules, and references therein, in *Controlled Release Polymeric Formulations*, Paul, D. R. and Harris, F. W., Eds., American Chemical Society, Washington, D.C., 1976, 195.
152. **Luzzi, L. A., Zoglio, M. A., and Moulding, H. V.,** Preparation and evaluation of the prolonged release properties of nylon microcapsules, *J. Pharm. Sci.*, 59, 338, 1970.
153. **Bottari, F., di Colo, G., Nannipieri, E., Saettone, M. F., and Serafini, M. F.,** Evaluation of a dynamic technique for studying drug-macromolecule interactions, *J. Pharm. Sci.*, 64, 946, 1975.
154. **Lowell, J. R. Jr., Culver, W. H., and de Savigny, C. B.,** Effect of wall parameters on the release of active ingredients from microencapsulated insecticides, presented at the Controlled-Release Pesticide Symp. American Chemical Society, New Orleans, March, 1977.
155. **Sourirajan, S., Ed.,** Reverse Osmosis and Synthetic Membranes, National Research Council of Canada, Ottawa, 1977.
156. **Theeuwes, F. and Yum, S. I.,** Principles of the design and operation of generic osmotic pumps for the delivery of semi-solid or liquid drug formulations, *Ann. Biomed. Eng.*, 4, 343, 1976.
157. **Short, M. P., Abbs, E. T., and Rhodes, C. T.,** Effects of nonionic surfactants on the transport of testosterone across a cellulose acetate membrane, *J. Pharm. Sci.*, 59, 995, 1970.
158. **Short, M. P. and Rhodes, C. T.,** Some investigations of the effect of a non-ionic surfactant on the diffusion of hydrocortisone across a cellulose acetate membrane, *J. Pharm. Pharmacol.*, 23, 2393, 1974.
159. **Short, M. P. and Rhodes, C. T.,** Effects of surfactants on diffusion of drugs across membranes, *Nature (London) New Biol.*, 236, 44, 1972.
160. **Withington, R. and Collett, J. H.,** The transfer of salicylic acid across a cellophane membrane from micellar solutions of Polysorbates 20 and 80, *J. Pharm. Pharmacol.*, 25, 273, 1973.
161. **Gardner, D. L., Fink, D. J., Patanus, A. J., Baytos, W. C., and Hassler, C. R.,** Steroid release via cellulose acetate butyrate microcapsules, in *Controlled Release Polymeric Formulations*, Paul, D. R. and Harris, F. W., Eds., American Chemical Society, Washington, D.C., 1976, 171.
162. **Kiraly, R. J. and Nosé, Y.,** Natural tissue as a biomaterial, *Biomater. Med. Devices Artif. Organs*, 2, 207, 1974.
163. **Ratner, B. D. and Hoffman, A. S.,** Synthetic hydrogels for biomedical applications in *Hydrogels for Medical and Related Applications*, Andrade, J., Ed., American Chemical Society, Washington, D.C., 1976, 1.

164. Andrade, J., Ed., *Hydrogels for Medical and Related Applications,* American Chemical Society, Washington, D.C., 1976.
165. **Wichterle, O. and Lim, D.**, Hydrophilic gels for biological use, *Nature (London),* 185, 117, 1960.
166. **Refojo, M. F. and Yasuda, H.**, Hydrogels from 2-hydroxyethyl methacrylate and propylene glycol monacrylate, *J. Appl. Polym. Sci.,* 9, 2425, 1965.
167. **Wichterle, O. and Chromecek, R.**, Polymerization of ethylene glycol monomethacrylate in the presence of solvents, *J. Polym. Sci. Part C.,* 16, 4677, 1969.
168. **Gouda, J. H., Povodator, K., Warren, T. C., and Prins, W.**, Evidence for a micro-mesomorphic structure in poly(2-hydroxyethyl methacrylate), *Polym. Lett.,* 8, 225, 1970.
169. **Ratner, B. D. and Miller, I. F.**, Interaction of urea with poly(2-hydroxyethyl methacrylate) hydrogels, *J. Polym. Sci. Part A,* 10, 2425, 1972.
170. **Nierzwicki, W. and Prins, W.**, Hydrogels of crosslinked poly(1-glyceryl methacrylate) and poly(2-hydroxypropyl methacrylamide), *J. Appl. Polym. Sci.,* 19, 1885, 1975.
171. **Chen, R. Y. S.**, Diffusion coefficients and swelling behavior of crosslinked poly(2-hydroxyethyl methacrylate), *Polym. Prepr. Am. Chem. Soc. Div. Polym. Chem.,* 15(2), 387, 1974.
172. **Ratner, B. D. and Miller, I. F.**, Transport through crosslinked poly(2-hydroxyethyl methacrylate) hydrogel membranes, *J. Biomed. Mater. Res.,* 7, 353, 1973, and references therein.
173. **Cowsar, D. R., Tarwater, O. R., and Tanquary, A. C.**, Controlled release of fluoride from hydrogels for dental applications, in *Hydrogels for Medical and Related Applications,* Andrade, J., Ed., American Chemical Society, Washington, D.C., 1976, 180.
174. **Abrahams, R. A. and Ronel, S. H.**, Biocompatible implants for the sustained zero-order release of narcotic antagonists, *J. Biomed. Mater. Res.,* 9, 355, 1975.
175. **Drobnik, J., Spacek, P., and Wichterle, O.**, Diffusion of antitumor drugs through membranes from hydrophilic methacrylate gels, *J. Biomed. Mater. Res.,* 8, 45, 1974.
176. **Anderson, J. M., Koinis, T., Nelson, T., Horst, M., and Love, D. S.**, The slow release of hydrocortisone sodium succinate from poly(2-hydroxyethyl methacrylate) membranes, in *Hydrogels for Medical and Related Applications,* Andrade, J., Ed., American Chemical Society, Washington, D.C., 1976, 167.
177. **Good, W. R.**, Diffusion of water soluble drugs from initially dry hydrogels, *In Polymeric Delivery Systems, Midland Macromolecular Mongraph,* Vol. 5, Gordon & Breach, New York, 1978.
178. **Levowitz, B. S., Laguerre, J. N., Calem, W. S., Gould, F. E., Scherrer, J., and Schoenfeld, H.**, Biologic compatibility and applications of Hydron, *Trans. Am. Soc. Artif. Intern. Organs,* 14, 82, 1968.
179. **Lazarus, S. M., Laguerre, J. N., Kay, H., Weinberg, S., and Levowitz, B. S.**, A hydrophilic polymer-coated antimicrobial urethral catheter, *J. Biomed. Mater. Res.,* 5, 129, 1971.
180. **Tollar, M., Stol, M., and Kliment, K.**, Surgical suture coated with a layer of hydrophilic Hydron gel, *J. Biomed. Mater. Res.,* 3, 305, 1969.
181. **Majku, V., Horakora, F., Vymola, F., and Stol, M.**, Employment of Hydron polymer antibiotic vehicle in otolaryngology, *J. Biomed. Mater. Res.,* 3, 443, 1969.
182. **Kaufman, H. E., Votila, M. H., Casset, A. R., Wood, T. O., and Varnell, E. D.**, Medical uses of soft contact lenses, in *Soft Contact Lenses,* Gassett, A. R. and Kaufman, H. E., Eds., C. V. Mosby, St. Louis, 1972, chap. 22.
183. **Podos, S. M., Becker, B., Asseff, C., and Hartstein, J.**, Pilocarpine therapy with soft contact lenses, *Am. J. Ophthalmol.,* 73, 336, 1972.
184. **Scott, H., Kronick, P. L., May, R. C., Davis, R. H., and Balin, H.**, Construction and properties of hydrogel-graft-coated copper-bearing intrauterine devices for rabbits, *Biomater. Med. Devices Artif. Organs,* 1, 681, 1973.
185. **Oster, G. K.**, Chemical reactions of the copper intrauterine device, *Fertil. Steril.,* 23, 18, 1972.
186. **Hagenfeldt, K.**, Intrauterine contraception with the copper-T device. I. Effect on trace elements in the endometrium, cervical mucus and plasma, *Contraception,* 6, 37, 1972.
187. **Davis, B. K., Noske, I., and Chang, M. C.**, Reproductive performance of hamsters with polyacrylamide implants containing ethinylestradiol, *Acta Endocrinol.,* 70, 385, 1972.
188. **Davis, B. K.**, Diffusion in polymer gel implants, *Proc. Natl. Acad. Sci. U.S.A.,* 71, 3120, 1974.
189. **Balin, H., Halpern, B. D., Davis, R. H., Akkapeddi, M. K., and Kyriazis, G. A.**, Prostaglandin delivery by cervical dilator, *J. Reprod. Med.,* 13, 208, 1971.
190. **Luttinger, M. and Cooper, C. W.**, Improved hemodialysis membranes for the artificial kidney, *J. Biomed. Mater. Res.,* 1, 67, 1967.
191. **Bradley, W. G. and Wilkes, G. L.**, Some mechanical property considerations of reconstituted collagen for drug release supports, *Biomater. Med. Devices Artif. Organs,* 5, 159, 1977.
192. **Meares, P., Ed.**, *Membrane Separation Processes,* Elsevier, Amsterdam, 1976.

193. **Michaels, A. S.**, Therapeutic systems for controlled administration of drugs: a new application of membrane science, in *Permeability of Plastic Films and Coatings,* Hopfenberg, H.P., Ed., Plenum Press, New York, 1975, 409.
194. **Graybiel, A., Knepton, J., and Shaw, J.**, Prevention of experimental motion sickness by scopolamine absorbed through the skin, *Aviat. Space Environ. Med.*, 47, 1096, 1976.
195. **Shaw, J. E., Schmitt, L. G., McCauley, M. E., and Royal, J. W.**, Transdermally administered scopolamine for prevention of motion sickness in a vertical oscillator, *Clin, Pharmacol. Ther.*, 21, 117, 1977.
196. **Chandrasekaran, S. K., Michaels, A., Campbell, P., and Shaw, J.**, Scopolamine permeation through human skin *in vitro, AIChE J.*, 22, 828, 1976.
197. **Chandrasekaran, S. K. and Shaw, J.**, Design of transdermal therapeutic systems, in *Contemporary Topics in Polymer Science,* Vol. 2, Pearce, E. M. and Schaefgen, J. R., Eds., Plenum Press, New York, 1977, 291.
198. **Shaw, J. E., Chandrasekaran, S. K., Michaels, A. S., and Taskovich, L.**, Controlled transdermal delivery, *in vitro* and *in vivo,* in *Relevance of Animal Models to Human Dermatopharmacology and Toxicology,* Maibach, H., Ed., Churchill-Livingston, London, 1975. 138.
199. **Scheuplein, R. J.**, Permeability of the skin: a review of major concepts and some new developments, *J. Invest. Dermatol.*, 67, 672, 1976.
200. **Michaels, A. S., Chandrasekaran, S. K., and Shaw, J.**, Drug permeation through human skin: theory and *in vitro* experimental measurement, *AIChE J.*, 21, 985, 1975.
201. **Zaffaroni, A.**, Bandage for the Administration of Drug by Controlled Metering Through Microporous Materials, U.S. Patent 3,797,494, 1974.
202. **Zaffaroni, A.**, Microporous Drug Delivery Device, U.S. Patent 3,993,072, 1976.
203, **Zaffaroni, A.**, Novel Drug Delivery Device, U.S. Patent 3,993,073, 1976.
204. **Sternberg, S., Bixler, H. J., and Michaels, A. S.**, Microporous Encapsulating Particles for Controlled Release or Immobilization of Reactants, U.S. Patent 3,639,306, 1972.
205. The DIFFUCAP® — CHRONODRUG® Method, EURAND, s.p.a., Milano, 1973.
206. **Sciarra, J. J. and Gidwani, R.**, Release of various ingredients from aerosols containing selected film-forming agents, *J. Soc. Cosmet. Chem.*, 21, 667, 1970.
207. **Sciarra, J. J. and Patel, S. P.**, In vitro release of therapeutically active ingredients from polymer matrixes, *J. Pharm. Sci.*, 65, 1519, 1976.
208. **Donbrow, M. and Friedman, M.**, Permeability of films of ethylcellulose and PEG to caffeine, *J. Pharm. Pharmacol.*, 26, 148, 1974.
209. **Donbrow, M. and Friedman, M.**, Enhancement of permeability of ethylcellulose films for drug penetration, *J. Pharm. Pharmacol.*, 27, 633, 1975.
210. **Donbrow, M. and Samoelov, Y.**, Controlled release of tripelennamine and other drugs dispersed in ethyl cellulose PEG films, *J. Pharm. Pharmacol.*, 28(Suppl.), 23P, 1976.
211. **Donbrow, M. and Samoelov, Y.**, Zero order release of drugs from polymeric films, *J. Pharm. Pharmacol.*, 28(Suppl.), 21, 1976.
212. **Yates, F. E., Benson, H., Buckles, R., Urquhart, J., and Zaffaroni, A.**, Engineering development of therapeutic systems: a new class of dosage forms for the controlled delivery of drugs, in *Advances in Biomedical Engineering,* Vol. 5, Brown, J. H. and Dickson, J. F., III, Eds., Academic Press, New York, 1975.
213. **Pharriss, B. B.**, Steroid delivery systems for contraception, *J. Reprod. Med.*, 17, 91, 1976.

Chapter 5

MULTILAYERED LAMINATED STRUCTURES

Agis F. Kydonieus and Alberto R. Quisumbing

TABLE OF CONTENTS

I. INTRODUCTION

To survive, man must protect himself, his crops, and his possessions from pest attack. At the same time, he has the important responsibility to utilize measures that are not detrimental to the environment. Failure to heed this responsibility, although inadvertent, has caused unforeseen disruptive effects and engendered an increased awareness of the danger of contaminating our environment. Thus, the public has often supported the environmentalists in their efforts to convince the government to enact legislation regulating the use of chemicals or industrial pollutants that may endanger man and his surroundings.

A. Hazards Associated with Use of Pesticides

Pesticides are often the target of environmentalists. The sole reliance upon and, at times, excessive use of these chemicals have caused serious problems, such as the development of pesticide-resistant strains of insects, increased outbreaks of secondary pests, population imbalances of beneficial insects, nontarget effects of fish and wildlife, off-flavors in foods, and, at times, long-time contamination of soils. Also, fluorocarbon gases in aerosol formulations of pesticides and in such common household items as hair sprays and room fresheners are believed to attack and diminish the ozone layer of the upper atmosphere. A reduction of the ozone layer would permit greater penetration of the ultraviolet rays of the sun to the earth and cause damage to ourselves and our ecosystem.

Ironically, this dependence on chemicals was the result of progress. Less than four decades ago, synthetic insecticides were introduced, and the public was led to believe that an almost pest-free environment was possible. With DDT and the revolutionary synthetic chlorinated hydrocarbons and organophosphates, pest control practices were reduced largely to a single system: chemical control.

In the late 1940s, the basic functions of an entomologist were to determine the killing potential of the new synthetic compounds and to fit them into a pest control program. Largely overlooked in this research was the effect of the new chemicals on the biology of the pest and on the ecosystem, which interacted with the insect and affected its control.

However, despite the opposition and limitations to the use of pesticide chemicals, there are often few practical alternatives to their use for avoiding serious damage, especially when an immediate reduction of pest populations becomes necessary. Currently, most responsible authorities agree that reliance on pesticides must continue, perhaps not as the only measure, but as an important part of an integrated pest management program.

There is also a need to judge the use of each chemical independently. It would be a gross mistake to replace chemical control with sophisticated, but untested or ineffective, pest management techniques. More effort must be exerted to find new means of increasing the efficiency of existing tools for pest control as well as to find new approaches. The traditional methodology of utilizing pesticides must be modified to make it both effective and safe to the user and to nontarget biota.

B. The Need for Alternative Products

The number of available organochlorine insecticides is steadily dwindling as a result of restrictions imposed by the U.S. Environmental Protection Agency on the use of the very persistent and supposedly carcinogenic compounds. Former "old reliables" — DDT, aldrin, dieldrin, chlordane, and heptachlor — are being replaced with the less persistent organophosphates, carbamates, and synthetic pyrethroids. This trend was prompted by the great stability of the organochlorine residues in the environment

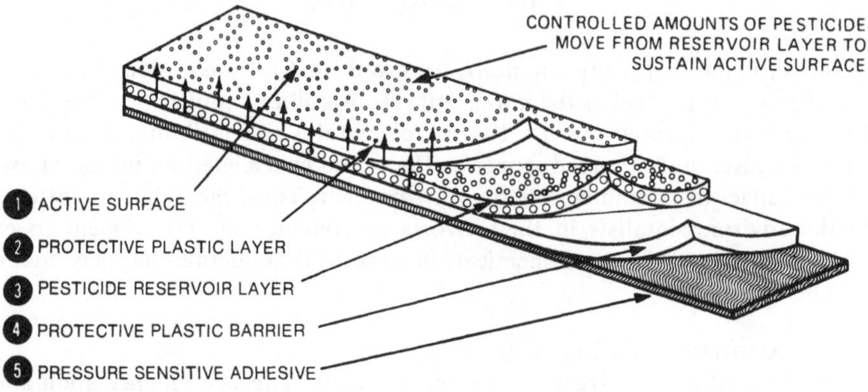

FIGURE 1. Schematic illustration of Hercon® multi-layered controlled release dispenser.

and the inability of many organisms to metabolize or degrade these materials; thus, causing residues to be found in air, food, water, soil, wildlife, and even in our bodies.

Unfortunately, the replacements are generally more toxic, and they are effective for a much shorter period of time, thereby necessitating multiple applications of the toxicant to effect an adequate degree of control. These shortcomings have given added impetus to a search for formulations that can extend the effectiveness of these materials, reduce the need for reapplications, facilitate handling, and provide greater safety to formulators and applicators, as well as to the environment. Clearly, as a means of meeting these goals, only optimal amounts of insecticide should be dispensed in combating target species.

C. Objectives

The application of a system that controls the release of active chemical to a steady optimum level and thereby minimizes or eliminates the usual problems associated with conventional methods of dispensing toxic chemicals seems to be a logical approach. This chapter will describe a system for the controlled release of chemicals through multilayered laminated structures and show how this knowledge is being utilized to improve pest management and other practices.

II. THE HERCON® SYSTEM: A GENERAL DESCRIPTION

The Hercon® controlled release system of dispensing active ingredients consists of several layers of laminated polymeric material. Basically, the Hercon® dispenser has the active ingredient implanted and protectively sealed in a layer between outer plastic layers. A schematic illustration of a typical Hercon® delivery system is shown in Figure 1.

The specially formulated inner layer serves as a reservoir for the active ingredient. An almost unlimited range of active agents can be implanted in this sealed reservoir, e.g., insecticides, insect pheromones or sex attractants, fragrance oils, room air fresheners, and antibacterial agents. Some of these multilayered dispensers have now been marketed for several years.

Essentially, the active agent (whether it be pesticide, fragrance oil, or insect pheromone) migrates continually, due to an imbalance of chemical potential, from the reservoir layer through one or more initially inert outer layers to the exposed surface, which is thereby rendered biologically or physiochemically active. At the surface, the active ingredient is removed by volatilization, thermal or ultraviolet degradation, alkaline or acid hydrolysis, or mechanical contact by humans, insects, rainfall, wind, or other agents.

Construction and composition of the dispenser varies with the active agent used and with the release rate and effective lifespan desired. One or both outer surfaces of the dispenser may be made active by having one or both layers permeable to the active ingredient.

The form of the dispenser may also be varied to aid in dispensing the active material. Thus, the original sheets may be cut into strips, ribbons, wafers, flakes, confetti, or even sprayable granules or powders.

III. MATHEMATICS AFFECTING TRANSPORT

The concentration of the implanted or stored chemical (i.e., diffusant) and the composition and/or construction of the plastic layer components control the release rate of the chemical.

The nonporous, homogeneous polymeric films, usually referred to as solution-diffusion membranes, are used in Hercon® dispensing systems. Typical examples are silicone rubber, polyethylene, polyvinyl chloride (PVC), and nylon films.

The diffusant is able to pass through the membrane material in the absence of pores or holes by absorption, solution and diffusion down a gradient of thermodynamic activity, and desorption. The permeation process is governed primarily by Henry's law and Fick's first law.[1,2] Thus, Fick's first law states that the rate of transfer of diffusing substance through unit area of a section is proportional to the concentration gradient measured normal to the section, i.e.,

$$J = -D \frac{dC_m}{dx} \tag{1}$$

where J is the flux in g/cm^2-sec, C_m is the connection of diffusant or permeant in the membrane in g/cm^3, dC_m/dx is the gradient in concentration, and D is the diffusion coefficient of the diffusant in the membrane in cm^2/sec.

When Henry's law applies, as with diffusing gases, the concentration of the dissolved gas is proportional to the pressure.

$$C_m = kp \tag{2}$$

where C_m is the concentration of the gaseous diffusant, k is Henry's law solubility constant, and p is pressure.

A. The Steady State

A schematic illustration of the concentration gradient across a three-layer laminate is shown in Figure 2. For purposes of the illustration, it has been assumed that the distribution coefficient is less than unity for Barrier Membrane I and more than unity for Membrane II.

The concentration just inside the membrane surface (C_m) can be related to the concentration in the reservoir (C) by the expressions,

$$\begin{aligned} C_{m(o)} &= KC_{(o)} \text{ at the upstream surface } (x = o) \\ C_{m(\ell)} &= KC_{(\ell)} \text{ at the downstream surface } (x = \ell) \end{aligned} \tag{3}$$

where K is a distribution coefficient and analogous to the familiar liquid-liquid partition coefficient.

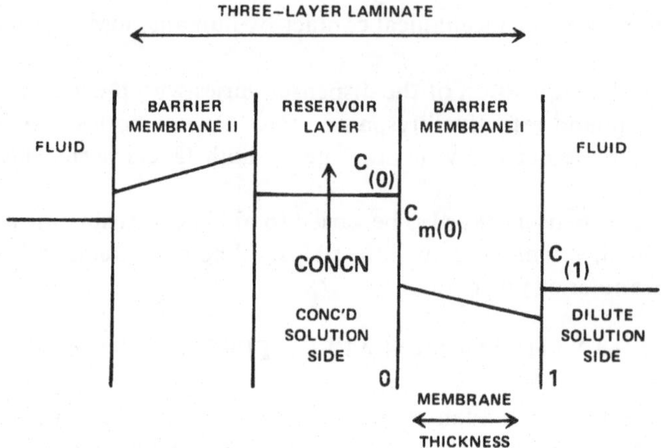

FIGURE 2. Schematic representation of the concentration gradient across a three-layer laminate.

Throughout the following, diffusion coefficients and distribution coefficients are assumed to be constant. This is a safe assumption for most polymer-diffusant (permeant) systems. Thus, in the steady state or when there is a constant gradient across the membrane, Equation 1 can be integrated to give

$$J = D\,\frac{C_{m(o)} - C_{m(\ell)}}{\ell} = D\,\frac{\Delta C_m}{\ell} \tag{4}$$

where ℓ is the thickness of the membrane. Since the concentration within the membrane is usually not known, Equation 5 is frequently written as

$$J = \frac{dM_t}{dt} = \frac{A\,D\,K\,\Delta C}{\ell} \tag{5}$$

where M_t is the mass of agent released, dM_t/dt is the steady state release rate at time t, A is the surface area of the barrier membrane, DK is the membrane permeability, and ΔC is the difference in concentration, i.e., $C^{(o)} - C(\ell)$, between the reservoir concentration and the fluid concentration adjacent to the barrier membrane.

It is noteworthy that the rate of release is proportional to diffusivity (a kinetic constant) and to the distribution coefficient (a thermodynamic constant). Equation 6 can be integrated between the limits,

$$M_t = o \qquad t = o$$

$$M_t = M_t \qquad t = t$$

to give

$$M_t = \frac{A\,D\,K\,\Delta C}{\ell}\,t \tag{6}$$

When the distribution coefficient between the reservoir layer and the barrier membrane is much smaller than unity, as is the case of Membrane I in Figure 2, the system will have excellent release kinetics, and the release rate can be maintained constant for extended periods of time. This situation is termed pseudo-zero order delivery. Equa-

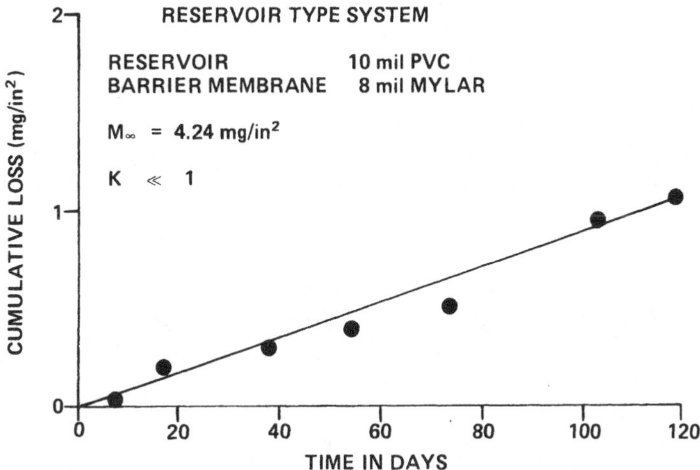

FIGURE 3. Release rate of MGK R-874 fly repellent from a Hercon® PVC-polyester multi-layered dispenser. (From Kydonieus, A. F., in *Controlled Release Pesticides,* ACS Symp. Series 53, Scher, H. B., Ed., American Chemical Society, Washington, D.C., 1977. With permission.)

tion 7 governs the process, and a straight line is obtained when M_t, the mass of agent released, is plotted against time t. This situation is exemplified in Figure 3, which shows the linear release rate of MGK R-874, a fly repellent formulation, from a PVC-polyester Hercon® dispensing system.

B. The Unsteady State

When the distribution coefficient between the reservoir layer and the barrier membrane is approximately unity, or larger than unity, as is the case with Membrane II of Figure 2, the Hercon® system will approximate the "dissolved system", i.e., the reservoir-membrane system forms a single homogeneous polymeric film. The concentration in the reservoir will not remain constant but will fall continuously with time. When the reservoir nears depletion or when the reservoir initially contains a less-than-saturated solution of permeant, the system remains continuously under unsteady state conditions, and the mass of agent released varies as a function of time. This is described as first order delivery.

The transport equations have been described by several investigators.[3-7] Two useful equations are the early time approximation, which holds over the initial portion of the release curve,

$$\frac{M_t}{M_\infty} = 4\left[\frac{D_t}{\pi \ell^2}\right]^{1/2} \qquad 0 \leqslant \frac{M_t}{M_\infty} \leqslant 0.6 \tag{7}$$

and the late time approximation, which holds over the final portion of the curve,

$$\frac{M_t}{M_\infty} = 1 - \frac{8}{\pi^2}\exp\left[\frac{-\pi^2 D_t}{\ell^2}\right] \qquad 0.4 \leqslant \frac{M_t}{M_\infty} \leqslant 1.0 \tag{8}$$

Equation 8 suggests that a plot of mass of agent released vs. time will give a parabolic curve. This was the case for the PVC-PVC system which dispensed the MGK R-

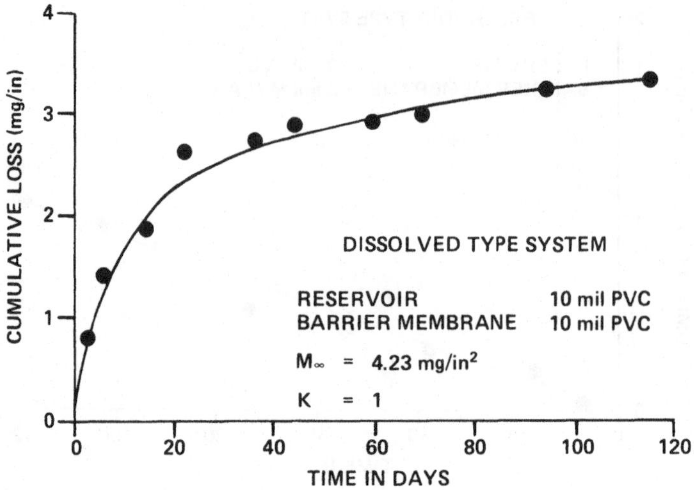

FIGURE 4. Release rate of MGK R-874 fly repellent from a Hercon® PVC-PVC multi-layered dispenser (loss vs. time). (From Kydonieus, A. F., in *Controlled Release Pesticides,* ACS Symp. Series 53, Scher, H. B., Ed., American Chemical Society, Washington, D.C., 1977. With permission.)

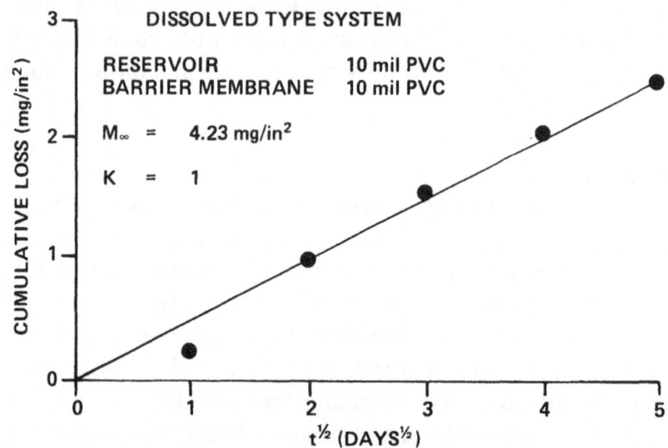

FIGURE 5. Release rate of MGK R-874 fly repellent from a Hercon® PVC-PVC multi-layered dispenser (loss vs. time$^{1/2}$). (From Kydonieus, A. F., in *Controlled Release Pesticides,* ACS Symp. Series 53, Scher, H. B., Ed., American Chemical Society, Washington, D.C., 1977. With permission.)

874 fly repellent[8] (Figure 4). When the same release rate data are plotted against $t^{1/2}$ (Figure 5), a linear curve is obtained[8] in accordance with Equation 8.

IV. FACTORS AFFECTING RELEASE RATE

Molecular and structural factors control the release of active ingredients from Hercon® laminated membrane structures. For a given combination of polymer structure and active agent, where energy to free rotations, free volume, and intermolecular attractions are constant, two parameters that play an important role in regulating the rate of transfer are reservoir concentration and membrane thickness. Their effect is quantified in Equations 7, 8, and 9.

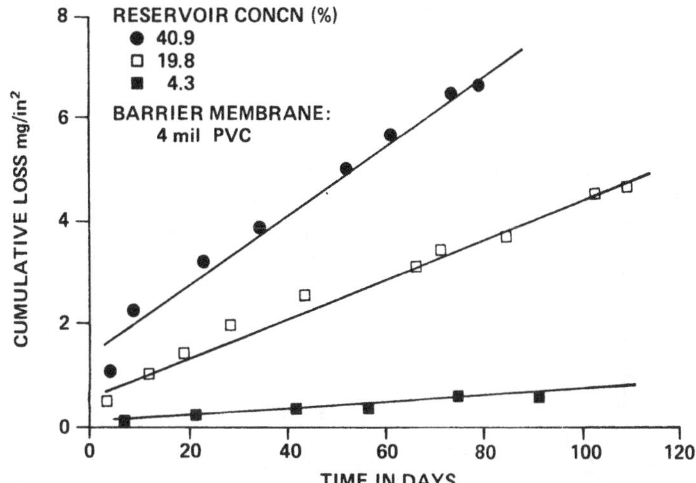

FIGURE 6. Effect of varying reservoir concentrations on the release rate of the insecticide chlordane from a Hercon® multi-layered dispenser. (From Kydonieus, A. F., in *Controlled Release Pesticides,* ACS Symp. Series 53, Scher, H. B., Ed., American Chemical Society, Washington, D.C., 1977. With permission.)

Other related factors affecting the transport of active ingredients through Hercon® membranes include polymer stiffness, co-diffusants, molecular weight of diffusant, and chemical functionality.

Diffusivity, D, and reservoir/membrane distribution coefficient, K, are directly proportional to the permeation rate. In polymers, diffusivity is strongly influenced by the molecular weight of the diffusant and by the stiffness of the backbone of the polymeric membrane. Simply speaking, the diffusant molecule will have to reorient several segments of polymer chain to allow its passage from site to site. The higher the molecular weight, the more the segments have to be reoriented to permit passage; the stiffer the polymer (such as those that are glassy or highly crystalline), the more difficult it is for the segments to undergo large reorientations. Thus, variables or additives that affect polymer membrane stiffness, such as codiffusants that soften, plasticize, or partially dissolve the membrane, will also affect diffusivity and permeation rate.

The reservoir/membrane distribution coefficient can be estimated from the solubility parameter of the diffusant, which can be calculated using Hildebrand's solubility theory. Solubility parameters and dissolution are strongly affected by molecular weight and the chemical functionality of the molecule, i.e., hydrogen bonding and polarity. When the solubility parameters for the diffusant and polymer membrane are similar, the polymer will be soluble in the diffusant. "Like dissolves like" is a good rule of thumb.

The various factors influencing the release rate are discussed below. For purposes of discussion, the reservoir of the laminated multilayered dispensers in the following examples is made of flexible PVC.[8]

A. Reservoir Concentration

The release rates of the insecticide, chlordane, and the repellent, deet (*N,N*-diethyl-*m*-toluamide), are illustrated in Figures 6 and 7. In both cases, zero order release rates were obtained. A closer look indicates that doubling the concentration in the reservoir does not double the mass of agent released. This deviation is more pronounced at higher concentrations, presumably because intermolecular attractions of the diffusant

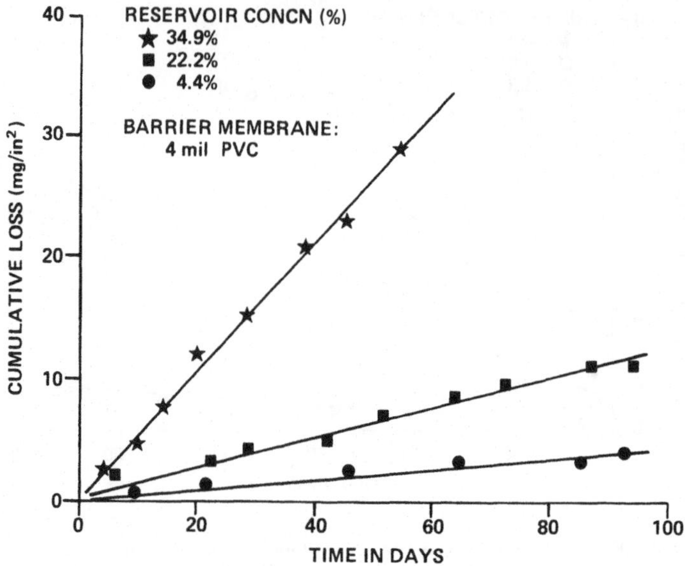

FIGURE 7. Effect of varying reservoir concentrations on the release rate of the repellent deet from a Hercon® multi-layered dispenser. (From Kydonieus, A. F., in *Controlled Release Pesticides*, ACS Symp. Series 53, Scher, H. B., Ed., American Chemical Society, Washington, D.C., 1977. With permission.)

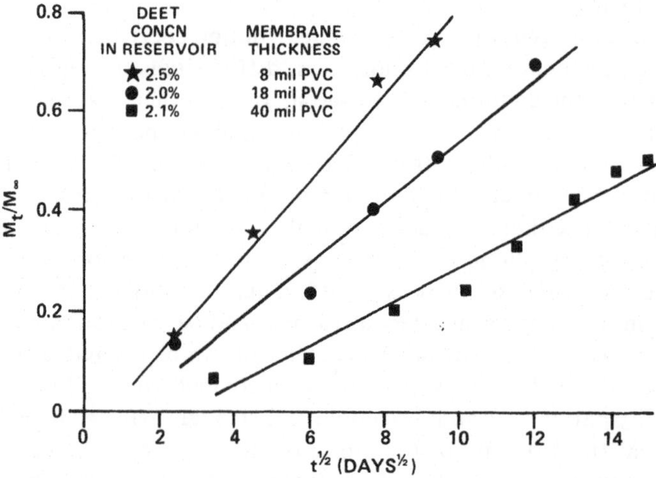

FIGURE 8. Effect of varying barrier film thickness on the release rate of the repellent deet from a Hercon® multi-layered dispenser. (From Kydonieus, A. F., in *Controlled Released Pesticides*, ACS Symp. Series 53, Scher, H. B., Eds., American Chemical Society, Washington, D.C., 1977. With permission.)

molecules increase exponentially with concentration. With the exceptions of these minor differences, the data, in general terms, follow Equations 6 and 7.

B. Membrane Thickness

Both Equations 6 and 7 indicate that the mass of active ingredient released or emitted is inversely proportional to the thickness of the barrier membrane. Figures 8 and 9

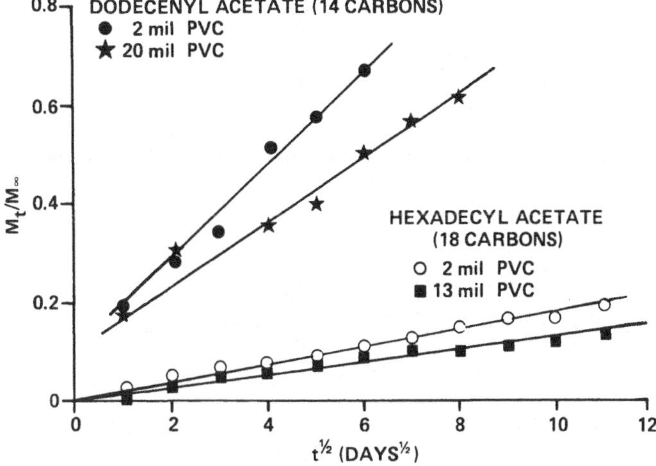

FIGURE 9. Effect of varying barrier film thickness on the release rate of insect pheromones dodecenyl acetate and hexadecyl acetate from a Hercon® multi-layered dispenser. (From Kydonieus, A. F., in *Controlled Release Pesticides,* ACS Symp. Series 53, Scher, H. B., Ed., American Chemical Society, Washington, D.C., 1977. With permission.)

TABLE 1

Amount of Active Agent Transported from the Flexible PVC Reservoir Layer to Polymers of Increasing Backbone Stiffness[a,8]

		Amount transported after 20 weeks (ppm)				
Active agent	Initial amount in reservoir (ppm)	Flexible PVC	Rigid PVC	Polypropylene	Nylon	Polyester (Mylar)
Captan	500	250	109	36	3	0
Malathion	24,000	12,000	9,700	498	23	8
Zineb	4,000	1,600	568	67	62	9

[a] All polymeric films were 5-mil thick.

From Kydonieus, A. F., *Controlled Release Pesticides,* ACS Symp. Series 53, Scher, H. B., Ed., American Chemical Society, Washington, D.C., 1977, 152. With permission.

show the release rates of deet and the insect pheromones, dodecenyl acetate and hexadecenyl acetate, respectively, through dispensers with membranes of varying thickness. In Figure 8, there was a faster release rate of deet through 8-mm PVC than through the thicker 40-mm PVC barrier membrane. A similar trend is shown in Figure 9, where greater amounts of insect attractants were released through the 2-mm PVC barrier films than through the 20- and 13-mm PVC films.

C. Polymer Stiffness

The distribution coefficient, K, in several polymer membranes of different backbone stiffness was studied by adhering large reservoir layers of PVC to these membranes. The transport of active agent from the reservoir to the membrane was determined by separating the layers and determining the amount of active agent by chemical analysis. Table 1 shows the amount of the three active agents (captan, zineb fungicide, and malathion insecticide) transported from the flexible PVC to the various polymers. For

TABLE 2

Amount of Active Agent Transported from the Flexible PVC Reservoir Layer to Controlling Polymeric Barrier (i.e., Rigid PVC, Nylon, and Polypropylene) as a Function of Time[a,8]

		Amount transported to polymers in time (ppm)								
		Rigid PVC			Nylon			Polypropylene		
Active agent	Initial amount in reservoir (ppm)	2	7 (weeks)	20	2	7 (weeks)	20	2	7 (weeks)	20
Malathion	24,000	6,300	9,000	9,700	6	12	23	387	334	498
Captan	500	29	55	109	—	—	—	1	26	36
Zineb	4,000	303	619	568	67	60	62	—	—	—

[a] All polymeric films were 5-mil thick.

From Kydonieus, A. F., *Controlled Release Pesticides,* ACS Symp. Series 53, Scher, H. B., Ed., American Chemical Society, Washington, D.C., 1977, 152. With permission.

TABLE 3

Property Improvement Imparted by the Diffusion of Active Agent from the Reservoir Layer to Polymers of Increasing Backbone Stiffness[a,8]

		Performance of activated polymers					
Active agent	Property	Flexible PVC	Rigid PVC	Acrylic	Polypropylene	Nylon	Polyester
Vinazene	Germicidal	7.0[b]	12	—	15	0	0
Captan	Antibacterial	99.9+[c]	99.9+	88.2	97.8	42.3	48.2
Ethoquad	Antistatic	10,000[d]	4,200	2,100	1,100	420	73

[a] All polymeric films were 5-mil thick.
[b] Zone of inhibition (mm).
[c] Reduction of bacteria over untreated control (NYS-63)(%).
[d] Surface resistivity (ohms/in.^{2})

From Kydonieus, A. F., *Controlled Release Pesticides,* ACS Symp. Series 53, Scher, H. B., Ed., American Chemical Society, Washington, D.C., 1977, 152. With permission.

these three agents, the amount transported into the membranes became progressively smaller as the material was varied from flexible PVC to rigid vinyl, polypropylene, nylon, and Mylar®, which are in the order of increasing backbone stiffness. Distribution coefficients were not calculated for the data in Table 1 because it was not certain that equilibrium had been established after 20 weeks when the experiment was terminated. The amounts of active chemical transported at several time intervals are shown for a few of the membranes in Table 2.

The diffusion into films of different backbone stiffness was also monitored by the activity imparted to these films (Table 3). Active agents used were captan, vinazene germicide, and Ethoquad® antistatic agent.

Generally, the activity imparted by the active agents was highest with flexible PVC, followed in decreasing order by rigid PVC, acrylic, polypropylene, nylon, and polyester. This order is similar to the results observed in tests discussed in Table 1, i.e., product effectiveness was reduced as backbone stiffness increased, undoubtedly owing to the migration of lesser amounts of chemicals to the exposed surface.

TABLE 4

The Effect of Co-diffusants on the Antistatic Properties of Carpets[8]

Carpet fiber	Antistatic agent	Co-diffusant	Active agent conc (% in reservoir)	Volume resistivity (ohms)
Nylon	Advastat 50	Phenol	14.7	7.0×10^{8}
Nylon	Advastat 50	None	14.8	2.2×10^{9}
Nylon	None	None	None	1.1×10^{11}
Polyester	Advastat 50	Dowanol EPh	17.6	7.0×10^{8}
Polyester	Advastat 50	None	16.9	8.0×10^{10}
Polyester	None	None	None	1.1×10^{11}

From Kydonieus, A. F., *Controlled Release Pesticides*, ACS Symp. Series 53, Scher, H. B., Ed., American Chemical Society, Washington, D.C., 1977, 152. With permission.

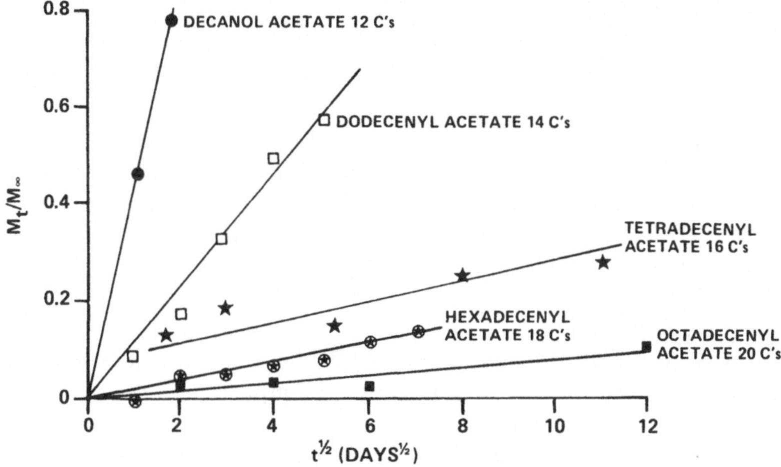

FIGURE 10. Effect of molecular weight on mass of insect pheromones released from Hercon® multi-layered dispensers. (From Kydonieus, A. F., in *Controlled Release Pesticides*, ACS Symp. Series 53, Scher, H. B., Ed., American Chemical Society, Washington, D.C., 1977. With permission.)

D. Co-Diffusants

Chemical agents capable of altering the polymer structure (e.g., stiffness) have a pronounced effect on the diffusion of active chemicals. These carrier chemicals must have the ability to swell, soften, and/or aid in the dissolution of the polymer matrix to facilitate the transport of the active chemicals through the Hercon® membranes.

The effect of co-diffusants on the rate of chemical transfer is demonstrated in Table 4 by the improvement of the antistatic properties of Advastat 50 on carpets through the addition of the co-diffusants, phenol and ethylene glycol phenyl ether (Dowanol EPh).[9] Three- and 100-fold improvements in resistivity were obtained with the co-diffusants in nylon and polyester carpet fibers, respectively.

E. Molecular Weight

The molecular weight of the diffusant is also an important factor in its rate of transfer because molecular weight is inversely related to diffusivity. Graham's law states that diffusion is inversely proportional to the square root of the molecular weight of the diffusant, i.e., as the molecular weight of the diffusant increases, the release rate decreases.

Figure 10 illustrates the effect of molecular weight on the transport of five insect

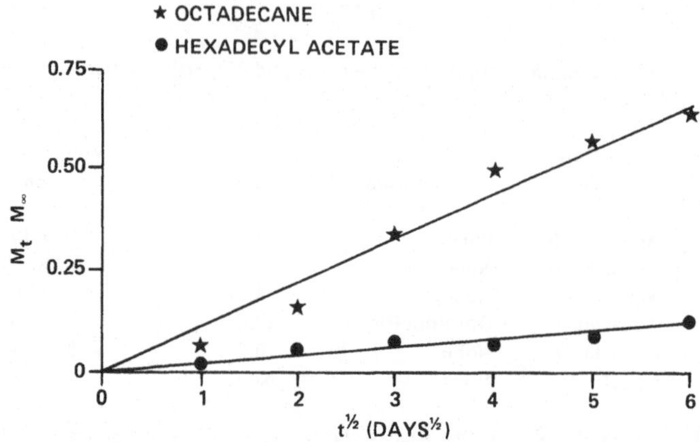

FIGURE 11. Effect of chemical functionality on release rate of 18-carbon chemicals, octadecane and hexadecyl acetate, from Hercon® multilayered dispensers. (From Kydonieus, A. F., in *Controlled Release Pesticides*, ACS Symp. Series 53, Scher, H. B., Ed., American Chemical Society, Washington, D.C., 1977. With permission.)

pheromones through a 2-mm flexible PVC membrane. Although not of the same homologous series, the pheromones studied were all acetates ranging from 12-carbon decanol acetate (lowest molecular weight) to 20-carbon octadecenyl acetate (highest molecular weight). The emission rates of the lower molecular weight pheromones were greater than those of their higher molecular weight analogs.

F. Chemical Functionality

The "like dissolves like" rule is equally applicable in the polymer area. Dissolution of the polymer matrix by the diffusing molecules is important in the transport process because it increases the distribution coefficient.

The ability of two sets of chemicals with 16- and 20-carbon atoms to dissolve the flexible PVC reservoir of the dispenser has been investigated. It was shown that the characteristic functional group in a chemical had varying effects on the release rates (Figures 11 and 12). With 18-carbonatom chemicals, hexadecenyl acetate had a slower release rate than octadecane (Figure 11). A similar trend was observed with 16-carbon-atom chemicals; epoxyhexadecane with its epoxy group had a much faster release rate than hexadecanone and hexadecenal (Figure 12).

V. APPLICATION OF THE HERCON® SYSTEM

The Hercon® dispensing system is utilized in a number of products that are now either marketed or in advanced stages of product development.[8,10,11] Herculite Staph-Chek®, Insectape® or Roach-Tape® insecticidal strips, Luretape™ pheromone dispensers for use in monitoring, trapping, and mating suppression of insects, and Scentstrip® air fresheners are now commercially available. Three other insecticidal products, the Lure 'N Kill™ Flytape contact-action fly killers, granules for soil insects, and fabrics for protecting stored products from insects, are in the developmental stage.

The advantages and benefits derived from multilayered laminated systems are described in the following discussion of specific controlled release commercial and developmental products.

A. Staph-Chek® Antibacterial Fabrics

Herculite Products, Inc. (formerly Herculite Protective Fabrics Corporation), the

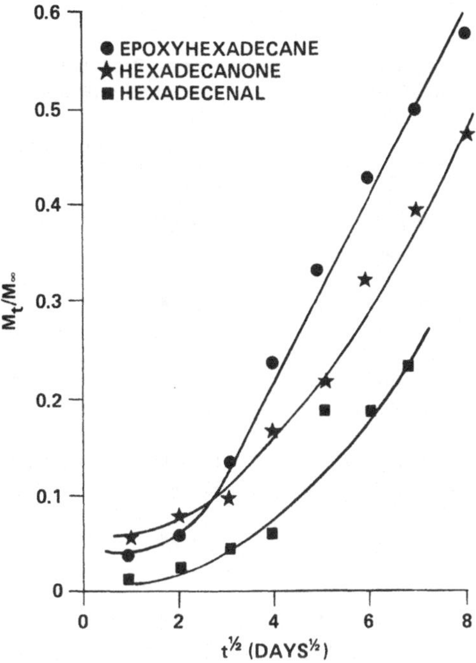

FIGURE 12. Effect of chemical functionality on release rate of 16-carbon chemicals, epoxyhexadecane, hexadecanone, and hexadecenal, from Hercon® multi-layered dispensers. (From Kydonieus, A. F., in *Controlled Release Pesticides*, ACS Symp. Series 53, Scher, H. B., Ed., American Chemical Society, Washington, D.C., 1977. With permission.)

parent firm of the Hercon® Group, initiated its controlled release technology when Staph-Chek® antibacterial nylon-reinforced vinyl fabrics were introduced in 1962. The product, sold under the Herculite trademark, has a reservoir layer containing two fungicides that give the product its antibacterial qualities. With the protected reservoir system, mattress coverings made of Staph-Chek® fabrics effectively control *Staphylococcus* and *Klebsiella* bacteria throughout the life of the fabric.

The development of Staph-Chek® fabrics was a direct result of the growing attention given to improving health care in this country. In addition to providing better and safer hospital products, the fabric also helped cut the skyrocketing costs of health care by providing longer lasting products. Thus, Staph-Chek® long-life fabrics on mattresses also provided protection from cross-infection and better flame resistance. The fabric also has an antistatic surface and is self-deodorizing to a degree previously unobtainable.

1. Longer Lasting Fabrics

On the average, mattresses made of Staph-Chek® last three times longer than the old-style cotton-ticked mattresses. The use of a nylon-reinforced fabric resulted in reduced costs by providing a more durable product, one resistant to wear and tear. This has meant fewer replacements, easier maintenance, and lower housekeeping costs.

Since 1962, Herculite has supplied leading manufacturers with enough Staph-Chek® fabrics for well over one million mattresses, and demand for the fabric has been increasing. After acceptance in hospitals as mattress tickings, Staph-Chek® fab-

TABLE 5

Bacteriostatic Evaluation of Multilayered Antibacterial Fabrics: Product Effectiveness as Measured by Percent Reduction in Bacterial Count of *Staphylococcus aureus* and *Klebsiella pneumoniae* cultures

Sample	Bacterium	Initial inoculation	Bacterial count (organisms/in²)	Reduction over inert control (%)
Inert control	*S. aureus*	126,000	4,990,000	—
Antibacterial vinyl wall cover (Hercon® process)	*S. aureus*	126,000	<1,000	99.9
Standard vinyl wall cover[a] (two samples)	*S. aureus*	126,000	1,040,000 1,235,000	79.1 75.2
Inert control	*S. aureus*	212,000	6,675,000	—
Staph-Chek® fabric	*S. aureus*	212,000	9,500	99.9
Inert control	*K. pneumoniae*	156,000	4,080,000	—
Antibacterial vinyl wall cover (Hercon® process)	*K. pneumoniae*	156,000	<1,000	99.9
Standard vinyl wall cover[a] (two samples)	*K. pneumoniae*	156,000	6,435,000 7,020,000	0 0
Inert control	*K. pneumoniae*	230,000	5,680,000	—
Staph-Chek® fabric	*K. pneumoniae*	230,000	31,000	99.5

[a] Standard vinyl wall cover did not contain any antibacterial agent.

rics were also used to fabricate pillows, cubicle and shower curtains, and for wall and floor covering in burn-treatment rooms.

The addition of an antistatic agent to the Staph-Chek® product offers an added dimension to its utility. The Staph-Chek® Anstat™ fabrics reduce explosion hazards by preventing the buildup or retention of dangerous electrostatic charges.

2. Antibacterial Effectiveness

The multilayered antibacterial fabrics were evaluated to determine their bacteriostatic activity against gram-positive *Staphylococcus* and gram-negative *Klebsiella* (Table 5). Bacterial counts were reduced in Staph-Chek® fabrics and in a white bacteriostatic vinyl covering prepared by the Hercon® process, while counts remained unchecked in the inert control.[11] Both *Staphylococcus* and *Klebsiella* species failed to grow on the treated multilayered fabrics (over 90% reduction compared to the control).

The bacteriostatic evaluation was done by the NYS-63 method, which involves the inoculation of samples with a broth culture of the organism. The samples were allowed to incubate at 37°C for 24 hr after inoculation and for an additional 24 to 48 hr after selecting the samples for the bacterial counts.

The NYS-63 tests also showed that cultures of bacteria, such as *Brevibacterium ammoniagenes, Streptococcus pyogenes. Alcaligenes fecalis, Escherichia coli* and *Proteus mirabilis,* failed to grow on Staph-Chek® material.[12]

B. Insectape® Insecticidal Strips

Hercon Insectape® insecticidal strip is a cockroach control product designed for use in residences and in commercial, institutional, and industrial establishments. The active agent in the reservoir of the 1-in. by 4-in. strip may be one of three insecticides long recognized as effective cockroach killers, i.e., propoxur, 2(1-methylethoxy)phenol methylcarbamate (trade name Baygon®), Diazinon®, *O,O*-diethyl *O*-(2-isopropyl-6-

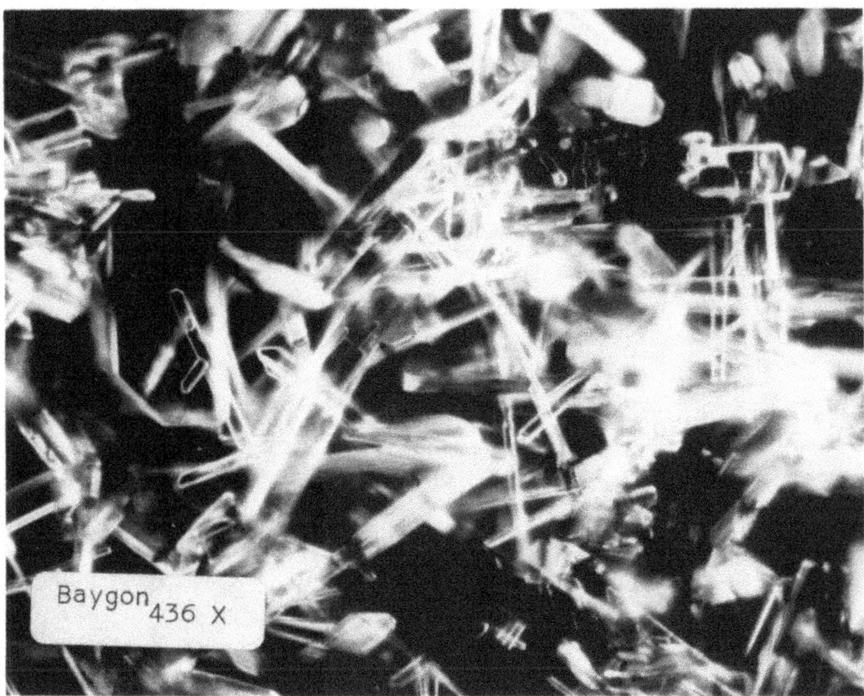

FIGURE 13. Crystals of propoxur insecticide on active surface of Hercon® Insectape® insecticidal strip (Magnification ×436).

methyl-4-pyrimidinyl)-phosphorothioate, and chlorpyrifos, *O,O*-diethyl *O*-(3,5,6-trichloro-2-pyridyl)-phosphorothioate (trade name Dursban®).

The three insecticides are also effective in controlling pests other than cockroaches. Field studies showing good Insectape® efficacy against various species of flies, ants, and spiders have been conducted.[12]

Benefits of the Hercon® multilayered Insectape® include (1) increased duration of effectiveness, which reduces the need for frequent reapplication, (2) elimination of toxic fumes through the use of involatile contact insecticides, (3) judicious placement of the strips to exploit known cockroach behavior, (4) reduction in toxicity of active ingredient in the dispenser compared to conventionally applied insecticide, and (5) increased safety to the user because toxicants are premeasured, thus eliminating spills and mixing errors.

1. Nonfumigant Action

As a departure from insecticidal products that release toxic fumes even when target pests are absent, Insectape® was designed to kill the insects by contact; thus, the insecticide is used only as needed. The cockroaches are killed after they pick up a lethal dose of the toxicant by walking across the active surface of the strip. Figure 13 is an electron microscope photograph showing the propoxur crystals on the active surface of the strip. Since there are no toxic fumes, and the insecticide remains inside or on the surface of the strip, the possibility of contamination by gas or spray drift is nonexistent with Insectape®.

The barrier films used in Insectape® were selected to maintain a very low rate of diffusion of the pesticide under normal-use conditions, thus minimizing the exposure of pesticides to pets and humans. Figure 14 shows the release rate of chlorpyrifos at three different temperatures from an Insectape® formulation.[13] At room temperature

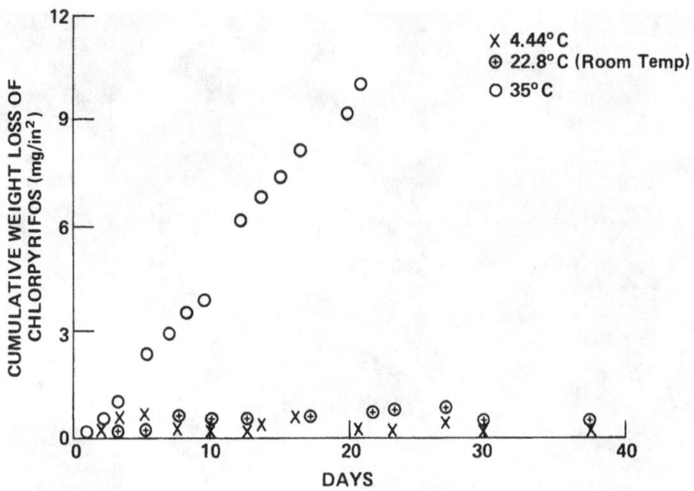

FIGURE 14. Release rate of chlorpyrifos from Hercon® Insectape® insecticidal strip at three different temperatures.

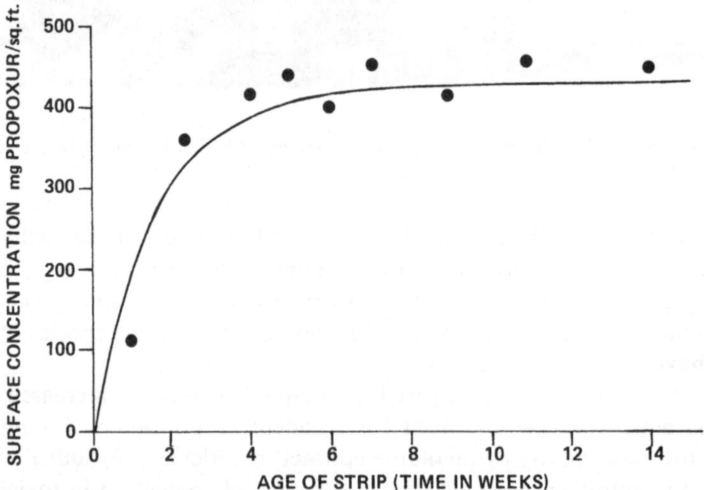

FIGURE 15. Release of propoxur from Hercon® Insectape® 4% Propoxur insecticidal strip at room temperature.

or less, the insecticide remains on the strip and is not released into the immediate environment. This nonfumigating delivery system makes feasible the servicing of pet shops, hospitals, nursing homes, and similar locations where sprays are restricted or impractical.

The nonfumigating characteristic is related to the release rate of the insecticide from the dispenser. In the Insectape® product with 4% propoxur (commercially available in supermarkets and grocery stores as Johnson Wax® Raid® Roach-Tape), the concentration of the insecticide on the surface remained approximately constant, about 400 to 450 mg of active ingredient per square foot, shown in Figure 15. The amount on the surface was determined by scraping the insecticide from the surface of several strips of varying age. The surface concentration equilibrates 2 to 3 weeks after production and is expected to remain at this level until the amount in the reservoir is substantially depleted.

TABLE 6

Results of Tests to Determine Whether Hercon® Insectape® Insecticidal Strips Containing Propoxur Emitted Toxic Fumes

Treatment	Percent knockdown + dead cockroaches after specified exposure time (days)		
	1	7	10
Cockroaches permitted to contact strip	62	73	75
Cockroaches not permitted to contact strip	2	7	8
Control	0	0	0

From Moore, R. C., *Pest Control*, 44(6), 37, 1976. With permission.

In separate studies, Smith[14] and Moore[15] determined that the fumigant action of Insectape® containing propoxur and Diazinon® against German cockroaches, *Blatella germanica* (L.), was negligible. In a laboratory study, Smith placed cockroaches with food and water in a 6.2-1 desiccator with an insecticidal strip at the bottom. The insects were not allowed to come in contact with the insecticidal surface of the strip by placing a screen 2 in. above the strip. The cockroaches remained active 1 week later, indicating an absence of vapor toxicity.

Moore used choice boxes to determine whether the Hercon® strips emitted toxic fumes. Choice boxes are used to detect repellency as well as the relative insecticidal potency of potential cockroach control products. The insects are given a choice between a dark compartment with the toxicant and a lighted compartment with food and water. In one study, Moore allowed the cockroaches to move normally from the lighted to the dark compartments. In another, a fine mesh screen blocked the hole between the two compartments; thus, cockroaches could not contact the strip in the darkened area. As shown in Table 6, the percentage of knockdown among cockroaches not permitted to contact the insecticidal strip was not significantly different from the control (wherein cockroaches were allowed to move freely in a choice box where no insecticide was used). If the product had fumigant action, cockroach mortality would have been high in the compartment separated from the treated area by the screen.

2. Pesticide Residue Studies

Lack of fumigant action was also demonstrated by residue analyses made following application of the chlorpyrifos-containing insecticidal strips.[16] Although the safety of the active ingredient had already been established, a test was conducted to demonstrate that the use of Insectape® strips does not result in a harmful residue in air, water, or food products, such as milk, meat, bread, and tomatoes. Gas-liquid chromatography analyses showed that the amounts of chlorpyrifos in the exposed samples were less than the minimum detectable values (0.010 ppm in air and 0.025 ppm in water and the food products).

The results of the residue analyses correlated with data obtained from release rate studies; only an insignificant amount of chlorpyrifos was lost from the dispenser held for 155 days at 21.1 to 26.7°C. More specifically, after 5 months, a single Insectape®

strip lost only 10 mg (2.5 mg/in.^{2}) or 6% of the original 143 mg chlorpyrifos. The low release rate is very important if the product is to be used in food processing plants and similar sensitive areas.

3. Toxicology

Since the insecticide in the strip is enclosed in the reservoir layer, the Hercon® insecticidal strips are susbstantially less toxic than the pure toxicant. An acute oral toxicity study in rats was performed with strips containing 10% Diazinon® by weight. The strips were ground to a 100-μm powder and administered by oral intubation as a 10% weight/volume suspension in aqueous 1% methylcellulose. Ten rats were dosed at levels of 1500, 3000, and 5000 mg/kg. Observations for mortality were made 1 and 6 hr after dosing and daily thereafter for 2 weeks. Although reported oral LD_{50} values for Diazinon® were 300 to 850 mg/kg,[17] no deaths were observed at any of the three dose levels.[13]

4. Insectape® Effectiveness

Because propoxur, Diazinon®, and chlorpyrifos have been proven to be effective products, they were evaluated in laboratory and field trials as an active ingredient of Insectape®. Accordingly, laboratory and field studies were conducted to prove that the Hercon® system can be used to effectively dispense, for prolonged periods, the three insecticides.

a. Laboratory Timed-Contact Studies

Tests were conducted to determine the contact time lethal to cockroaches for each Insectape® formulation.[15] With the propoxur-containing Insectape®, a 2-sec contact was sufficient to knock down and kill all German cockroaches within 50 min. Longer contact, i.e., 10 sec or more, resulted in faster knock-down rates. As observed with conventional sprays,[18] the propoxur formulation was faster acting than Insectape® with Diazinon® or chlorpyrifos. Moore[15] reported that Insectape® formulations of Diazinon® and chlorpyrifos needed an exposure greater than 20 to 60 min to produce significant cockroach moribundity and mortality.

b. Laboratory Forced-Contact Studies

In a replicated study conducted by USDA entomologists[19] and reported by Kydonieus, Smith, and Hyman,[13] the efficacy of Insectape® prototype formulations containing 21% Diazinon® was compared with that of plywood panels sprayed with an acetone solution of Diazinon® at the rate of 100 mg of Diazinon® per square foot by exposing cockroaches to their surfaces for 30 min. After exposure, the cockroaches were transferred to clean petri dishes, and knock down-kill observations were recorded after 0, 24, and 48 hrs. Insectape®-treated and sprayed panels were then allowed to age and retested at various intervals to determine residual effectiveness of each treatment.The results (Table 7) showed that the sprayed panels became totally ineffective after 2 weeks, while Insectape® strips with Diazinon® were 100% effective for 52 weeks. Similar results were obtained with propoxur and chlorpyrifos Insectape® formulations.

c. Laboratory Choice-Box/Mock-Up Closet Studies

In a series of tests done by University of California entomologists, additional evidence of the residual effectiveness of Insectape® was obtained. In one study conducted by Ebeling and Reierson and reported by Kydonieus, Smith, and Hyman,[13] the university researchers applied Diazinon®-containing strips at the rate of 19.5 in.^{2}/0.5 ft^2 to floorwall intersections of a choice box; the vertical intersections and cover were

TABLE 7

Efficacy of Hercon® Controlled Release Dispenser of Diazinon® Determined by Residue Testing at USDA ARS Insects Affecting Man Research Laboratory[a]

Treatment	30-min. exposure time + 48-hr observation, percent mortality after specified aging time[b] (weeks)					
	Start	1	2	8	36	52
Hercon® dispenser with 21% Diazinon®	100	100	100	100	100	100
Diazinon-acetone spray residue on plywood (100 mg/ft^2)	100	90	0	—	—	—

[a] Evaluations made by D. E. Weidhaas, S. Burden, and L. R. Swain, Jr. (USDA).
[b] Percent mortality was obtained using 20 adult male German cockroaches as observed 48 hours after a 30-min exposure to Hercon® strip or spray residues.

From Kydonieus, A. F., Smith, I. K., and Hyman, S., *Proc. 1975 Int. Controlled Release Pesticides Symp.*, Harris, F. W., Ed., Wright State University, Dayton, 1975, 60. With permission.

TABLE 8

Performance of Hercon® Insectape® Containing Diazinon® in Choice Boxes for Up to 258 Days of Aging at Ambient Atmospheric Conditions

Age of strip (days)	Cockroach mortality within specified time (days) after start of test (%)						
	1	2	3	4	5	6	7
Fresh	5	68	85	93	95	97	97
102	63	93	98	98	98	98	100
131	92	95	98	98	100		
154	65	ND[a]	100				
183	95	98	100				
258	ND[a]	ND[a]	100				

[a] No data collected.

From Kydonieus, A. F., Smith, I. K., and Hyman, S., *Proc. 1975 Int. Controlled Release Pesticides Symp.*, Harris, F. W., Ed., Wright State University, Dayton, 1975, 60. With permission.

not treated. Thirteen hr after strip placement, 20 adult male cockroaches were put into each of three choice boxes, and cockroach mortality was recorded daily for a week. Afterwards, the dead cockroaches were removed, and the choice box covered; the product was again tested in the same way after 102, 131, 154, 183, and 258 days of aging under ambient conditions.

Table 8 shows the performance of fresh and aged Insectape® strips containing Diazinon®. Fresh Insectape® gave 97% cockroach mortality within 1 week. The aged strips (up to 258 days) gave 100% kill within 1 week.

In another series of tests, conventional cockroach control spray materials were applied to the dark compartments of choice boxes and tested after periods of aging up

TABLE 9

Performance of Deposits of Various Insecticides After Aging for up to 85 Days Under Ambient Atmospheric Conditions

Conventional formulation[a] treatment	Age of Deposit (days)	Total cockroach knockdown within specified time (days) after start of test (%)						
		1	2	5	6	10	18	25
Diazinon® 1%	Fresh	57	98	100				
	20	75	93	98	98	98	98	100
	85	ND[b]	0	4	4	62	95	100
Chlorpyrifos 0.5%	Fresh	18	60	98	98	100		
	20	80	97	100				
	85	ND[b]	32	39	58	87	97	97
Propoxur 1%	Fresh	17	50	83	100			
	20	80	97	98	98	98		
	85	ND[b]	7	13	18	22	28	62
Untreated	Fresh	0	0	2	2	2	2	ND[b]
	20	2	3	8	8	17	21	26
	85	2	3	5	5	7	10	15

[a] Applied 3 ml dilute emulsion to the masonite floor of the dark compartment with a pipette. Brushed deposit over the substrate with a damp camel hair brush.

[b] No data collected.

From Kydonieus, A. F., Smith, I. K., and Hyman, S., *Proc. 1975 Int. Controlled Release Pesticides Symp.*, Harris, F. W., Ed., Wright State University, Dayton, 1975, 60. With permission.

to 85 days. The liquids were prepared in water from commercial products, and 3 ml of 1% Diazinon® as active ingredient, 0.5% chlorpyrifos as active ingredient, or 1% propoxur as active ingredient were applied; 3 ml approximates the amount applied when the material is sprayed to runoff under normal conditions. The deposits were allowed to dry for 24 hr prior to the start of standard choice-box evaluations. The same deposits were again tested after 20 and 85 days of aging under ambient conditions.

Table 9 shows that, unlike the controlled release Insectape® products, percent knock down obtained by the liquid deposits gradually decreased with time. Insectape® products gave 100% mortality within 1 week despite aging the strips for 258 days, while knock down by conventional formulations decreased when deposits aged for just 85 days.

Insectape® formulations were also compared with conventional materials under simulated field conditions using mock-up closets in the laboratory.[20] The closets were 90 × 35 × 30.5-cm wooden boxes with an 18-V DC electric barrier near the top of each box to prevent the insects from escaping; cockroaches in the closets were also provided with food and water. Inverted 0.95-l cardboard cartons were also placed in the closet as a harborage for the insects.

A significantly greater mortality over untreated controls was obtained by placing Insectape® directly inside a harborage carton.[20] Data also showed that stimuli which increase cockroach movement increased kill and that Insectape® could be significantly influenced by food, harborage, light, and the amount of area treated. These observations emphasized that for satisfactory cockroach control, proper placement of strips

TABLE 10

Summary of Periodic Testing to Determine Duration of Residual Effectiveness of Hercon® Insectape® containing 10% Propoxur, by Weight, in Controlling Adult German Cockroaches

Age of strips[a] (months)	No. of tests	Average knockdown and kill (%) within specified time (min)								
		15	30	45	60	75	90	120	180	250
Fresh	11	19	92	100						
3	11	10	63	89	98	100				
4	11	0	38	85	96	99	99	99	99	100
5	11	4	57	90	98	100				
6	11	47	95	100						
8	11	1	15	56	81	91	95	98	100	
11	11	19	85	100						
12	10	4	32	53	72	100				

Note: Each test consisted of ten German cockroaches placed in an 80-in^3 glass cage with food, water, and a single 1 in. × 4 in. insecticidal strip.

[a] Age of strips is the number of months strips were exposed under actual apartment conditions.

From Kydonieus, A. F., Quisumbing, A. R., Smith, I. K., Baldwin, S., and Conroy, R. A., Hercon Technical Bulletin 26, Herculite Protective Fabrics Corporation, New York, 1977. With permission.

in harborages and across travel pathways is critical in maximizing insect contact with the strip surface.

d. Laboratory Residual Action Studies

Strips exposed in actual home apartments were also bioassayed in the laboratory to determine the duration of residual control. Table 10 shows that propoxur strips, exposed for one year, continued to provide 100% cockroach mortality.[21] Knockdown rates were similar regardless of the duration that strips were exposed in the field.

e. Field Tests: Insectape® vs. Sprays

Field studies also demonstrated the long-term effectiveness and superiority of Insectape® over conventional spraying. Table 11 compares residual spray treatments and application of strips containing propoxur, Diazinon®, or chlorpyrifos.[21] From 71 to 86% reduction in cockroach infestation was achieved with the strips after 1 month and from 87 to 93% after 5 months. Conventional spraying produced a 65% reduction after 2 to 3 months and only 34% after 5 months, indicating that cockroaches were numerous again.

Insectape® performance is directly affected by the size of the area treated. This has been observed in mock-up closet studies as well as in field tests.[20] Since Insectape® use involves strategic placements of the strips, field effectiveness is also affected by the level of cockroach infestation, the abundance or lack of cockroach harborages, construction of the infested site, and the degree of sanitation.

In field studies conducted in public housing facilities in Memphis, Tenn., apartment construction or the presence of cracks and crevices appeared to be a more significant factor relative to the level of infestation than sanitation.[22] Thus, there was no significant correlation between sanitation and German cockroach precounts; however, kitch-

TABLE 11

Summary of Field Tests Comparing Efficacy of Conventional Residual Spray Treatment Only and Hercon® Insectape® Only, with Either Propoxur (Baygon®), Diazinon®, or Chlorpyrifos As the Active Ingredient of the Insecticidal Strips

		Insectape® insecticidal strips only Active ingredients		
Variables	Spray only	Baygon	Diazinon®	Chlorpyrifos
No. sites tested	19	38	27	28
No. cockroaches/site	47.74	30.92	36.33	12.18
Average Infestation Reduction (%)				
After 1—1.5 months	43.31	85.78	71.10	78.70
After 2—3 months	65.28	92.07	92.59	88.42
After 3—5 months	34.37	86.69	90.18	92.91

From Kydonieus, A. F., Quisumbing, A. R., Smith, I. K., Baldwin, S., and Conroy, R. A., Hercon Technical Bulletin 26, Herculite Protective Fabrics Corporation, New York, 1977. With permission.

ens with cracks and narrow spaces between furniture and walls generally had more cockroaches than sites with few or no observable nesting sites.

With heavy infestations, Insectape® is more effective when used in conjunction with general spraying or dusting treatments that quickly reduce pest population levels.[15,22] Frequently unsatisfactory control observed in sites with poor structural condition and a general lack of good maintenance cleaning results from use of an insufficient number of strips. Treating each harborage is possible, but may be impractical, especially if treatment sites are of poor construction and the number of nesting sites numerous. In a combined Insectape®-plus-spray program, strips placed in preferred harborages maintain cockroach counts at low or zero levels for periods longer than ordinarily obtained by sprays.

f. Field Tests: Insectape® in Localized Infestations

Use of Insectape® is ideal when a localized cockroach problem exists and spraying is not advisable. In tests conducted at a hospital in Essex County, New Jersey, the strips maintained excellent control throughout the testing period.[23] The application of one strip per nightstand or three strips per desk sufficed to rid the furniture of cockroaches for more than 5 months (Table 12). With this kind of protection, harborages in hospital rooms can be effectively treated without causing major disruptions or discomfort to patients.

g. Field Tests: Cockroach Harborages

To ensure proper placement of strips, which is essential in obtaining sufficient cockroach-insecticidal surface contact, insect behavior was studied and preferred harborages specified in a directions guide. Figure 16 shows the recommended strip locations and common nesting sites in a residential kitchen. The illustrated guide is part of the label directions.

Generally, the Insectape® strips are placed in warm, dark, and moist out-of-sight locations, where the insects are known to crawl or hide. The common treatment sites include waterpipes, where they enter walls or floors, corners of bases of cabinets, motor compartments, and behind kickplates of such appliances as refrigerators, washers, and stoves. Field data show that control of cockroach populations can be more easily obtained when strips are placed in every spot suggested in the directions guide.[21,22]

TABLE 12

Effectiveness of Hercon® Insectape® Containing 10% Propoxur, by Weight, in Controlling Localized German Cockroach Infestations in Hospital Room Furniture

Treated furniture	No. tests	No. strips/test	Precount	No. cockroaches at specified time (months) after start of test		
				1	2	5.5
Insecticidal strips applied during precount						
Nightstand	6	1	42	1	2	0
Desk	5	3	26	0	1	3
Insecticidal strips applied 1 month after start of test						
Nightstand	3	1	20	24	0	2
Desk	2	3	14	24	0	0

From Mampe, C. D., personal communication.

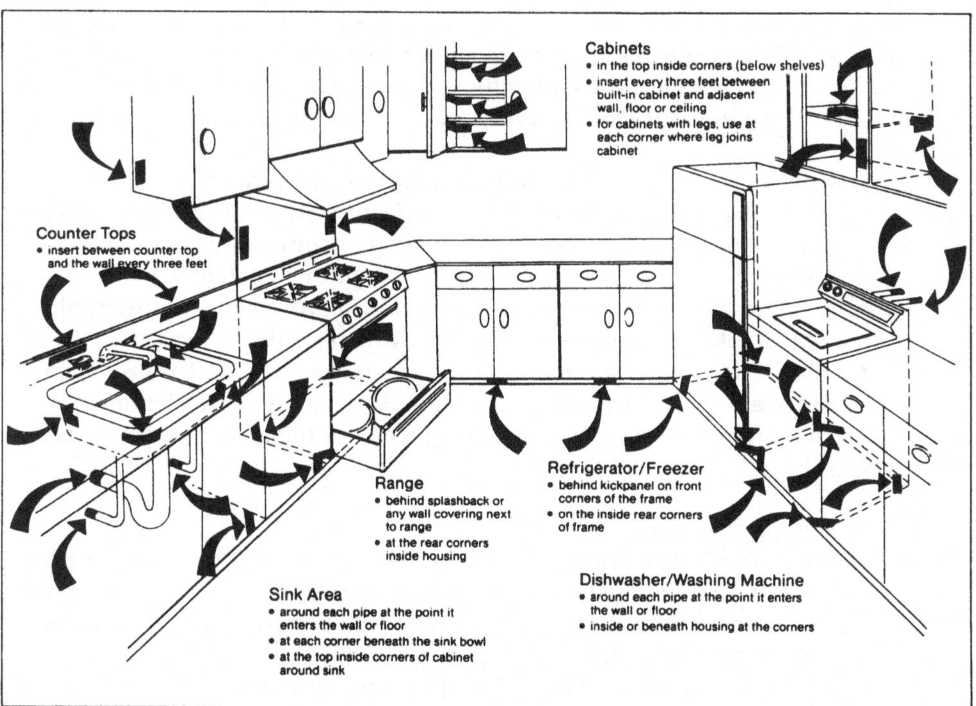

FIGURE 16. Illustration of typical residential kitchen showing recommended insecticidal strip locations for optimum cockroach control.

5. Economic Advantages

Obviously, extending the duration of insecticidal activity enables commercial applicators to utilize personnel and equipment more efficiently. The time intervals between service calls are increased, while clients are provided with continuous cockroach control, and the pest control operator or exterminator is able to both reduce his overall labor costs and expand his market coverage.

C. Luretape™ Insect Attractant Dispensers

The Hercon® technology has also been used to dispense insect attractants or pheromones, a technique useful in pest management. The utilization of these innocuous chemical attractants signals a marked departure from sole reliance on insecticides for agricultural pest control, and interest in sex attractants as an important tool in the effort to develop alternative methods of insect pest control has been steadily mounting.[24,25] With the highly specific action of these compounds, damage to nontarget species can be avoided, and contamination of the environment by insecticides reduced.

1. Background Information

Initially the U.S. Department of Agriculture preferred testing a great variety of chemicals to find a synthetic insect attractant instead of trying to isolate and identify an attractive chemical from an insect. A half dozen attractants were found in this way and put to good use.

For example, since the early 1960s, traps baited with synthetic lures for the Mediterranean fruit fly, *Ceratitis capitata* (Wiedemanno), oriental fruit fly, *Dacus dorsalis* Hendel, and melon fly, *D. cucurbitae* Coquillett, have been deployed about ports of entry to the U.S. to detect any accidental importation of these serious pests. Infestations of all three species were found, and the pests were eradicated quickly with the aid of traps which showed where and when to apply the appropriate counter measures. The USDA stated that the traps saved them millions of dollars in potential eradication costs, but saved the agricultural community and the public far more in terms of the increased prices of food that would have resulted if it became necessary to live with these pests.

With the great value of insect attractants established, the possibility of similar use of the compounds that the insects themselves manufactured (the so-called pheromones) became most appealing. For a long time, the isolation, identification, and synthesis of these materials were hampered by the minute amounts of these potent substances present in insects. In the late 1960s, however, remarkable improvements in spectrophotometric, chromatographic, instrumental, and specialized techniques facilitated the isolation and identification of minute amounts of pheromones. Progress has been so rapid in the past five years that pheromones or chemicals believed to be pheromones are now known for hundreds of insect species,[26] and many of these are of great economic importance.

2. Protection of Pheromones

The Luretape™ pheromone dispenser is usually manufactured as a three-layer plastic strip with the insect attractant concentrated in the inner layer. As in other Hercon® dispensers, the two outer plastic layers act to regulate the release of the pheromone, i.e., the pheromone gradually diffuses from the reservoir through the outer layers to the exposed surface from which the pheromone evaporates into the atmosphere. At the same time, the outer layers protect the inner layer and the bulk of the pheromone from the degradative action of the weather. Sensitive compounds, e.g., aldehydes or unsaturated chemicals, are amply protected. Thus, the dispenser not only regulates release of the pheromone, it also protects the compound from degradation by sun and weather.

In addition to protecting unstable and volatile chemicals and controlling their release rate, the Hercon® pheromone dispenser offers economic benefits in terms of long-term effectiveness, elimination of reapplication costs, and reduction in the amount of active ingredient used.

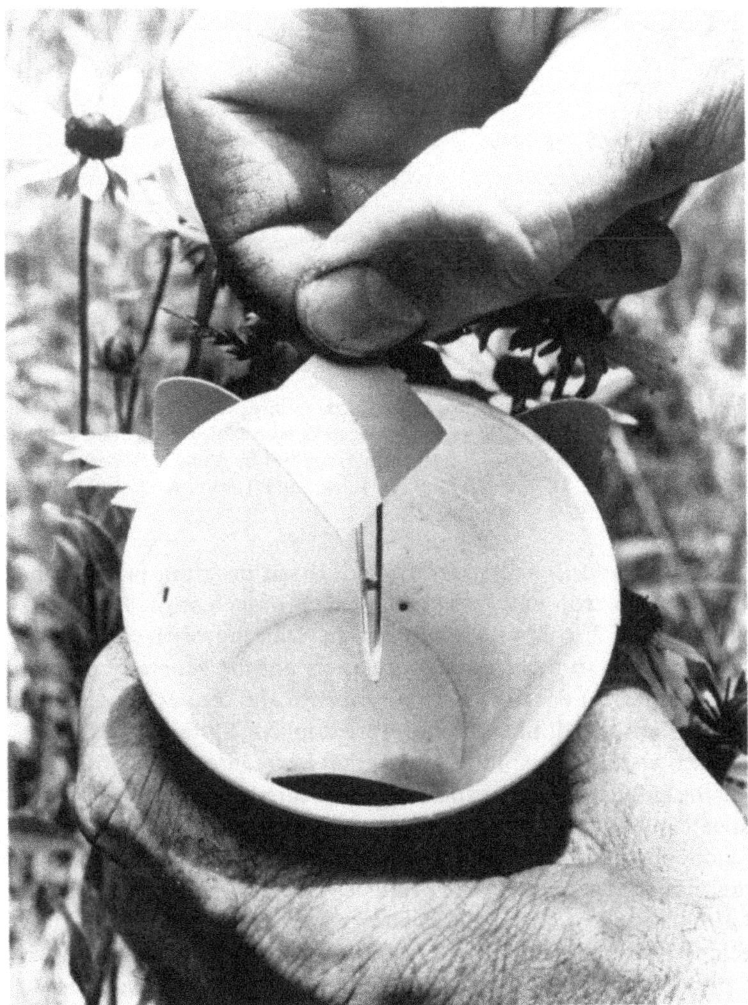

FIGURE 17. A cylindrical trap is baited by a Hercon® Luretape® pheromone dispenser which is held in place by a clip. This trap was used in gypsy moth trapping programs. The male moths are lured by the pheromone into entering the trap where they adhere to the glue on the inside of the trap (U.S. Department of Agriculture photo).

3. Field Application of Luretape™ Dispensers

In an integrated pest management program, pheromones may be used in lure-baited traps to signal the presence of a specific insect pest. This high specificity makes them valuable in monitoring pest populations. After the level of infestation is detected and estimated, control measures may be applied, either to prevent the spread of the insect or only to those areas where the pest is found. The traps are also used to time the application of control measures and thereby cut down on the number of treatments needed.

In addition to survey and detection, pheromones may also be used as direct control measures by (1) mating suppression through insect disorientation, and (2) mass trapping. Figure 17 shows a cylindrical paper trap baited by a 1-in.^{2} Hercon® pheromone dispenser. Pheromones may also be used with "trap crops"; the bait (or attractant dispenser) is applied to plants (the trap crop) in a specific area, which is then sprayed with an insecticide after the target insects have been lured to the trap crop area.

The trap crop technique was evaluated by the USDA as part of the cotton boll wee-

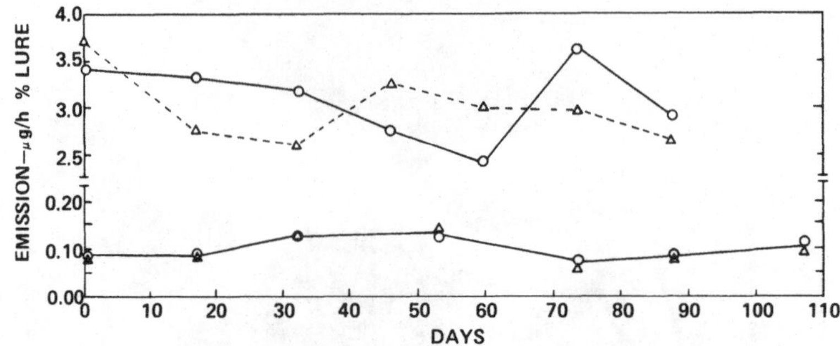

FIGURE 18. Rate of emission of Hercon® Luretape® dispensers of disparlure (lower graph) and percentage lure found in these dispensers (upper graph) after aging at room temperature. Points for laminates with sealed ends designated by circles; points for laminates with open ends designated by triangles. (From Beroza, M., Paszek, E. C., DeVilbiss, D., Bierl, B. A., and Tardif, J. G. R., *Environ. Entomol.*, 4, 712, 1975. With permission.)

vil, *Anthonomus grandis* Boheman, trial eradication program in 1972 and 1973.[27] A four-row strip of trap crop cotton was planted in every cotton field under test. The trap crop was planted 2 to 3 weeks prior to the planting of the producer's cotton in order for the trap crop to be larger, fruit earlier, and be generally more attractive to the insect. To further enhance the attractiveness of the trap crop, dispensers of grandlure, the cotton boll weevil pheromone, were applied. Systemic and foliar insecticide treatments were then made to kill the weevils that aggregated on the trap crop.

Results of the pilot trap crop experiment indicated that a combination of the bait and the systemic insecticide treatment effectively controlled overwintering boll weevils. The researchers observed that almost all of the weevils that entered the field before the appearance of flower buds on the producer's cotton moved onto the trap crop and were killed after feeding on the systemic insecticide-treated plants.[28]

The Luretape™ dispenser has already been tested for trapping and/or mating disruption of the cotton boll weevil, the gypsy moth, *Lymantria dispar* (L.), the pink bollworm, *Pectinophora gossypiella* (Saunders), the oriental fruit moth, *Grapholita molesta* (Busck), the peachtree borer, *Synanthedon exitiosa* (Say), the lesser peachtree borer, *Synanthedon pictipes* (G. & R.), the soybean looper, *Pseudoplusia includens* (Walker), the cabbage looper, *Trichoplusia ni* (Hubner), the smaller European elm bark beetle, *Scolytus multistriatus* (Marsham), the tobacco budworm, *Heliothis virescens* (Fabricius), the fall armyworm, *Spodoptera frugiperda* (J. E. Smith), the Douglas-fir tussock moth, *Orgyia pseudotsugata* (McDunnough), the eastern spruce budworm, *Choristoneura fumiferana* (Clemens), the Japanese beetle, *Popillia japonica* Newman, and many other insect species. The performance of the Hercon® dispensers against these insects has been reviewed.[29,30]

a. Luretape™ with Gypsy Moth Pheromone, Disparlure

The USDA gypsy moth survey program was the first to use the Hercon® controlled release pheromone dispensers in their traps.[31] The active ingredient for such dispensers is the sex pheromone emitted by the virgin female to attract the male, a single compound identified as *cis*-7,8-epoxy-2-methyl-octadecane and called disparlure.[32]

In 1974 and again in 1975, at least 100,000 of the traps were strategically placed in the eastern half of the U.S. to monitor the presence of the gypsy moth, probably the most important defoliating insect of hardwoods, especially the oak, in the northeastern states. The rate of emission of disparlure from the dispensers was ample and fairly constant, and the dispensers were effective for an entire season. Figure 18 shows the

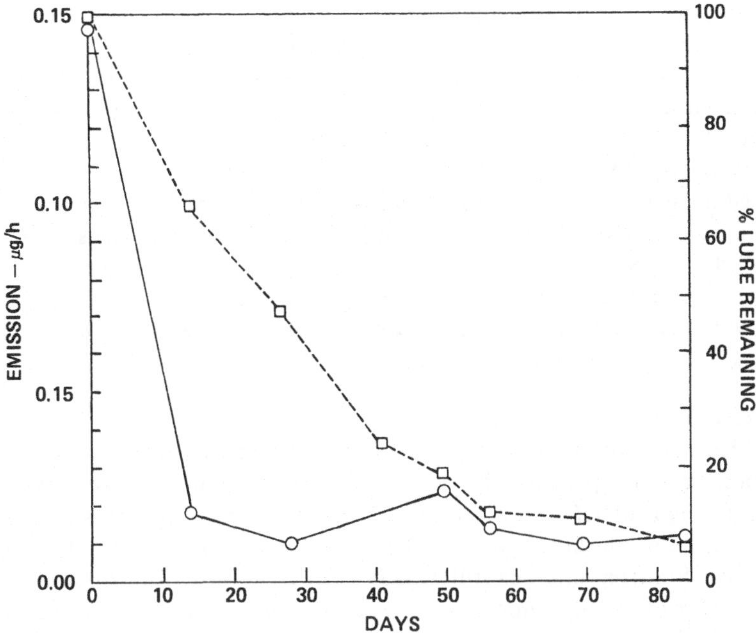

FIGURE 19. Rate of emission of cotton wicks, each initially containing 300 μg disparlure and 2 mg trioctanoin (solid line, ordinate on left), and percentage lure remaining (dashed line, ordinate on right, lure initially = 100%) after aging up to 84 days at room temperature. (From Beroza, M., Paszek, E. C., DeVilbiss, D., Bierl, B. A., and Tardif, J. G. R., *Environ. Entomol.*, 4, 712, 1975. With permission.)

emission rate of the Hercon® Luretape™, while Figure 19 gives the emission rate of the previously used cotton wick dispenser.[32]

i. Mass Trapping Studies

Mass trapping with a new, inexpensive, and highly efficient trap baited with a Luretape™ showed potential for direct control of the gypsy moth in light infestations.[33] Earlier demonstrations that mass trapping (using formulations other than the Luretape™) can prevent a build-up of low-level populations have been reported.

It is important to note that the design of an optimum trapping system, whether for protection or control, requires consideration of many parameters. In addition to pheromone emission rate, lure stability and quantity, and duration of effectiveness (which are related to the design of the pheromone dispenser), consideration must also be given to trap design and color, trap height, trap placement, trapping means (e.g., use of adhesive or insecticide), the position of lure dispenser in the trap, effect of host crop, effective distance of attraction, time of insect response, cost of trap, and ratio of ingredients if the lure is multicomponent.

ii. Effective Emission Rates

Permeating the atmosphere with disparlure to disorient males trying to find the nonflying female gypsy moth has resulted in up to 98% suppression in mating.[33,34] Webb et al.[35] also reported that in tests conducted in Cecil County, Maryland, during the summer of 1976, disparlure emitted from Hercon® dispensers significantly suppressed mating in sparse populations of the gypsy moth. They reported that when Luretape™ dispensers were placed at densities of 1600 and 200/ha both treatments reduced mating by 96% compared to control plots.

Success in mating disruption hinges on the ability of the dispenser to provide an adequate pheromone concentration in the atmosphere throughout the gypsy moth's flight period.[36] By trapping and measuring released pheromone, the emission rate of disparlure from a Hercon® dispenser, under controlled conditions, has been shown to be relatively constant.[31,37] Bierl, DeVilbiss, and Plimmer[36] reported that, relative to the amount of disparlure originally present, the emission rate from 1-in.^{2} 21-mm dispensers used in field tests in 1974 was very low (0.1 μg/hr) and remained constant over extremely long periods. Approximately 90% of the pheromone was still in Hercon® Luretape™ dispensers, although these were used in traps for an entire season. The Luretape™ in 1975 was 6 mm thick and contained 6 mg of lure; it released 0.24 μg/hr at 26.7°C with a 100 ml/min passage of dry air; an increase in the release rate of 1.8 μg/hr/6.3°C rise in temperature was observed.

iii. Air Permeation Experiments

One-in.^{2} Hercon® dispensers containing approximately 20 mg disparlure were tested in an air permeation experiment in 1976. Two dispensers were hung on trees at 10-m intervals at a height of 1.5 m for a total of 200 strips per hectare. Also evaluated were eight other formulations at the rate of 20 g disparlure per hectare.[38,39] These were four NCR and one Stauffer microcapsules, one Conrel hollow-fiber dispenser, and one McLaughlin-Gormley-King impregnated-wax formulation. Efficacy of the treatments was determined by placing virgin nonflying females in the test areas (including an untreated control area) every 3 days during the flight season and determining the number of females mated in each area at the end of each 3-day period of exposure.

The best result, 98% reduction in mating relative to mated females in the untreated control area, was obtained with an NCR microcapsule formulation with 2% disparlure. The next best, 96% mating reduction, was obtained with the Hercon® Luretape.™[35] The remaining formulations were less effective. The 96% reduction obtained in the Hercon® dispenser-treated area indicates that the Luretape™ emitted the pheromone and functioned effectively despite the wind, rain, sun, and humidity during the entire season. The amount of pheromone present in Luretape™ was equivalent to 4 g/ha, of which, from other studies,[31] about 90% still remained after exposure for one season. Since this amount is significantly less than the 20-g/ha rate used in other formulations (including the NCR microcapsules, which gave 98% mating reduction), excellent field performance was obtained with much lesser amounts of the expensive pheromone.

The long-lasting quality of the Luretape™ also makes possible their application well in advance of the gypsy moth flight. Furthermore, it was found that the Hercon® dispensers were effective for more than 1 year; thus, Luretape™ put out one year will be operative in the succeeding year.

b. Luretape with the Cotton Boll Weevil Pheromone, Grandlure

The Hercon® pheromone dispenser has also been used successfully by the USDA in an integrated control program against the cotton boll weevil, considered the worst pest of cotton. Luretape™ was used in traps contained the boll weevil sex and aggregating pheromone, a four-component mixture of chemicals,[40] two of which are aldehydes: *cis* -3,3-dimethyl-Δ^1,α-cyclohexaneacetaldehyde and *trans* -3,3-dimethyl-Δ^1,α-cyclohexaneacetaldehyde; and two of which are alcohols: (+-*cis* -2-isopropenyl-1-methylcyclobutaneethanol and *cis*-3,3-dimethyl-Δ^1,β-cyclohexaneethanol.

i. Mass Trapping Studies

The use of grandlure previously has been limited because the field effectiveness of available formulations — gel-type solutions, a polyethylene glycol mixture impregnated on a cigarette filter and enclosed in a glass vial — had been less than 14 days. Hardee, McKibben, and Huddleston[41] field-tested early Hercon® dispensers of grandlure and found them to be equal in effectiveness to their standard formulations and attractive for 14 days or more. Johnson et al.[42] reported that in field tests of an improved formulation, Luretape™, applied once, was as effective or more effective for 8 weeks than the standard filter-in-vial formulation which was changed weekly. The tests were conducted during the winter of 1974-1975 throughout the cotton belt in the U.S. and Mexico; a summary of the comparison between Luretape™ and two standard formulations is shown in Table 13.[42] Other formulations, such as impregnated cigarette filters in plastic tubes or gel inside a plastic straw, hollow fiber dispensers, and grandlure-containing vials with a flexible foam barrier, were not as effective as the Luretape™ or the standards.

The longer-lasting effectiveness of Luretape™-baited traps is assured by the linear release rate of grandlure (Figure 20); this was the result of the outer polymeric layers of the dispenser controlling the rate of release and preventing the fast degradation of the aldehyde components of the pheromone. Davich and McKibben[43] stated that the Hercon® dispenser, with its 8 to 9 weeks of effectiveness, holds a slight edge over filter-in-vial formulations (effective for 6 to 7 weeks only) in terms of longevity, handling, and storage convenience.

ii. Mating Disruption Studies

Although grandlure is used primarily in mass trapping, Mitchell[44] has shown that cotton boll weevils can be disoriented in a field with Luretape™ acting as evaporators of the pheromone. In one study, 1-in.-square dispensers were arrayed in a checkerboard pattern in a test plot with a pheromone-baited trap at its center. Failure of the insects to find the trap served to indicate disorientation of the weevil, owing to the pheromone emitted by the dispensers. Over a 20-day period, trap captures of boll weevils in the plot with Luretape™ were suppressed 83% compared to captures in similar size control plots that did not contain dispensers. Suppression in captures was also obtained with dispensers containing only two of the four grandlure components.

c. Luretape™ with the Pink Bollworm Pheromone, Gossyplure

In addition to the selective use of pesticides, pheromones will be valuable in a pest control program against the pink bollworm, considered the second most injurious cotton pest in the U.S. Serious outbreaks of destructive secondary pests, such as beet armyworm, *Spodoptera exigua* (Hubner), cotton leafperforator, *Bucculatrix thurberiella* Busck, and spider mites, *Tetranychus* spp., have occurred in California because of the heavy chemical treatment schedule used to control the pink bollworm on cotton. The reliance on broad-spectrum insecticides allowed secondary pests to multiply unchecked after their natural enemies were eliminated.

Gossyplure will be an important addition to the cotton farmer's arsenal of weapons against insect pests. With its high specificity, this pheromone may be employed in a pink bollworm control program without harming the predators and parasites of other pests. Reduction of larvae-infested cotton bolls was obtained in the field after gossyplure applications; the frequency of insecticidal applications was also reduced when the pheromone was used to disrupt mating.[45] In this study, gossyplure was released by Conrel hollow fibers which were attached to cotton plants at 3-week intervals (see Volume II, Chapter 9).

i. Gossyplure-Baited Traps

Hercon® dispensers of the pheromone were used in the USDA pink bollworm survey program. The baits, 1 cm^2 in size and containing about 1.5 mg of the lure, a 1:1

TABLE 13

Response of Native Boll Weevils in Sinaloa, Mexico to Grandlure Formulations in Tests by the Boll Weevil Research Laboratory

Formulation	Grandlure (mg)	$\overline{X}$ no. weevils/trap at indicated week[a]								Average captures/week
		1	2	3	4	5	6	7	8	
HERCON®										
L 64-43-7	40	5.8a	3.1a	10.1a	7.1a	4.5a	9.9a	3.4a	1.6a	5.69
Standard I	40	2.8a	1.0ab	4.3bc	3.4ab	1.4bcd	2.6cd	0.3c	0.3a	2.01
Standard II (replaced weekly)	3	1.3a	0.9ab	3.8bc	2.5bc	2.1abc	7.3ab	4.0a	1.1a	2.88

[a] Means followed by same letter not significantly different at 5% level of confidence. Data on less effective formulations deleted from table.

From Johnson, W. L., McKibben, G. H., Rodriguez, J., and Davich, T. B., *J. Econ. Entomol.*, 69, 263, 1976. With permission.

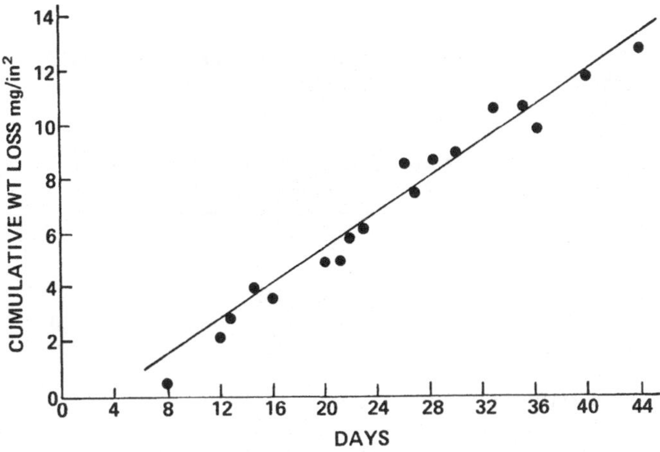

FIGURE 20. Rate of emission of Hercon® Luretape® dispensers of grandlure, initial amount of grandlure in formulation, 4.28% a.i. or 38.19 mg/in. (From Kydonieus, A. F., Smith, I. K., and Beroza, M., in *Controlled Release Polymeric Formulations,* ACS Symp. Series 33, Paul, D. R. and Harris, F. W., Eds., American Chemical Society, Washington, D.C., 1976. With permission.)

TABLE 14

Mean Number of Male Pink Bollworm Moths Caught in Five Traps Baited with Standard Cotton Wick and Hercon® Luretape™ Dispensers of Gossyplure, the Pink Bollworm Pheromone

	Mean no. moths caught at specified time (days)						Cumulative total no. caught
Dispenser	3	7	13	22	27	34	
Standard cotton wick	3.0	11.4	11.6	2.8	2.0	7.0	189
Luretape™ #1	8.0	70.0	35.8	26.4	4.8	19.6	823
Luretape™ #2	8.6	90.0	54.4	36.2	7.8	29.0	1130
Luretape™ #3	10.2	86.4	48.2	36.6	7.2	28.8	1080
Luretape™ #4	20.6	36.4	54.2	25.6	1.2	32.2	911

From Staten, R. T., personal communication.

mixture of *cis, cis* - and *cis,* trans-7,11-hexadecadien-1-ol acetates, have been highly effective in capturing male pink bollworm moths during field tests in different parts of the country.

Table 14 shows the number of moths caught in traps baited with Luretape™ and a cotton dental roll wick containing the lure.[46] The cotton wick, which used to be the standard bait, was generally replaced at 2-week intervals to guard against erratic behavior which often occurred after 18 to 22 days. The Luretape™ dispensers contained approximately 1 mg of gossyplure per square centimeter. The data show that the Hercon® formulations were more effective than the cotton wicks in luring the moths to the traps.

d. Luretape™ in the Pest Management of Other Insects

i. Smaller European Elm Bark Beetle

This insect has been responsible for the decline of the American elm tree, *Ulmus*

TABLE 15

Catches of Smaller European Elm Bark Beetle in Traps Baited with Hercon® Dispensers of Multilure, May 22 to August 20, 1975

	Mean no. beetles trapped in 20 traps within specified time (weeks)						
Dispenser	1	3	5	7	9	11	13
Hercon®	391	368	476	162	95	515	567
Check[a]	342	283	477	128	294	518	820
% of Check	114	130	100	127	33[b]	100	69

[a] Fresh dispenser every two weeks.
[b] Probably anomalous due to high value for check.

From Peacock, J. W., *Proc. 1975 Int. Controlled Release Pesticides Symp.*, Harris, F. W., Ed., Wright State University, Dayton, 1975, 216.

americana L. It is the major vector of the virulent and destructive parasitic fungus, *Ceratostomella ulmi* Schwarz, of the Dutch elm disease. After the ingredients of the beetle's multicomponent pheromone were identified to be 4-methyl-3-heptanol, cubebene, and multistriatin,[47] these chemicals were formulated into Hercon® dispensers, which proved to be highly attractive to the insects.[48] A production run of the dispensers was tested over a prolonged period by comparing captures of traps with the same dispenser, but freshly baited every 2 weeks. For the first 11 weeks, captures with the aged and freshly baited dispensers were not too different. If a 70% level of captures by aged vs. fresh dispensers is considered acceptable, the Luretape™ formulation gave a total of 13 weeks of acceptable performance without the need of rebaiting (Table 15). These results indicate that only one rebaiting would be needed to retain trap effectiveness for the 150- to 180-day season of the beetle.

ii. Oriental Fruit Moth

Hercon® dispensers containing *cis* -8-dodecenyl acetate, the synthetic sex pheromone of *Grapholitha molesta*, were compared with NCR plastic-coated gelatin-based microcapsules to evaluate their efficiency in disorienting the male moths seeking females for mating.[49] Tests were conducted from 1973 to 1975 at the Southeastern Fruit and Tree Nut Research Station in Byron, Ga. The results of the test are shown in Table 16. The Hercon® dispensers, one on every other tree of a 2-acre peach orchard, at a height of 1.2 m, suppressed the response of the male moths 100% to traps baited with virgin females for 12 weeks. Twig damage was reduced from 78 to 100% (compared to an untreated orchard) during the same 12-week period, and the lure in the strips decreased from 2.26 to 1.17%. Though the experiment was continued for 4 more weeks, the captures of males in the untreated control were too low to attach significance to the data of the last 4 weeks. Microencapsulated formulations were effective for only 5 weeks after treatment. The workers also tested Hercon® dispensers containing the pheromone with a dodecyl alcohol synergist; performance of this modified formulation was not as good as with pheromone alone.

iii. Peachtree Borers

The peachtree borer and the lesser peachtree borer have been successfully disoriented in peach orchards.[50] In an experiment conducted by C. R. Gentry and associates, 1-

TABLE 16

Effect of Air-Permeation Trials in 2-acre Orchards with Hercon® Luretape® Dispensers of *cis*-8-dodecenyl acetate on Captures of Native Oriental Fruit Moth Males in Traps (2 per Orchard) Baited with Ten Virgin Females Each, and Associated Twig Damage Byron, Georgia, 1975

Weeks[a]	With pheromone strips[b]	Untreated	Lure remaining in strips (%)
	Male captures		
0			2.26
0—4	0	0	1.58
5—8	0(100)	5	1.24
9—12	0(100)	24	1.17
13—16	0(100)	1	
	Twig damage		
0—4	0(100)	9	
5—8	4(89)	38	
9—12	19(78)	88	
13—16	25(51)	51	

[a] Started June 5, 1975; cumulative weeks after treatment.

[b] Values in parentheses are percent reductions compared to untreated.

From Gentry, C. R., Bierl, B. A., and Blythe, J. L., *Proc. 1976 Int. Controlled Release Pesticides Symp.*, Cardarelli, N. F. Ed., University of Akron, 1976, 3.22. With permission.

in.-square dispensers of *cis*, -3,13-octadecadien-1-ol acetate, the peachtree borer pheromone, were hung on trees in two peach orchards; another orchard was left untreated as a check. In comparison with the check, captures of the lesser peachtree borer in the pheromone-baited traps in orchards with the dispensers were suppressed by 100% for 9 weeks; then results became erratic. Fresh Hercon® dispensers restored almost complete suppression of captures for another 10 weeks (Table 17). Similar results were recorded for the peachtree borer, but significance of the results is questionable because the insect population was extremely low.

McLaughlin, Mitchell, and Tumlinson[51] stated that the Hercon® system provides an "excellent means for baiting pheromone traps" for the two peachtree borer species because the Hercon® Luretape^{TM} "can be stored and handled with relative ease, and the rate of release of active compound is readily adjusted." Diminishing or extending the duration of effectiveness of the bait can be easily obtained by decreasing or increasing the pheromone reservoir. Emission rates can be easily adjusted by cutting a suitable size area of the dispenser. A proper adjustment of these parameters will effect savings in the use of the sex attractant and assure optimum activity for the desired time interval.

iv. Douglas-Fir Tussock Moth

Daterman and Sower[52] evaluated the Hercon® dispenser for possible use in a trap-bait system to monitor the presence of the moth. Traps are baited with -6-heneicosen-11-one, the Douglas-fir tussock moth sex pheromone, and catches used to predict possible serious outbreaks, thus giving workers time to prepare measures to control the larval population of the next season.

TABLE 17

Effect of Air-Permeation Trials on Captures of Lesser Peachtree Borer Males in Pheromone Traps[a]

	No. Lesser Peachtree Borers Trapped		
	Treated		Untreated
Week	Orchard 1[b]	Orchard 2	Orchard 3
1	1 (86)	0 (100)	7
2	0 (100)	2 (93)	27
3	2 (92)	0 (100)	24
4	1 (93)	0 (100)	15
5	0 (100)	0 (100)	2
6	1 (90)	0 (100)	10
7	0 (100)	1 (92)	12
8	0 (100)	0 (100)	4
9	0 —	0 —	0
10	0 (100)	2 (83)	12

[a] Each orchard had five traps baited with 100 mg of pure E, Z.
[b] Value in parentheses is percent reduction compared to the untreated orchard.

From Mitchell, E. R. and Tumlinson, J. H., personal communication.

For survey trapping, it is important that trap catches correctly reflect the pest population size. In previous studies, problems were encountered when traps quickly filled to capacity even at low population levels. The researchers are currently attempting to determine the lowest pheromone concentration that can lure enough moths to serve as a valid indicator of field populations.

In 1976, the lowest pheromone dosage used, which nevertheless captured moths at the very lowest density populations encountered, was formulated in a Hercon® dispenser. The Luretape™ dispenser, 2 mm by 2mm, contained a maximum of 900 ng of the pheromone and had a release rate of only 1 to 7 pg/min during the flight season. This is significantly less than the amount released by a female tussock moth; on the average, the insect contains about 40,000 pg, which is released within a few days of its emergence and then only for a few hours-each day.

Daterman and Sower[52] stated that in the survey-trapping system for the tussock moth, it is important that dispensers must not only be long-lasting but release rates of the chemical must be controlled to avoid excessive trap captures.

4. *Potential of Luretape™ Products*

With pheromones of economically important insect species rapidly becoming available, the Hercon® dispensing system appears to have excellent potential for use in trapping, air permeating, and in lure-insecticide combinations. The flexibility of the system makes it easy to adjust the rate of emission of the behavior chemical to values approaching optimum and greatly facilitates their use as dispensers in traps or as evaporators of the sex attractant in the air-permeation mating-inhibition technique. Ease of handling and storage, protection of a broad variety of chemicals from degradation by light and air, and excellent weathering qualities are other unique features of the Hercon® multilayered polymeric dispensers.

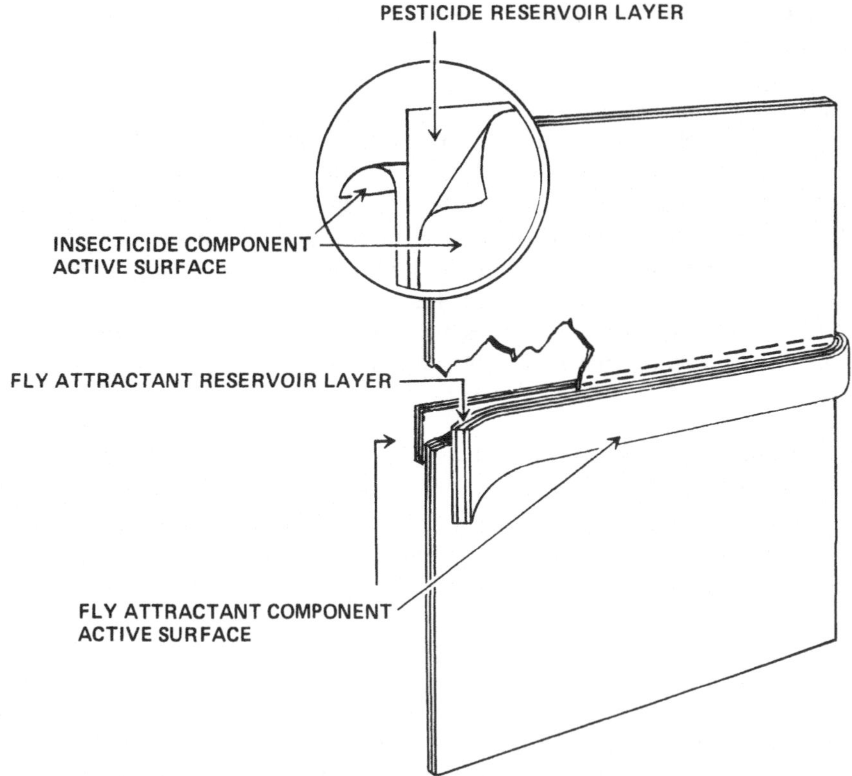

FIGURE 21. Schematic diagram of Hercon® Lure 'N Kill™ Flytape showing insecticide dispenser and fly food attractant dispenser components.

D. Lure 'N Kill™ Flytape Nonfumigant Fly Killer

An alternative has been developed to replace the widely used fly strips that utilize the fumigant DDVP or Vapona® insecticide, and this new product is now under EPA review for pesticide registration. In view of the criticism directed at DDVP, the development of this alternative may be timely. DDVP, *O,O*-dimethyl-2,2-dichlorovinylphosphate, is one of 45 pesticides listed by the EPA as potentially too hazardous for continued use.[53] A Rebuttable Presumption Against the Reregistration (RPAR) of DDVP is expected; thus, benefits and risks of the continued use of the insecticide can be weighed. Pursuant to existing EPA policies, a RPAR may be issued against any pesticide for any of a number of reasons, including indications that the compound is carcinogenic, embryotoxic, or because its continued use would cause a population reduction in nontarget organisms.

1. Description of Lure 'N Kill™ Flytape

Unlike conventional fumigant-action fly strips, the Hercon® Lure 'N Kill™ Flytape[54] utilizes a contact insecticide, thus eliminating potential problems associated with volatile toxicants. Flies drawn by a food attractant come in contact with the Flytape and thereby acquire a lethal dose of the insecticide.

The Lure 'N Kill™ product consists of two separate components: (1) the insecticide dispensing system, and (2) the attractant dispensing system (Figure 21). The release of chemical from both dispensers is based on the Hercon® multilayered structure. Both outer surfaces of the 6-in. by 4-in. insecticide dispenser are active, and the product is suspended to allow access by the flies to both surfaces. As shown in Figure 21, a relatively thin attractant strip, 0.05 in. by 8 in., is fastened about the middle of the insecticide dispenser.

a. Insecticide Dispensing System

Resmethrin (SBP-1382®), a relatively nonvolatile, highly potent solid synthetic pyrethroid, is the active ingredient used. This insecticide, (5-benzyl-3-furyl)methyl 2,2-dimethyl-3-(2-methyl-propenyl)cyclopropane carbolxylate, is widely used in homes, where safety of humans and pets is of paramount importance. Since resmethrin kills by contact and does not release toxic vapors, the Flytape may be used in kitchens and nurseries. Furthermore, with the attractant luring the flies to the insecticide, Flytape may be used in ventilated rooms, porches and open garages. Air movement will not impair effectiveness because the relatively stable insecticide is protected by the Hercon® structure.

b. Fly Food Attractant Dispensing System

The attractive ingredient located in the 0.05-in. by 8-in. dispenser is vanillin (3-methoxy-4-hydroxybenzaldehyde). This chemical was selected after screening dozens of candidate fly food attractants. Vanillin is made synthetically from waste of the wood pulp industry and is used primarily as a flavoring agent in confections, beverages, and food products.

Although several bait-type fly controls products are available, the Hercon® product is the first to employ a controlled release mechanism. Conventional fly baits usually are sugar-baited formulations containing insecticides such as DDVP, methomyl, and ronnel. Muscalure, *cis* -9-tricosene, the synthetic house fly pheromone, has also been incorporated into fly baits with sugar, propoxur, and DDVP, but field effectiveness is not long lasting.

2. Flytape Evaluations

In order to determine the fly-killing effectiveness of Flytape, CSMA F58WT flies, *Musca domestica* (L.) were released into a 1,000 ft.^{3} chamber with the product. The CSMA F58WT was developed from DDT-resistant field-collected strains and reared under procedures developed by the Chemical Specialties Manufacturers Association. Water and food (milk and sugar) were supplied throughout the test. Product effectiveness was expressed in terms of fly knock down and mortality.

The mortality of caged flies placed in the test chamber indicated whether or not the test product released toxic fumes. Significant mortality among the caged insects could only suggest fumigant action since the flies were unable to contact the tested product.

a. Attractiveness of Flytape

Table 18 shows that fly knock down and mortality values with the two sample strips without lure (IA and IIA), while appreciable, were significantly less than the corresponding values obtained with the vanillin-containing Flytape. In the 2-day test, the lure increased fly mortality between 24 and 42%.

b. Effect of Renewing Lure on Aged Flytape

Test data have also shown that aged fly strips were "revitalized" by the addition of a fresh lure dispenser (Table 19). Two formulations of the product (samples I and II) showed excellent fly control (90.7 and 88.4%, respectively) when tested fresh, but after two months fly control decreased 30%. The addition of a new attractant dispenser to the 4½-month-old strip restored the 48-hr fly mortality to levels close to those of the fresh fly strips.

TABLE 18

Attractiveness of Hercon® Lure 'N Kill™ Flytape: Increase in Mortality Due to the Addition of Fly Food Attractant

Sample no.	Type	No. flies	Knockdown 7 hr (%)	Mortality, 48 hr (%)
IA	without lure	144	18.1	47.2
IB	with lure	75	96.0	89.3
IIA	without lure	101	42.6	73.3
IIB	with lure	166	77.7	98.2

From Quisumbing, A. R., Kydonieus, A. F., Calsetta, D. R., and Haus, J. B., *Proc. 1976 Int. Controlled Release Pesticides Symp.*, Cardarelli, N. F., Ed., University of Akron, 1976, 3.40. With permission.

TABLE 19

Influence of Lure on Flytape Efficacy: Change in Fly Mortality After Application of Fresh Attractant Strip to Aged Fly Strip

Sample no.	Active ingredient (%)	Date of test	No. flies	Mortality 48 hr (%)
I	11.95% SBP-	12/08/74	108	90.7
	1382® + 10%	1/06/75	87	62.1
	Starch	2/04/75	126	57.1
IA	+ 10% Attractant	4/16/75	279	89.6
II	7.71% SBP-1382[a]	12/09/74	215	88.4
		1/08/75	59	76.3
		2/06/75	226	52.7
IIA	+ 10% Attractant	4/16/75	359	81.9

[a] From SBP-1382® PGXY (45% active ingredient).

From Quisumbing, A. R., Kydonieus, A. F., Calsetta, D. R., and Haus, J. B., *Proc. 1976 Int. Controlled Release Pesticides Symp.*, Cardarelli, N. F., Ed., University of Akron, 1976, 3.40. With permission.

c. *Nonfumigant Action of Flytape*

Table 20 shows the results of tests (four replicates) in which flies, both caged and free flying, were exposed in the 1000-ft^3 test chamber. A high mortality was obtained with the free flying insects (61.5 and 95.3% mortality after 48 hr) and virtually no mortality with the caged flies. These results indicate that toxic fumes were not present; if they were, mortality of the caged flies would have been noted.

3. *Release Rates*

Figure 22 shows the release rates of resmethrin from the Lure 'N Kill™ Flytape and DDVP from the Shell® No-Pest Insect Strip, the conventional fly killer available in supermarkets. The cumulative weight loss data show that in 3 months at room temperature approximately 7 g of DDVP were volatilized from one insect strip. This is an appreciable release of DDVP fumes, and it could cause a contamination problem, especially in or near a food area.

With the Hercon® multilayered dispenser, there was practically no loss of resmeth-

TABLE 20

Effect of Lure 'N Kill™ Flytape on Free-Flying (Released) and Caged Flies

Test no.	Active SBP-1382® (%)	No. flies[a]	Mortality, 48 hr (%)
I-Released	11.82	296	61.5
I-Caged	11.82	(281)	0.7
II-Released	12.06	284	62.3
II-Caged	12.06	(302)	5.3
III-Released	11.56	406	73.6
III-Caged	11.56	(421)	0.9
IV-Released	11.78	516	95.3
IV-Caged	11.78	(497)	0

Note: Test area: 1,000-ft^3 room.

[a] Number in parentheses is number of flies inside cage.

From Quisumbing, A. R., Kydonieus, A. F., Calsetta, D. R., and Haus, J. B., *Proc. 1976 Int. Controlled Release Pesticides Symp.*, Cardarelli, N. F., Ed., University of Akron, 1976, 3.40. With permission.

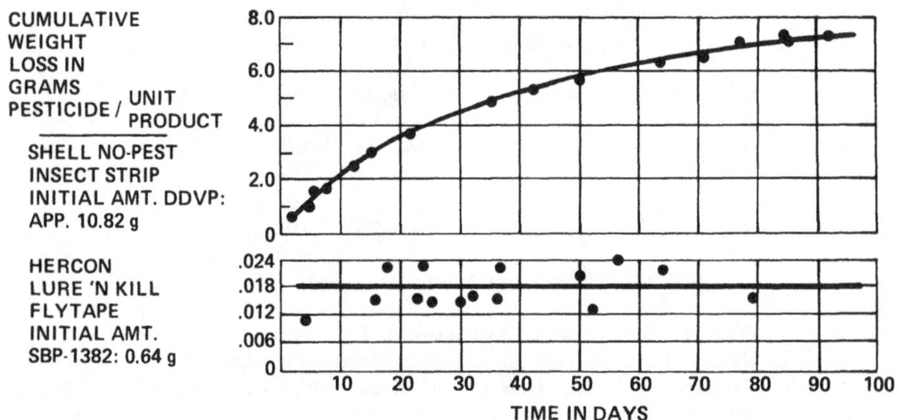

FIGURE 22. Release rate of SBP-1382® insecticide from Hercon® Lure 'N Kill™ Flytape and DDVP insecticide from Shell No Pest® Insect Strip. (From Quisumbing, A. R., Kydonieus, A. F., Calsetta, D. R., and Haus, J. B., in *Proc. 1976 Int. Controlled Release Pesticides Symp.*, Cardarelli, N. F., Ed., University of Akron, Ohio, 1976. With permission.)

rin. The emission rate shown in Figure 22 suggests that the Flytape will be effective for at least 3 months. The insecticide, when formulated into the Hercon® dispensing system, has already been shown to be active for more than 1 year.[55] In another study of residual action, 420-day-old Hercon® dispensers containing SBP-1382® gave 100% mortality of German and oriental, *Blatta orientalis* (L.), cockroaches.

The release of vanillin from its dispenser, shown in Figure 23, is linear; 39% of the original 54.87 mg of vanillin per square inch was volatilized or otherwise lost after 3 months of exposure.

4. *Flytape vs. Another Contact-Action Product*

The Lure 'N Kill™ Flytape was compared with commercially available Black Flag®

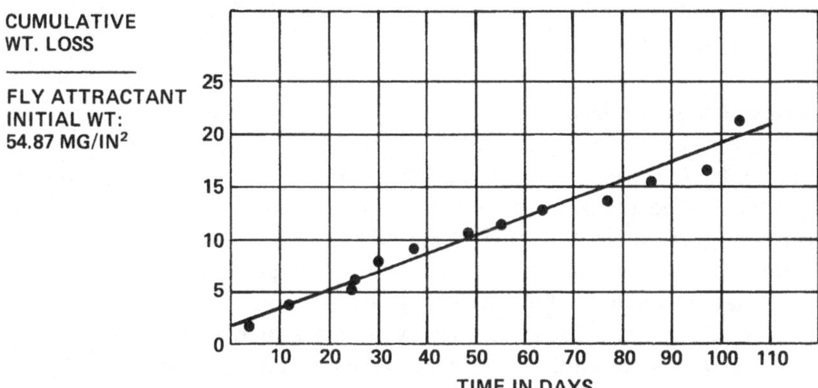

FIGURE 23. Release rate of vanillin fly food attractant from Hercon® Lure 'N Kill™ Flytape at 21.1° to 26.7°C. (From Quisumbing, A. R., Kydonieus, A. F., Calsetta, D. R., and Haus, J. B., in *Proc. 1976 Int. Controlled Release Pesticides Symp.*, Cardarelli, N. F., Ed., University of Akron, Ohio, 1976. With permission.)

TABLE 21

Comparison in Effectiveness of Hercon® Lure 'N Kill™ Flytape and Black Flag® Flyded on Free-Flying and Caged Flies

Product	Active ingredient(s)	Number of flies[a]	Down and/or dead (%) 1	2	3	4
			(days)			
Hercon®	SBP-1382®	5	0	20	40	60
Lure 'N	plus fly	5	40	80	80	80
Kill	food	46	50	65.2	89.1	89.1
flytape	attractant	(302)	0.3	0.3	1.3	2.6
Black	Propoxur	5	0	20	0	0
Flag®	(Baygon®)	40	10.0	20	15	20
Flyded	only	(290)	0.7	0.3	0.7	1.0

Note: Test area: 1,000-ft³ room.

[a] Number in parentheses is number of flies inside cage.

From Quisumbing, A. R., Kydonieus, A. F., Calsetta, D. R., and Haus, J. B., *Proc. 1976 Int. Controlled Release Pesticides Symp.*, Cardarelli, N. F., Ed., University of Akron, 1976, 3.40. With permission.

Flyded® product. The latter was used as a standard in tests submitted to the EPA because Flyded® also killed flies by contact; its active ingredient is propoxur. Flyded® has been registered by the EPA, although the label does not specify nor claim any fly attractant ingredient.

The Hercon® and Black Flag® products were tested identically. Table 21 shows that both products did not kill by fumigant action. The extremely low and practically negligible mortality among caged flies after 4 days (i.e., 2.6 and 1.0%) indicated the absence of toxic fumes in rooms treated with either Flytape or Flyded®. However, mortality was rather high when the insects were allowed to fly freely. Table 21 also shows that the addition of vanillin to the Hercon® formulation significantly improved its performance compared to that of Flyded®, which has no fly attractant. After 4 days, 60 to 89% of the test flies were killed with the Hercon® Lure 'N Kill™ Flytape and only 20% with Flyded.

E. Scentstrip® Fragrance Dispensers

In addition to the multilayered products for pest control, Hercon® has employed its patented controlled release system to dispense fragrances, deodorizers, and malodor counteractants. Called the Scentstrip®, this basic Hercon® dispenser contains the scent as the active ingredient in the inner reservoir layer.

The product is now commercially available and competes in the varied deodorant-air freshener market against solid gels, wicks, aerosol sprays, saturated products, and deodorizing disinfectants. Deodorants act by covering or masking offensive or undesirable odors with an acceptable fragrance, while air fresheners function by releasing a refreshing, pleasant scent.

1. The Air Freshener Market

According to a market review, solid gels have captured 55% of the grocery store volume. This gain in sales is a consequence of the aerosol-ozone controvery and the resulting expanded promotion of the solid gels. Aerosols now account for only 24% of the grocery store market, a significant reduction from 95% in 1970. Liquid wicks account for 4% and sachets, pads, and other methods for about 7% of the sales. Fragrance-impregnated filter paper products make up the remaining 10% of the air freshener market.

Air fresheners are also used by janitorial companies, but this industry prefers mechanical aerosols and fan-powered dispensers. The latter units utilize a small fan that blows across a can of fragrance in solid gel or liquid wick form. They are generally serviced monthly, on a regular contractural basis, as are the disinfectant deodorizer wall units, the oldest devices in use. The most popular unit by far utilizes the fan blowing over a canned solid gel.

2. Limitations of Present Methods

Strong resistance to aerosol products developed after claims were made that fluorocarbon propellants were damaging the upper-atmospheric ozone layer. The high cost of these units was also a major factor in its decline from favor.

Liquid wick dispensers have other drawbacks. They require increasingly expensive containers and packaging. Little success has been achieved in reducing their size and bulk as well as the built-in problems of spills and general untidiness.

Solid gel air fresheners promise 4 or more weeks of life; however, fragrance-perception studies indicate that effective life is 1 to 2 weeks at best. Similar results are noted with liquid wick systems.

With industrial and institutional dispensers, spills in handling containers of liquid fragrance can cause problems. The placement of the wick in the can is vital to proper dispensing of the scent. Wick placement, which differs for AC and battery-operated units, is left to the judgment of the individual service operator.

3. Scentstrip® Marketing Flexibility

The Hercon® Scentstrip® is a unique way of dispensing fragrances and other odor chemicals into the air, automatically and continuously. These thin plastic strips can be fabricated in virtually any size or shape. They require no expensive containers and, within limits, can easily be "programmed' to dispense a wide array of chemical compounds at given release rates, and they can remain effective for extended periods.

The Hercon® fragrance dispenser is produced in flat continuous sheets up to 60 in.

FIGURE 24. Scentcoil air-freshener spools may be used with various AC or battery-operated room deodorizing units such as table-top model (left) or industrial wall models (center and right). Scentcoil may be installed in the units with a plastic "T" holder or within a plastic cylindrical frame.

wide, making it adaptable to many die-cut forms and inexpensive packaging. The dispensers may be produced in strip, squares, ovals, stars, or designer shapes, such as pine trees, flowers, fruits, animals, or cartoons. The strips may also be packaged singly or joined on a common backing sheet.

Scentstrip® can be manufactured such that one or both sides of the strip are active in dispensing fragrance molecules. Typically, such units are designed for hanging, insertion into air conditioning units, or inclusion in a perforated package. Dispensers with one side impervious to chemical passage can be given a coating of pressure-sensitive adhesive for easy stick-on use. In this form, the dispenser is affixed to unobtrusive surfaces, such as beneath tables, behind picture frames, under clothes hampers, garbage can lids, or toilet seats. They can be applied almost anywhere near sources of offensive odors and can be equally or more long lasting than the solids or liquids in common use.

4. Hercon® Scentcoil® Version

The Scentcoil™, a modified version of the Scentstrip®, is essentially a long dispenser rolled into a spool (Figure 24). It is designed to directly replace cans of solid gels or liquid wick fragrances used in industrial air freshener units. A decorative, tabletop unit with a battery-operated fan to circulate the fragrance from the Scentcoil™ is available for use in living rooms, dens, bathrooms, and offices. The combination of Scentcoil™ and fan provides a long-lasting product with the fragrance circulated much more effectively than from the solid gels in passive plastic holders with holes.

TABLE 22

Relative Consumer Acceptance Tests: Comparison Between Hercon® Scentstrip® Fragrance Dispenser and Leading National Brand Solid Gel Air Freshener

	Solid gel	Hercon® Scentstrip®			
	Fragrance released				
	Lemon	Rose	Powder Room	Lemon	Herbal
Number of respondents	34	36	36	33	32
	Acceptance Ratings				
Amount of awareness of fragrance upon entering	13.6	45.0	43.5	36.5	46.3
Appeal of fragrance	35.2	43.2	42.5	39.6	52.5
Suitability	33.8	38.3	50.7	41.7	45.2
Naturalness of fragrance	34.8	43.3	45.0	37.0	48.3
Clean-smelling	35.2	45.0	50.1	44.1	50.6
Fresh fragrance	35.8	44.3	47.9	41.5	51.1
Irritating fragrance	13.9	22.9	22.6	18.9	19.4
Overall acceptance	36.1	44.9	48.2	44.1	52.7

5. Product Evaluation

An independent test panel was asked to evaluate the relative consumer acceptance and effectiveness of several Scentstrip® dispensers containing a number of scents. The specially trained panel was composed of 32 to 40 women. To allow circulation of the fragrance, the Scentstrip® was placed on the wall of a room measuring 8 ft by 7.5 ft by 8 ft 2 hr before starting the tests. Panel members then evaluated the product for awareness of fragrance on entering the room, appeal, suitability, and other qualities of the fragrance.

In order to provide a relative quantitative measurement of current consumer preferences, the leading solid gel product — lemon scented — was also rated following the same procedures by the same test panel. Ratings for the leading brand product were then used as the standard in scoring relative consumer acceptance. Products that were evaluated contained similar amounts of the fragrance.

Table 22 shows the results of the relative consumer acceptance tests. While there was considerable variation in performance among the four Scentstrip® formulations (rose, powder room, lemon, and herbal fragrances), the Hercon® dispensers were all rated superior to or more acceptable than the leading U.S. product. The data suggest that Scentstrip® is capable of attaining consumer acceptance levels greater than those enjoyed by the leading national product.

The duration of effectiveness of Hercon® formulations for industrial and institutional applications was also determined in a separate study. The 3-in. by 3-in. dispensers were checked for 30 days and rated in comparison to the 2-day performance of the leading solid gel product. Table 23 shows that even after 30 days the three Scentstrip® products (mint, rose, and neutral) were still more perceptible than a fresh, 2-day-old solid gel. Similar tests have indicated that the lifespan of gels generally, in terms of actual fragrance perception, rarely exceed 8 to 10 days. The Hercon® multilayered dispenser, with its protected fragrance reservoir, extends this effectiveness period significantly.

In a similar test, a panel of seven women evaluated the effectiveness of the Scent-

TABLE 23

Duration of Effectiveness Evaluation: Comparison Between three 30-day-old Hercon® Scentstrip® Fragrance Dispensers and the Leading National Brand 2-day-old Solid Gel Air Freshener

	Solid gel	Hercon® Scentstrip®		
	Fragrance released			
	Lemon	Mint	Rose	Neutral
Number of respondents	34	40	38	36
	Test Panel Ratings			
Amount of awareness of fragrance upon entering	13.6	31.7	34.9	32.8
Appeal of fragrance	35.2	38.9	40.2	35.6
Suitability	33.8	37.9	39.8	35.3
Naturalness of fragrance	34.8	35.1	43.0	35.0
Clean-smelling	35.2	41.5	36.7	38.1
Fresh fragrance	35.8	40.3	39.4	38.4
Irritating fragrance	13.9	11.4	12.8	13.6
Overall acceptance	36.1	39.5	40.4	38.7

coil® on a weekly basis. The results showed that the fragrance continued to be perceptible, even after 1 month. In terms of reduced service calls alone, the long-term duration of effectiveness can save labor costs and thereby provide a significant advantage over short-lived gels.

F. Insecticidal Fabrics for Stored-Food Protection

Multilayered fabrics have been evaluated for the protection of stored-food products. Such protection will help increase the food supply, which is dwindling as the exploding population of the world is rapidly outpacing agricultural production.

Protection of stored-food products from deterioration is made difficult by many interacting physical, chemical, and biological variables. With stored grain, the quality of the product is affected by temperature, moisture, oxygen, local climate, granary structure, and physical, chemical, and biological properties of grain bulks, as well as attack by microorganisms, insects, mites, rodents, and birds.[56] These variables seldom act alone or all at once, but their presence can significantly diminish the food supply.

Controlling insect pests will not necessarily win the war against stored products deterioration; however, it will be a significant victory because grain injury and organic litter due to insect feeding will be minimal, and much of the growth of fungi and bacteria can be checked.

There are a number of insecticides effective against such stored-product insects as the confused flour beetle, *Tribolium confusum* duVal, the rusty grain beetle, *Cryptolestes ferrugineus* (Stephens), and larvae of the black carpet beetle, *Attagenus megatoma* (F.), but the rapid degradation or volatilization of these materials under actual use conditions greatly curtails their effective lifespan. Since stored products require protection from harvest through storage, transport, and processing to consumption, repeated insecticidal treatments, which are often difficult or impractical, are needed.

Fumigants and a few residual insecticides have been widely used to control general insect infestations.[57] However, several of these chemicals, e.g., DDVP, ethylene dibromide, carbon tetrachloride, hydrogen cyanide, and lindane, are encountering regulatory difficulties and may have to be withdrawn from use because they are potentially hazardous. Other problems, such as off flavors, pesticide residues, and the development of resistance among stored-product insects to insecticides, also exist, and they have spurred the search for improved or new pest control products and methods.

Fabrics with insecticides incorporated into and protected by the Hercon® multilayered structure have been evaluated to determine their potential usefulness as toxicants, repellents, or attractants against stored-product insects.[58] Such fabrics are potentially useful as packaging containers or insect-resistant barriers for stored-food products. A film containing pyrethrins and piperonyl butoxide is now used successfully to protect packaged dried fruits.[59]

1. Effectiveness of Hercon® Fabrics

The Stored-Product Insects Research and Development Laboratory (USDA) at Savannah, Ga. has tested multilayered fabrics reinforced with a nylon scrim for durability and for containing the insecticide in the inner reservoir layer.[58] Ten Hercon® formulations were evaluated, namely, fabrics with 5% carbaryl, 5 and 19% chlordane, 1 and 15% chlorpyrifos, 8.5% stirofos, 5 and 16% malathion, 5% methoxychlor, and 5% Pyrenone®. Test insects, from laboratory cultures, consisted of 7- and 14-day-old adults of the confused flour beetle and 3- to 5-month-old larvae of the black carpet beetle. The residual toxicity was determined as described by McDonald, Guy, and Spiers[60] The test areas were prepared by cutting panels from the fabric and placing glass rings on the surface of the panels. Insects were placed in the test arenas for 24 hr, removed to clean petri dishes, and observed for mortality (expressed as dead + moribund). The repellency-attractancy tests were also conducted using a similar procedure, except that the test arena consisted of treated and untreated surfaces.[60] The repellency of the chemical in the fabric was based on the number of live insects on the untreated surface; because toxic materials were used, only tests in which the insects remained alive were considered valid.

The test chamber was maintained at 27 ± 1°C and 60 ± 5% relative humidity. The materials were tested when fresh, after 6-months, and annually thereafter for 4 years. Multilayered fabrics without insecticides were used as controls. After testing, test panels were stored at room temperature in separate compartments of a cabinet having forced air ventilation.

a. Mortality of T. confusum

The effectiveness of the ten Hercon® fabrics and the control is shown in Table 24. Four formulations, 15% chlorpyrifos, 8.5% stirofos, and 5 and 16% malathion, killed all confused flour beetles exposed during the 4th year of testing; of these four, only the 5% malathion dispenser killed less than 100% at any time during the study. The 5 and 19% chlordane formulations were effective for 6 months (80% mortality) and 2 years (97%), respectively. The 1% chlorpyrifos fabric showed only a moderate, but consistent, effectiveness for 4 years; its performance suggests a higher concentration of the toxicant is needed for greater mortality. Those materials formulated with carbaryl, methoxychlor, and Pyrenone® showed little or no toxicity to the flour beetles.

b. Mortality of A. megatoma

Except for the two chlordane products, the fabrics effective against the confused flour beetles were also effective against larvae of the black carpet beetle. As shown in Table 25, the 19% chlordane formulation showed an 87% effectiveness when 1 year old; the 5% chlordane product was ineffective. The 1% chlorpyrifos dispensers were highly effective against the black carpet beetle larvae throughout the test period. Other materials were ineffective.

TABLE 24

Mortality of Adult Confused Flour Beetles, *Tribolium confusum*, 5 Days After a 24-hr Exposure to the Surfaces of Hercon® Insecticidal Dispensers

			Mortality (%) at indicated test period (date)					
Insecticide	Content (% by weight)	Data fabricated (1972)	0 (1-2-73)	6 months (7-10-73)	1 year (1-12-74)	2 years (1-7-75)	3 years (1-6-76)	4 years (1-5-77)
Carbaryl	15	11-28	0	0	10	0	3	0
Chlordane	5	8-19	83	80	67	37	0	3
	19	11-28	97	100	100	97	23	24
Chlorpyrifos	1	8-19	50	40	63	37	53	38
	15	11-28	100	100	100	100	100	100
Stirofos	8.5	11-28	100	100	100	100	100	100
Malathion	5	4-15	97	100	100	100	93	100
	16	11-28	100	100	100	10	100	100
Methoxychlor	5	4-15	7	3	0	0	0	0
Pyrenone	5	4-15	3	0	10	0	0	0
Control	0		0	0	0	0	0	0

From McDonald, L. L., Guy, R. H., and Spiers, R. D., *USDA Mktg. Res. Rept.*, 882, 1970.

TABLE 25

Mortality of Black Carpet Beetle Larvae, *Attagenus megatoma*, 14 Days After a 24-hr Exposure to the Surfaces of Hercon® Insecticidal Dispensers

Insecticide	Content (% by weight)	Date Fabricated (1972)	Mortality (%) at indicated test period (date)					
			0 (1-2-73)	6 months (7-10-73)	1 year (1-12-74)	2 years (1-7-75)	3 years (1-6-76)	4 years (1-5-77)
Carbaryl	15	11-28	0	3	20	0	0	2
Chlordane	5	8-19	0	0	0	0	0	0
	19	11-28	17	17	87	0	3	3
Chlorpyrifos	1	8-19	93	90	100	93	63	92
	15	11-28	100	97	100	97	90	100
Stirofos	8.5	11-28	27	97	93	100	77	96
Malathion	5	4-15	100	100	97	97	73	99
	16	11-28	100	100	97	97	97	99
Methoxychlor	5	4-15	0	3	0	0	0	0
Pyrenone	5	4-15	0	0	0	0	0	0
Control	0		0	0	0	0	0	0

From McDonald, L. L., Guy, R. H., and Spiers, R. D., *USDA Mktg. Res. Rept.*, 88.2, 1970.

c. Repellency and Attractancy

The five dispensers that were relatively nontoxic to the flour beetles were evaluated for their repellency or attractancy to these insects (Table 26). The chlordane and Pyrenone® fabrics showed appreciable levels of repellency during the study. The carbaryl and methoxychlor formulations were attractive throughout the test, but Gillenwater and McDonald theorized that this may have been caused by some other component in the Hercon® fabric.[60] No attraction was observed when the researchers assayed carbaryl-impregnated paper.

Repellency is undesirable in insecticides intended for use in insect control. However, repellent chemicals are potentially useful in preventing insect attack. Synergized pyrethrins are currently used to treat multilayered paper bags for holding flour and cereal products, and 9- to 10-months protection against insect infestations are now obtained with these bags.[61] There is, therefore, a great potential in using Hercon® fabrics with a repellent, e.g., Pyrenone®, within the sealed reservoir layer for longer periods of stored-product protection.

d. Control of Rusty Grain Beetle

The effectiveness of 14-month-old 5% malathion dispensers against the rusty grain beetle, *C. ferrugineus*, has been reported.[11] Discs of the Hercon® dispensers (2 and 4 cm in diameter) dispersed inside a container of wheat killed the beetles infesting the material. Increasing the diameter of the disc doubled the effectiveness. In stored wheat with a moisture content of 12 to 16%, the 2-cm discs gave 39.3 to 44.7% mortality, and the 4-cm discs killed 84.6 and 88%.

G. Hercon® Granules for Use Against Soil Insects

Hercon® granular formulations, which are ground-up laminated material, have been evaluated against soil insects and for other agricultural and turf applications. These experimental products have been tested against the northern corn rootworm, *Diabrotica longicornis* (Say), banded cucumber beetle larvae, *D. balteata* LeConte, southern corn rootworm, *D. undecimpunctata howardi* Barber, western potato leafhopper, *Empoasca abrupta* DeLong, cabbage maggot, *Hylemya brassicae* (Bouche), sweet corn borer, *Sesamia nonagrioides*, white grub larvae of European chafer, *Amphimallon majalis* (Razoumowski) and the Japanese beetle, *Popillia japonica* Newman.

Protecting the active ingredient of a formulation under field conditions is necessary when the local environment adversely affects the stability of the toxicant. Thus, some insecticides are superior as foliar sprays, but are ineffective below ground, i.e., against soil insects. Conventional granular formulations are likewise likely to be highly susceptible to hydrolysis, photodecomposition, or volatilization and, therefore, ineffective in protecting the active ingredient from degradation and a rapid loss of effectiveness.

Controlled release products generally offer definite economic advantages by prolonging the effectiveness of insecticides, eliminating costly overspraying, and reducing the toxic hazard of the compounds to the applicator by sealing them within a protective structure. These long-lasting products are also beneficial because they reduce or eliminate reapplications which can be impractical because of crop growth or poor weather.

1. Northern Corn Rootworm Control

Hercon® controlled release granules containing 8.3 and 11.3% diazinon were used to protect roots of field corn from rootworm feeding.[62] Such damage can reduce crop yields owing to lodging of plants in the field. Tests showed that plants treated with Hercon® 8.3% formulation or with commercially available 10% Furadan® and 10% Orthene® granules had the least root damage and the highest yields per acre. Plants

TABLE 26

Repellency (+) and Attractancy (−) of Five Hercon® Insecticidal Dispensers to Adult Confused Flour Beetles, *Tribolium confusum*

Insecticide	Content (% by weight)	Date fabricated (1972)	Repellency-attractancy (%) at indicated test period (date)				
			0(1-29-73)	1 year (1-28-74)	2 years (1-13-75)	3 years (1-5-76)	4 years (1-10-77)
Carbaryl	15	11-28	−63.0	−76.4	−29.0	−38.4	−10.5
Chlordane	5	8-19	53.0	49.4[a]	76.6	39.0	35.0
Chlorpyrifos	1	8-19	29.6	24.6	40.6	42.0	37.0
Methoxychlor	5	4-15	−22.4	−13.0	−9.0	−32.0	−28.0
Pyrenone	5	4-15	56.6	53.6	84.8	56.6	52.0
Control	0		±2.6	±0.6	±5.0	±3.2	±5.0

[a] Repellency percentage based on eight instead of ten regular readings.

From McDonald, L. L., Guy, R. H., and Spiers, R. D., *USDA Mktg. Res. Rept.*, 882, 1970.

TABLE 27

Percent Mortality of Banded Cumcumber Beetle Larvae, *Diabrotica balteata*, Held in Field Soil Treated with Diazinon in Hercon® and Commercial Granular Formulations

Treatment	Rate (ppm)	Mortality at indicated week (%) 1	3	5	7	9	11	13	15
Diazinon-Hercon® 6.0%	1	90	100	90	70	75	60	45	0
	2	75	100	90	100	80	95	80	35
	4	85	95	100	100	100	95	100	95
Diazinon 14G	1	75	90	80	60	30	25	25	0
	2	90	100	100	95	50	30	20	10
	4	100	85	70	95	100	85	90	30
Check #1	—	15	35	0	20	25	25	30	0
Check #2	—	10	25	15	15	5	25	20	5

From Kydonieus, A. F., Baldwin, S., and Hyman, S., *Proc. 1976 Int. Controlled Release Pesticides Symp.*, Cardarelli, N. F., Ed., University of Akron, September 13 to 15, 1976, 4.23. With permission.

receiving the commercial Diazinon® 14G formulation had roots pruned by larval feeding.

2. Banded Cucumber Beetle Larval Control

Against *D. balteata* larvae, Hercon® 6% Diazinon® granules outperformed the standard Diazinon® 14G product in terms of percent larval mortality and duration of effectiveness[63] (Table 27). In laboratory tests, soil treated with Hercon® granules (2 ppm) gave 80% larval mortality 13 weeks after treatment, while Diazinon® 14G (2 ppm) gave 50% mortality or less after 9 weeks. The 1-ppm rates of the Hercon® and standard products showed 60% mortality after 9 and 7 weeks, respectively.

The tests showed that the Hercon® granules protected young corn plants from injury by soil insects for at least 13 weeks, yet the insecticide concentration was substantially below that present in commercially available products providing the same duration of effectiveness.

3. Soil and Systemic Insecticidal Treatments

In a greenhouse study, Hercon® granular formulations of Thimet® systemic insecticide were compared with the standard Thimet® clay granules for effectiveness against the southern corn rootworm.[63] Thimet® efficacy as a soil treatment, in terms of feeding damage to roots, was measured using a rating system in which 0 = no feeding, 1 = slight feeding, 2 = moderate feeding, and 3 = severe feeding damage.

As shown in Table 28, the standard attapulgite clay granules reduced or eliminated feeding damage only after applying at least 30 mg of the toxicant. At lower rates (7.5 and 15 mg of active ingredient) effectiveness lasted only 73 days, and severe feeding damage was observed 94 days after treatment. The Hercon® formulation protected the plants throughout the 164-day study at all four rates.

The standard clay and Hercon® granules were further tested to determine the residual availability of Thimet® for systemic control of insects on corn plants. The presence of Thimet® in the soil was determined by bioassays with the western potato leafhopper.[63] Leaves in test were excised from greenhouse-grown, 15- to 18-cm-tall corn plants. Excellent leafhopper mortality throughout the 124-day experiment was obtained with all four Hercon® formulations (i.e., 7.5 mg, 15 mg, 30 mg, and 60 mg of active ingredient). The standard Thimet® formulation, on the other hand, gave

TABLE 28

Effectiveness of Thimet® Insecticide in Hercon® and Attapulgite Clay Granular Formulations Against the Southern Corn Rootworm, *Diabrotica undecimpunctata howardi*

Treatment system	Rate Al Thimet® (mg)	Average corn damage rating after indicated days of treatment[a]					
		17	42	73	94	122	164
Hercon®	—	3.0	3.0	3.0	3.0	3.0	3.0
Attapulgite clay	7.5	1.0	0.7	0.3	3.0	3.0	3.0
	15.0	1.0	0.0	0.0	2.3	3.0	3.0
	30.0	0.0	0.0	0.0	0.0	0.7	1.7
	60.0	0.0	0.0	0.0	0.7	0.0	0.0
Hercon®	7.5	0.7	0.3	0.0	1.0	0.3	0.0
	15.0	0.7	0.0	0.0	0.0	0.0	0.0
	30.0	1.0	0.0	0.0	1.3	0.7	0.0
	60.0	0.0	0.0	0.0	0.0	0.0	0.0
Control	—	3.0	3.0	3.0	3.0	3.0	3.0

[a] Feeding damage rating: 0 = no feeding, 1 = slight feeding, 2 = moderate feeding, 3 = severe feeding.

From Kydonieus, A. F., Baldwin, S., and Hyman, S., *Proc. 1976 Int. Controlled Release Pesticides Symp.*, Cardarelli, N. F., Ed., University of Akron, 1976, 4.23. With permission.

excellent control for 124 days only when the 30- and 60-mg rates were used; the 7.5- and 15-mg rates were effective for only 45 and 96 days, respectively.

Both the southern corn rootworm and leafhopper studies showed that the current standard formulation will require four to eight times more Thimet® to obtain the degree of effectiveness afforded by the Hercon® granules. This is added evidence that effectively protecting the active ingredient from degradation when in the soil can reduce the quantity of toxicant used without sacrificing plant protection and long-lasting performance. Possible problems of Thimet® residues and persistence in food products treated with controlled release formulations can be averted by selecting the proper protective layer and release rates.

4. Cabbage Maggot Control

Hercon® granular formulations of Diazinon® and chlorpyrifos were tested for cabbage root maggot control following in-furrow application.[62] Roots of plants treated in 1975 with Hercon 8.8 and 11.3% Diazinon® granules showed only 16.7 and 20% damage, respectively, while those treated with the commercial Diazinon® 14G product had 53.3% damage. The roots of the untreated check had 60% damage. In-furrow treatments with the Hercon® Diazinon® and chlorpyrifos formulations the following year gave 51.8 and 44.4% maggot control, respectively. The standard Diazinon® 14G corn cob-based granule had a 29.6% maggot control.

5. Sweet Corn Borer Control

A series of field tests were also conducted to compare the effectiveness against stem borers of 10% Diazinon® calcite and Hercon® granular formulations.[63] The granules were applied to the top of 3-week-old plants at a rate of 30 kg/ha. Thirty larvae were then placed at the leaf axils of each plant 1 day after insecticide application and at weekly intervals for 3 weeks. After allowing the larvae to remain undisturbed for 10 days, ten plants of each granular treatment were uprooted, examined, and the number of live larvae that bored into the plant recorded.

Both calcite and Hercon® granules were very effective when insects were placed on the plants 1 day after insecticidal applications, i.e., no live larvae were found. However, 1 week after granular application, live larvae were observed with the number being higher in plants that had been treated with the diazinon calcite granules. Plants treated with the Hercon granules gave 91.3, 86.6, and 62.2% control of the stem borer 7, 14, and 20 days, respectively, after granular application relative to the untreated check plants. The calcite-treated plants gave only 72.9, 50, and 38.5% control during these weekly intervals.

The tests showed that in spite of exposure to sunlight and high temperatures in the field, the eventual breakdown of Diazinon® was significantly slowed down by the protective layers of the granules.

6. White Grub Control in Turf

Spray and granular insecticidal materials were tested for the control of European chafer and Japanese beetle larvae.[62] In 1975, the European chafer white grubs were more easily controlled than the Japanese beetle larvae in turf sprayed with Diazinon®, Primicid®, Dyfonate®, and chlorpyrifos insecticides. The highest level of control was obtained with a spray treatment of Diazinon®, which gave 93% control of the Japanese beetle and 83% control of the European chafer. The other three sprays did not exceed 75% control for either insect species.

The Diazinon® granular formulations, both Hercon® and commercial (Spectracide® 6000), outperformed the non-diazinon® sprays and the Hercon® and commercial chlorpyrifos granules in the 1975 and 1976 tests. Since turf plots were examined for the presence of grubs only 40 days after treatment, a longer test would be required to ascertain the benefits of protecting the toxicant by encapsulation from fast degradation. It appears, however, that against white grubs the standard granular form is as good as an encapsulated product.

The effectiveness of insecticide granules, whether conventional type or encapsulated, depends on a number of factors, including insect species, host crop, and release rates. A quick-release granular formulation may be satisfactory against some insects, but the sustained slow-release types may be better if a longer duration of effectiveness is desired or if an easily degraded toxicant is used.

While controlled release formulations will definitely play an important role in agricultural pest management, this technology, in some situations, may be impractical and expensive. Each case must be considered individually, and for those controlled release products that turn out to be higher in cost, this cost must be weighed against desired benefits. Generally, a significant portion of conventionally applied pesticides is wasted because such products are prepared or used at concentrations far in excess of that actually required to compensate for the expected loss of active ingredient from degradation, evaporation, surface run-off, and leaching. Such overuse allows a sufficient amount to be present over a practical period of time. With proper attention to the particular application and the factors affecting the efficacy of the insecticide at the treatment site, controlled release can provide a basis for safer and more efficient insect control.

VI. LABORATORY PROCEDURE FOR PREPARATION OF MULTILAYERED LAMINATED STRUCTURES

The following description of the preparation of a multilayered laminated dispenser containing an antibacterial agent is based on an example given in U.S. Patent No. 3,705,938 covering the manufacture of polymeric articles containing antibacterial and/or antifungal properties.[64] In the following procedure, the active ingredient is the antibacterial agent, 10,10′-oxybisphenoxyarsine.

A sheet of PVC film, 13 mm thick, is coated on one side with a plastisol mixture containing about 3% of a solution of the antibacterial agent in epoxidized soybean oil. The plastisol is prepared by dispersing 100 parts of PVC resin and about 8 parts of calcium carbonate in about 100 parts of dioctyl phthalate and then dispersing the antibacterial agent into the PVC resin-calcium carbonate-dioctyl phthalate mixture. This mixture is stirred until uniform, and a coating about 2 mm thick is applied to one side of the PVC film.

The coated PVC film is then overlayed with a piece of nylon scrim and a second sheet of PVC film, 3.75 mm thick. This three-layered structure (PVC film/plastisol mixture/PVC film) is then allowed to set under suitable conditions of heat and pressure until an integral, firmly bonded product is obtained.

The outside surfaces of the laminated unit were shown by the NYS-63 method to have antibacterial properties. One-in.-square test samples of the laminates were inoculated with broth cultures of *Staphylococcus aureus* and incubated for 24 hr at 37°C along with an untreated check. Compared to the untreated sample, bacterial counts of *S. aureus* were reduced 99.6% on the 13-mm side and 99.7% on the 3.75-mm side.

REFERENCES

1. **Crank, J. and Park, G. S.,** Methods of measurement, in *Diffusion of Polymers,* Crank, J. and Park, G. S., Eds., Academic Press, New York, 1968, 1.
2. **Richards, R. W.,** The Permeability of Polymers to Gases, Vapours and Liquids, Explosive Research and Development Establishment ERDE (Ministry of Defense) Tech. Rept. No. 135, National Technical Information Service NTIS AD-767 627, U.S. Department of Commerce, Springfield, Va., March, 1973.
3. **Baker, R. W., Lonsdale, H. K., and Gale, R. M.,** Membrane-controlled delivery systems, in *Proc. Int. Controlled Release Pesticides Symp.,* Cardarelli, N. F., Ed., University of Akron, Ohio, 1974, 40.1
4. **Baker, R. W. and Lonsdale, H. K.,** Controlled release: mechanisms and rates, in *Controlled Release of Biologically Active Agents,* Tanquary, A. C. and Lacey, R. E., Eds., Plenum Press, New York, 1974, chap. 2.
5. **Harris, F. W.,** Theoretical aspects of controlled release, in *Proc. Int. Controlled Release Pesticides Symp.,* Cardarelli, N. F., Ed., University of Akron, Ohio, 1974, 8.1.
6. **Crank, J.,** *The Mathematics of Diffusion,* Oxford University Press, London, 1956.
7. **Barrer, R. M.,** Diffusion and permeation in heterogeneous media, in *Diffusion of Polymers,* Crank, J. and Park, G. S., Eds., Academic Press, New York, 1968, 165.
8. **Kydonieus, A. F.,** The effect of some variables on the controlled release of chemicals from polymeric membranes, in *Controlled Release Pesticides,* ACS Symp. Series 53, Scher, H. B., Ed., American Chemical Society, Washington, D.C., 1977, 152.
9. **Kydonieus, A. F. and Smith, I.,** U.S. Patent 3,961,117, 1976.
10. **Kydonieus, A. F., Quisumbing, A. R., and Hyman, S.,** Application of a new controlled release concept in household products, in *Controlled Release Polymeric Formulations,* ACS Symp. Series 33, Paul, D. R. and Harris, F. W., Eds., American Chemical Society, Washington, D.C., 1976, 295.
11. **Kydonieus, A. F., Rofheart, A., and Hyman, S.,** Marketing and economic considerations for Hercon consumer and industrial controlled release products, in *Chemical Marketing and Economic Preprints of Symposium on Economics and Market Opportunities for Controlled Release Products,* ACS Chemical Marketing and Economics Division, San Francisco, August 29 to September 3, 1976, 140.
12. Herculite Protective Fabrics Corporation, unpublished data.

13. **Kydonieus, A. F., Smith, I. K., and Hyman, S.,** A polymeric delivery system for the controlled release of pesticides: Hercon Roach-Tape, in *Proc. 1975 Intl. Controlled Release Pesticides Symp.*, Harris, F. W., Ed., Wright State University, Dayton, Ohio, 1975, 60.
14. **Smith, I. K.,** unpublished data.
15. **Moore, R. C.,** Efficacy of Hercon Roach Tape, *Pest Control,* 44(6), 37, 1976.
16. **Quisumbing, A. R. and Kydonieus, A. F.,** Advantages of using controlled release formulations for insect control in commercial food handling establishments, in *Proc. 1977 Int. Controlled Release Pesticides Symp.*, Goulding, Robert, Ed., Oregon State University, Corvallis, 1977, 216.
17. **Wiswesser, W. J., Ed.,** *Pesticide Index,* 5th ed., Entomological Society of America, College Park, Md., 1976, 72.
18. **Ebeling, W., Wagner, R. E., and Reierson, D. A.,** Influence of repellency on the efficacy of blatticides. I. Learned modification of behavior of the German cockroach, *J. Econ. Entomol.*, 59(6), 1374, 1966.
19. **Weidhaas, D. E., Swain, L. R., Jr., and Burden, S.,** personal communication.
20. **Reierson, D. A. and Rust, M. K.,** Utilization of insecticidal laminated tapes to control German cockroaches, *J. Econ. Entomol.*, 70(3), 357, 1977.
21. **Kydonieus, A. F., Quisumbing, A. R., Smith, I. K., Baldwin, S., and Conroy, R. A.,** Hercon Insectape: the PCO's New Weapon for the Ancient War on Roaches, Hercon Technical Bulletin 26, Herculite Protective Fabrics Corporation, New York, 1977.
22. **Quisumbing, A. R., Kydonieus, A. F., and Latwatsch, D. J.,** Field studies on the control of the German cockroach, *Blattella germanica* (L.), using Hercon "Roach-Tape" and standard sprays, in *Proc. 1975 Intl. Controlled Release Pesticides Symp.*, Harris, F. W., Ed., Wright State University, Dayton, Ohio, 1975, 247.
23. **Mampe, C. D.,** personal communication.
24. **Beroza, M. and Green, N.,** Synthetic chemicals as insect attractants, in *New Approaches to Pest Control and Eradication,* Advances in Chemistry Series 41, Gould, R. F., Ed., American Chemical Society, Washington, D.C., 1963, 11.
25. **Beroza, M., Green, N., and Knipling, E. F.,** Gypsy moth control with the sex attractant pheromone, *Science,* 177, 19, 1972.
26. **Mayer, M. S. and McLaughlin, J. R.,** An Annotated Compendium of Insect Sex Pheromones, Florida Agricultural Experiment Station Monograph Series 6, University of Florida, Gainesville, August, 1975.
27. **Mulhern, F. J.,** Administrator, Trial Boll Weevil Eradication Program Environmental Statement, USDA APHIS, (ADM)-75-1, Washington, D.C., 1975.
28. **Scott, W. P., Lloyd, E. P., Bryson, J. O., and Davich, T. B.,** Trap crops for the suppression of low density overwintered boll weevil populations, *J. Econ. Entomol.*, 67, 281, 1974.
29. **Kydonieus, A. F., Smith, I. K., and Beroza, M.,** Controlled release of pheromones through multilayered polymeric dispensers, in *Controlled Release Polymeric Formulations,* ACS Symp. Series 33, Paul, D. R. and Harris, F. W., Eds., American Chemical Society, Washington, D.C., 1976, 283.
30. **Kydonieus, A. F. and Beroza, M.,** Insect control with the multi-layered Luretape dispenser, in *Proc. 1977 Intl. Controlled Release Pesticides Symp.*, Goulding, R., Ed., Oregon State University, Corvallis, 1977, 78.
31. **Beroza, M., Paszek, E. C., DeVilbiss, D., Bierl, B. A., and Tardif, J. G. R.,** A 3-layer laminated plastic dispenser of disparlure for use in traps for gypsy moths, *Environ. Entomol.*, 4, 712, 1975.
32. **Bierl, B. A., Beroza, M., and Collier, C. W.,** Potent sex attractant of the gypsy moth, *Porthetria dispar* (L.): its isolation, identification and synthesis, *Science,* 170, 87, 1970.
33. **Beroza, M., Hood, C. S., Trefrey, D., Leonard, D. E., Knipling, E. F., and Klassen, W.,** Field trials with disparlure in Massachusetts to suppress mating of the gypsy moth, *Environ. Entomol.*, 4, 705, 1975.
34. **Schwalbe, C. P., Cameron, E. A., Hall, D. J., Richerson, J. V., Beroza, M., and Stevens, L. J.,** Field tests of microencapsulated disparlure for suppression of mating among wild and laboratory-reared gypsy moths, *Environ. Entomol.*, 3, 589, 1974.
35. **Webb, R. E., Dull, C. W., McComb, C. W., Bierl, B. A., and Plimmer, J. R.,** Gypsy moth mating suppressed by disparlure emitted from laminated plastic dispenser, submitted.
36. **Bierl, B. A., DeVilbiss, E. D., and Plimmer, J. R.,** Use of pheromones in insect control programs: slow release formulations, in *Controlled Release Polymeric Formulations,* ACS Symp. Series 33, Paul, D. R. and Harris, F. W., Eds., American Chemical Society, Washington, D.C., 1976, 265.
37. **Bierl, B. A. and DeVilbiss, E. D.,** Insect sex attractants in controlled release formulations: measurements and applications, in *Proc. 1975 Intl. Controlled Release Pesticides Symp.*, Harris, F. W. Ed., Wright State University, Dayton, Ohio, 1975, 230.
38. **Plimmer, J. R., Bierl, B. A., Webb, R. E., and Schwalbe, C. P.,** Controlled release of pheromone in the gypsy moth program, in *Controlled Release of Pesticides,* ACS Symp. Series 53, Scher, H. B., Ed., American Chemical Society, Washington, D.C., 1977, 168.

39. **Schwalbe, C. P., Paszek, E. C., Webb, R. E., McComb, C. W., Dull, C. W., Plimmer, J. R., and Bierl, B. A.**, Field evaluation of controlled release formulations of disparlure for gypsy moth mating disruption, *Ann. Entomol. Soc. Am.*, submitted.
40. **Tumlinson, J. H., Hardee, D. D., Gueldner, R. C., Thompson, A. C., Hedin, P. A., and Minyard, J. P.**, Sex pheromones produced by male boll weevils: isolation, identification and synthesis, *Science*, 166, 1010, 1969.
41. **Hardee, D. D., McKibben, G. H., and Huddleston, P. M.**, Grandlure for boll weevils: controlled release with a laminated plastic dispenser, *J. Econ. Entomol.*, 68, 477, 1975.
42. **Johnson, W. L., McKibben, G. H., Rodriguez, J., and Davich, T. B.**, Boll weevil: increased longevity of grandlure using different formulations and dispensers, *J. Econ. Entomol.*, 69, 263, 1976.
43. **Davich, T. B. and McKibben, G. H.**, Belt-Wide Grandlure Formulation Test, presented at the Entomological Soc. America Annual Meeting, New Orleans, November 30 to December 1, 1975.
44. **Mitchell, E. R.**, personal communication.
45. **Gaston, L. K., Kaae, R. S., Shorey, H. H., and Sellers, D.**, Control of the pink bollworm by disruption of adult moth sex pheromone communication, *Science*, 196, 904, 1977.
46. **Staten, R. T.**, personal communication.
47. **Pearce, G. T., Gore, W. E., Silverstein, R. M., Peacock, J. W., Cuthbert, R. A., Lanier, G. N., and Simeone, J. B.**, Chemical attractants for the smaller European elm bark beetle, *Scolytus multistriatus* (Coleoptera: Scolytidae), *J. Chem. Ecol.*, 1, 115, 1975.
48. **Peacock, J. W. and Cuthbert, R. A.**, Field and laboratory evaluations of controlled release dispensers for *Scolytus multistriatus* pheromone, in *Proc. 1975 Intl. Controlled Release Pesticides Symp.*, Harris F. W., Ed., Wright State University, Dayton, 1975, 216.
49. **Gentry, C. R., Bierl, B. A., and Blythe, J. L.**, Air permeation field trials with the oriental fruit moth pheromone, in *Proc. 1976 Int. Controlled Release Pesticides Symp.*, Cardarelli, N. F., Ed., University of Akron, Ohio, 1976, 3.22.
50. **Mitchell, E. R. and Tumlinson, J. H.**, personal communication.
51. **McLaughlin, J. R., Mitchell, E. R., and Tumlinson, J. H.**, Evaluation of some formulations for dispensing insect pheromones in field and orchard crops, in *Proc. 1975 Int. Controlled Release Pesticides Symp.*, Harris, F. W., Ed., Wright State University, Dayton, 1975, 209.
52. **Daterman, G. E. and Sower, L. L.**, Douglas-fir tussock moth pheromone research using controlled release system, in *Proc. 1977 Int. Controlled Release Pesticides Symp.*, Goulding, R., Ed., Oregon State University, Corvallis, 1977, 68.
53. **Anon.**, EPA lists pesticides that may be too dangerous to use, Chemical & Engineering News, American Chemical Society, Washington, D.C., June 14, 1976, 18.
54. **Quisumbing, A. R., Kydonieus, A. F., Calsetta, D. R., and Haus, J. B.**, Hercon "Lure 'N Kill"™ Flytape: a non-fumigant insecticidal strip containing attractants, in *Proc. 1976 Int. Controlled Release Pesticides Symp.*, Cardarelli, N. F., Ed., University of Akron, Ohio, 1976, 3.40.
55. **Brown, B. B., Haus, J. B., Calsetta, D. R., and Grassl, E. F.**, Residual Action of Synthetic Pyrethroids, presented at Int. Congr. of Pesticide Chemistry, Helsinki, Finland, July 1974.
56. **Sinha, R. N.**, Interrelations of physical, chemical, and biological variables in the deterioration of stored grains, in *Grain Storage: Part of a System*, Sinha, R. N. and Muir, W. E., Eds., AVI Publishing, Westport, Conn., 1973, 15.
57. **Bond, E. J.**, Chemical control of stored grain insects and mites, in *Grain Storage: Part of a System*, Sinha, R. N. and Muir, W. E., Eds., AVI Publishing, Westport, Conn., 1973, 137.
58. **Gillenwater, H. B. and McDonald, L. L.**, Toxicity, repellency and attractancy of slow-release insecticide dispensers, *J. Ga. Entomol. Soc.*, 12(3), 261, 1977.
59. **Anon.**, Piperonyl butoxide and pyrethrins in the adhesive of cellophane-polyolefin two-ply bags, Federal Register 39(210): 38224-38225, October 30, 1974.
60. **McDonald, L. L., Guy, R. H., and Spiers, R. D.**, Preliminary evaluation of new candidate materials as toxicants, repellents, and attractants against stored product insects, *USDA Mktg. Res. Rept.*, 882, 1970.
61. **Gillenwater, H. B. and Burden, G. S.**, Pyrethrum for the control of household and stored-product insects, in *Pyrethrum: The Natural Insecticide*, Casida, J. E., Ed., Academic Press, New York, 1973, 243.
62. **Gauthier, N. L.**, Field experiments with experimental controlled release granular insecticides, presented at the 1977 Int. Controlled Release Pesticides Symp., Oregon State University, Corvallis, 1977.
63. **Kydonieus, A. F., Baldwin, S., and Hyman, S.**, Hercon granules and powders for agricultural applications, in *Proc. 1976 Int. Controlled Release Pesticides Symp.*, Cardarelli, N. F., Ed., University of Akron, Ohio, 1976, 4.23.
64. **Hyman, S., Bernstein, B. S., and Kapoor, R.**, U.S. Patent No. 3,705,938, 1972.

Chapter 6

CONTROLLED RELEASE FROM ULTRAMICROPOROUS TRIACETATE

Arthur S. Obermayer

TABLE OF CONTENTS

I. INTRODUCTION

A new controlled release matrix that is beginning to receive broad commercial acceptance is an ultramicroporous open-celled form of cellulose triacetate.[1,2,3] When it is in a film or membrane configuration, it has the tradename Poroplastic®. When it is in a powder or microbead form for time release applications, the name Sustrelle® is applied. When used for pressure release, it is called Liqui-Powder®. In addition, it is possible to coat it on a variety of substrates, reinforce it, or make hollow fibers. Experiments which have been performed thus far on film and powder have shown only differences that could be explained by the difference in surface area-to-volume ratio between film and powder.

These products are essentially molecular sponges where the sponge matrix is cellulose triacetate. The pore dimensions of this sponge-like material are extremely small, ranging in diameter from one order of magnitude to three orders of magnitude smaller than the wavelength of light. This very fine porosity means that there is a large internal surface area and large quantities of liquid can be held very strongly within the material by capillary action. In fact, it can aptly be described as a solid composed mostly of liquid.

The properties of Poroplastic® film are somewhat similar to a typical membrane, some properties are in common with hydrogels, and some resemble a porous polymer of plastisol. In its powder form it has some characteristics in common with both microencapsulated products and solid absorbants, but it, nevertheless, is quite different.

II. GENERAL CHARACTERISTICS

Following is a list of the unique properties of Poroplastic.® The same list applies to its companion products, Sustrelle® and Liqui-Powder®. No other material has these six properties, and only a few materials combine even four of the listed properties.

A. Extremely Small Yet Variable Pore Size

Characteristic pore diameters range from 14 Å to over 250 Å as measured by selective ultrafiltration for globular proteins of well-defined dimensions.

B. High Liquid Content, Almost any Liquid

Liquid content can be varied from 70 to 98%. Poroplastic® can be impregnated with almost any liquid. The water in aqueous Poroplastic® can be replaced, for example, by mineral oil, alcohols, moisturizing oils, esters, ketones, essential oils, silicone oils, or almost any kind of solution. In fact, chemical reactions or precipitations can be made to take place within the Poroplastic® structure.

Generally, one liquid is replaced by another liquid through a diffusion exchange of miscible liquids. If they are not miscible, more than one step may be necessary. For example, because many organic compounds are not miscible with water but are miscible with alcohol, the first step is to exchange the water for alcohol, and then directly exchange the alcohol for the organic compound. The replacement of one liquid by another is usually quite rapid; for example, this diffusion controlled process normally is about 90% complete in less than 2 min for a film of 10 mil thickness. Sustrelle® or Liqui-Powder® whose particle diameter is typically about 75 mil, will undergo 90% replacement in less than 10 sec.

C. Irreversible Shrinkage on Drying

If the liquid in Poroplastic® is allowed to evaporate, the pore structure will progressively collapse. Shrinkage can be as great as a factor of 50. In contrast to crosslinked gels, it shrinks irreversibly on drying because the swollen state is not thermodynamically preferred over the partially or completely dry state. The shrinkage can be stopped at any stage by preventing further loss of liquid.

When an internal liquid, such as water, is partially or wholly removed, shrinkage appears to occur initially with a collapse of the smallest pores, and eventually, it also includes the larger pores. Thus, liquids or solids can initially be diffused into the Poroplastic® matrix; and then by a drying and shrinkage process, the pores can be partially collapsed so as to entrap or reduce the rate of active agent diffusion to the surface of the matrix. The collapsed structure is strong, thermally resistant, and will not reabsorb liquids.

D. Hydrophobic and Oleophilic Surface

The surface of Poroplastic® film containing as much as 90% water is hydrophobic and oleophilic. Water has a low contact angle and beads on its surface, while mineral oil will form a thin, continuous surface film.

E. Homogeneous and Transparent

Poroplastic® is distinct from thick skinned (assymetric) membranes in that it is a homogenous structure. The best evidence for this is that its hydrolic permeability and its diffusive permeability are inversely proportional to thickness. Because of the extremely fine dimensions of the pores and the lattice structure (typically 1/100 of the wavelength of light), Poroplastic® will not scatter light and is transparent.

F. Strong, Crystalline, and Noncrosslinked

The mechanical strength of Poroplastic® is due to its ordered crystalline lattice structure. Its strength does not depend on crosslinking which rigidizes most gels.

III. TRANSPORT PROPERTIES

The characteristic pore dimensions can be inferred from measurements of molecular weight cutoff in the filtration of globular proteins of well defined sizes (Table 1). A direct correlation between the water content of a Poroplastic® membrane and its molecular weight cutoff can be observed and is shown in Figure 1. The molecular weight cutoff can then be used to estimate a characteristic pore diameter. The pores have a reasonably broad size distribution, probably with a preponderance below the characteristic diameter.

The permeability characteristics of Poroplastic® film are consistent with the model of an open cell molecular spongelike structure made up of cylindrical micellular units. The Kozeny Equation[4] for viscous flow through a porous solid can be used to derive pore dimensions, micelle dimensions, and internal surface area, directly from hydraulic permeability information. Table 2 lists experimental hydraulic permeability data for water in terms of flux through membranes of various thickness and porosities at different applied pressures. In accordance with the model, flux is directly proportional to applied pressure and inversely proportional to thickness. The Kozeny Equation states that

$$Q = \frac{a^3}{3\eta S^2} \times \frac{\Delta P}{\Delta h}$$

where Q is the volume of transported liquid per unit area per unit time, a is the open area per unit cross-section, S is the internal surface area, η is the viscosity of the liquid,

TABLE 1

Retention Characteristics of Poroplastic®

Solute	Molecular weight	% Retention at 600 g/cm² on 0.1-mm films			
		MA-70	MA-85	MA-92	MA-97.5
Phenyl-alanine alanine	165	0	0	0	0
Sucrose	342	0	0	0	0
Vitamin B-12	1,355	70	0	0	0
Inulin	5,200	—	10	0	0
Cytochrome C	12,400	>99	87	7	0
Beta lactoglob-ulin	35,000	>99	—	—	0
Hemoglobin	64,000	>99	97	—	0
Albumin (bo-vine)	67,000	>99	95	25	0
Gamma globu-lin	153,000	>99	>99	>98	0
Apoferritin	480,000	>99	>99	>99	42
Blue dextran 2000	2,000,000	>99	>99	>99	93
Apparent pore diameter (Å)		14	25	60	>200
Apparent pore diameter (μm)		0.0014	0.0025	0.006	>0.02
Water content (wt %)		70	85	92	97.5

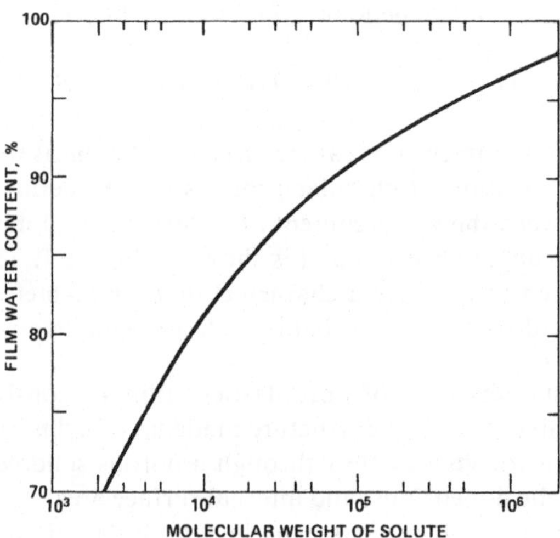

FIGURE 1. Effect of water content on molecular weight cutoff (90% retention, 0.004 in. film).

and $\Delta P/\Delta h$ is the pressure gradient through the film. The number 3 represents a reasonable assumed value for the geometry-dependent Kozeny constant.

Using experimental data,[5] we can derive for MA-85 Poroplastic® film (85% water) a pore diameter of 36 Å, a micellular diameter of 27 Å, and an internal surface area of 180 m²/cm³ of volume. Data based on Poroplastic® film of other water contents lead to values ranging from 22 to 32 Å for micellular diameters and pore diameters

TABLE 2

Hydraulic Permeability of Poroplastic® Film (Water Flux in mℓ/cm²/min)

Membrane	Water content (wt %)	Applied pressure		
		700 g/cm²	2100 g/cm²	3900 g/cm²
MA70	70	0.003	0.01	0.02
MA85	85	0.02	0.07	0.13
MA92	92	0.05	0.14	0.26
MA97.5	97.5	1.2	2.5—3.5	Compacts

Note: Tabular values of flux are for 0.1-mm film. Fluxes are divided by 2 for 0.2-mm film and multiplied by 2 for 0.05-mm film.

TABLE 3

Diffusive Permeability of Poroplastic® Membrane

Effect of Molecular Weight on Diffusion for MA-85 Membrane

Solute (%)	Mol wt	Transport method	Diff. coef.
0.3 NaCl	58	Film, 4 mil	4.9 cm²/sec × 10^{-6}
0.2 urea	60	Hollow fiber	1.6 cm²/sec × 10^{-6}
2.0 ionic chemical	400	Film, 4 mil	1.9 cm²/sec × 10^{-6}
0.005 vitamin B12	1355	Hollow fiber	0.25 cm²/sec × 10^{-6}

Effect of Membrane Water Content on Diffusion for 2% Ionic Chemical with mol wt of 400 through 4 mil Film at 32°C

Poroplastic®	Flux (mg/cm²/sec)	Diff. coef.
MA-92 film	0.00411	2.1 cm²/sec × 10^{-6}
MA-85 film	0.00378	1.9 cm²/sec × 10^{-6}
MA-70 film	0.00291	1.4 cm²/sec × 10^{-6}
Av. area corrected diff. coeff.		2.2 cm²/sec × 10^{-6}

Effect of Membrane Shrinkage on Diffusion For 2% Ionic Chemical with mol wt of 400 from MA-85 Film at 32°C

Total volume loss (%)	Water loss (%)	Area corrected diff. coef.
0	0	1.9 cm²/sec × 10^{-6}
33	38	1.8 cm²/sec × 10^{-6}
63	72	1.4 cm²/sec × 10^{-6}
71	82	1.6 cm²/sec × 10^{-6}
85 (dry)	100 (dry)	0.0 cm²/sec × 10^{-6}

Effect of Membrane Water Content and Thickness on Dialysance (0.3% NaCl through Membrane into Pure Water)

Poroplastic®	Dialysance of 4 mil film	Dialysance of 8 mil film
MA-92	500 mℓ/m²/min	270 mℓ/m²/min
MA-85	310 mℓ/m²/min	170 mℓ/m²/min
MA-70	200 mℓ/m²/min	110 mℓ/m²/min

TABLE 4

Typical Properties of Water and Oil Filled Poroplastic® Film

Property	MA-70	MA-85	MA-92
Water			
Composition			
Resin (%)	30	15	8
Liquid (%)	70	85	92
Apparent pore size			
Å	14	25	60
mils	0.00006	0.00010	0.00023
Specific gravity	1.09	1.04	1.02
Tensile strength (psi)	1,300	550	175
Elongation at break (%)	44	40	20
Tensile strength after drying (psi)	2,600	1,600	550
Elastic modulus (psi)	21,000	5,000	200
Elongation at elastic limit (%)	2.4	2.7	4.0
Mineral Oil			
Specific gravity	1.01	0.94	0.91
Tensile strength (psi)	2,300	0.900	400
Elongation at break (%)	41	35	32
Elastic modulus (psi)	41,000	10,000	3,400
Elastic limit (psi)	1,000	275	110
Elongation at elastic limit (%)	2.4	2.7	3.1
Plastic modulus (psi)	2,700	1,700	950

Courtesy Nichols, L. D., private communications.

ranging from 21 Å (MA70) to 145 Å (MA97.5). These values are in rough agreement with other data even though many assumptions were made in their derivation.

Diffusive permeability studies as shown in Table 3 have been carried out with various solutes for film and hollow fiber with film of different thickness, with film initially made with differing water contents, and with partially dried films. The most significant observation from this data is that the area-corrected diffusion coefficients can be as high as 50% of those for the pure liquid. The permeability of Poroplastic® film is so high that it is a good matrix for rapid solute transport and chemical reactions.

IV. MECHANICAL PROPERTIES

The mechanical properties are extremely good considering its high liquid content. Table 4 lists some of these properties for both water- and oil-filled film. When desirable, the film may be reinforced with numerous materials, e.g., woven fiberglass or nonwoven polypropylene or polyester. Alternatively, the film may be coated on a substrate, such as paper, cloth, or metal.

V. SOLVENT COMPATIBILITY

Cellulose triacetate is generally considered to have very good solvent resistant properties, and the solvent resistance of Poroplastic® film and Sustrelle® powder are accordingly quite good. It is adversely affected only by a few polar organic solvents, such as dimethyl sulfoxide, dimethyl formamide, acetone, methylene chloride, chloroform, phenol, pyridine, and glacial formic or acetic acid, but is compatible with equal weight mixtures of water with many of the above solvents. It can be readily loaded with solvents such as methanol, toluene, methyl isobutyl ketone, aromatic oils, aliphatic hydrocarbons, silicones, and most related compounds. Over long periods of time it is stable from pH 1 through pH 10.

VI. CONTROLLED RELEASE

Sustained release from ultramicroporous cellulose triacetate is basically diffusion-controlled, in which respect it differs from microencapsulation which depends primarily on rupture or dissolving of a thin shell for release. In general, a three-step process is involved in controlled release of the active agent. The steps are (1) impregnation, (2) fixation, and (3) diffusional release.

The impregnation step has already been generally described. It involves the series of diffusional exchanges of miscible liquids until the liquid within the polymer matrix is the active agent, or a solution of the active agent to be released. The fixation step utilizes some procedure to retard the rate of diffusion of active agent to the environment. In addition to the pore collapse fixation method described previously, there are a number of fixation methods based on the precipitation of active agent within the porous matrix. For example, when the porous matrix is impregnated with a saturated solution, evaporation of the solvent can cause the solute to precipitate within the porous matrix. In addition, the solvent evaporation causes partial pore collapse, and the precipitated active agent fills the remaining pore structure and prevents complete collapse. Internal precipitation of an active agent can also be accomplished by a solvent substitution method. In this case, after impregnation with a solution of the active agent, another liquid is added, with which the original solvent is miscible, but in which the active agent is insoluble. A third precipitation technique involves chemical reaction within the porous matrix to produce a precipitate.

Diffusional release may require a trigger mechanism for initiation. It differs from microencapsulation in that no outer shell is broken, and it differs from common absorbents, such as activated carbon, in that no chemical bonding to the substrate is involved.

VII. COMPARISON WITH MICROENCAPSULATION

Sustrelle®, the time release powder, and Liqui-Powder®, the pressure release powder, are commonly compared with microencapsulated products. Although the applications are similar, the mechanism for action is quite different. Microencapsulation involves a coating, whereas Liqui-Powder® or Sustrelle® is a molecular sponge. The former requires rupture or dissolution of a coating for action, whereas the latter involves squeezing or diffusion of a liquid out of a sponge. Microencapsulation usually requires harsher processing conditions for incorporation, whereas Liqui-Powder® or Sustrelle® can be loaded by a room temperature diffusional exchange process using a mild solvent.

Because payloads are generally about 90% for Liqui-Powder® and the matrix itself is crystalline, there is no gummy residue apparent when the liquid is squeezed out of the sponge-like powder. Also, because it is spongy and compliant, there is no gritty feeling when it is spread on the skin.

VIII. PRACTICAL APPLICATIONS

Many commercial applications for these materials have been developed and new ones are continuing to appear. They cover a broad range of consumer and industrial product areas. The types of applications fall into two categories: those based on its characteristics as a matrix or carrier and those related to controlled release. In the former category, the most important properties are its high diffusion coefficients combined with the ability to hold large quantities of liquid in a solid form as represented by the following examples:

- Electrical insulating oils
- Radiation absorbers used in solar energy collectors
- Flavor and fragrance modifiers
- Photochromic solutions for eye goggles
- Indicating chemical reactions for toxic substance monitoring
- Solvent extraction systems in hydrometallurgy

The controlled release applications usually depend on characteristics based on regulated pore size, shrinkage, deformation, and liquid or vapor expulsion. For example, it can be used as a controlled release material for:

- Pharmaceuticals
- Pesticides
- Air fresheners
- Dermatological products
- Toiletries
- Surfactants
- Shelf life extenders
- Cosmetics and perfumes
- Adhesives
- Analytical procedures, such as chromatography, ultrafiltration, enzymes or antibody immobilization, and electrophoresis

REFERENCES

1. **Obermayer, A. S. and Nichols, L. D.,** Controlled release from ultramicroporous cellulose triacetate, in *Controlled Release Polymeric Formulations,* ACS Symp. Ser. No. 33, Paul, D. R. and Harris, F. W., Eds., American Chemical Society, Washington, D.C., 1976, 303.
2. **Nichols, L. D.,** Poroplastic® and Sustrelle®, controlled release vehicles having broad compatibility with dissolved and precipitated pesticides, *Proc. Int. Controlled Release Pesticide Symp.,* Cardarelli, N. F., Ed., Engineering and Science Division, Community and Technical College, University of Akron, Ohio, 1974.
3. **Obermayer, A. S. and Nichols, L. D.,** Controlled release from Poroplastic®, Sustrelle®, and LiquiPowder®, Society of Cosmetic Chemists Annual Meeting, New York, December 6 to 7, 1976.
4. **Gray, D., Ed.,** *American Institute of Physics Handbook,* McGraw-Hill, New York, 1957, 2.
5. Technical Staff, Poroplastic® Description and Properties, Technical Bulletin, Moleculon Research Corp., Cambridge, Massachusetts.
6. **Nichols, L. D.,** private communications.
7. Technical Staff, *Sustrelle® Laboratory Manual,* Moleculon Research Corp., Cambridge, Mass., 1975.

INDEX

A

B

C

D

E

F

I

J

K

L

M

N

R

S

U

W

X

Z